Molecular Basis of Oxidative Stress

Molecular Basis of Oxidative Stress

Edited by Tiffani Damron

hayle
medical

New York

Hayle Medical,
750 Third Avenue, 9ᵗʰ Floor,
New York, NY 10017, USA

Visit us on the World Wide Web at:
www.haylemedical.com

ISBN: 978-1-64647-541-4

Cataloging-in-Publication Data

Molecular basis of oxidative stress / edited by Tiffani Damron.
    p. cm.
Includes bibliographical references and index.
ISBN 978-1-64647-541-4
1. Oxidative stress. 2. Oxidative stress--Molecular aspects. 3. Oxidation.
4. Oxidation-reduction reaction. I. Damron, Tiffani.
RB170 .M65 2023
616.07--dc23

Table of Contents

VI   Contents

Preface

Oxidative stress refers to an imbalance between antioxidants and free radicals in the body. Free radicals are oxygen-containing molecules with an unbalanced number of electrons. The body creates free radicals during regular metabolic processes. Oxidative stress happens naturally and contributes in the aging process. It can occur due to a variety of reasons including air pollutants, consumption of alcohol, radiation, cold, drugs, toxins, and trauma. In the body, numerous pathophysiological conditions such as chronic fatigue syndrome, atherosclerosis, heart attack, cancers and gene mutations are caused by oxidative stress. It may also lead to neurodegenerative diseases such as Alzheimer's disease and Parkinson's disease. Certain dietary and lifestyle adjustments may aid in the reduction of oxidative stress. These include exercising regularly, maintaining a healthy body weight, and eating a well-balanced nutritious diet rich in vegetables and fruits. This book contains some path-breaking studies related to the molecular basis of oxidative stress. It will serve as a reference to a broad spectrum of readers.

Significant researches are present in this book. Intensive efforts have been employed by authors to make this book an outstanding discourse. This book contains the enlightening chapters which have been written on the basis of significant researches done by the experts.

Finally, I would also like to thank all the members involved in this book for being a team and meeting all the deadlines for the submission of their respective works. I would also like to thank my friends and family for being supportive in my efforts.

Editor

Could Oxidative Stress Regulate the Expression of MicroRNA-146a and MicroRNA-34a in Human Osteoarthritic Chondrocyte Cultures?

Sara Cheleschi [1], Anna De Palma [1,2], Nicola Antonio Pascarelli [1], Nicola Giordano [3], Mauro Galeazzi [1], Sara Tenti [4] and Antonella Fioravanti [1,*] (iD)

[1] Rheumatology Unit, Azienda Ospedaliera Universitaria Senese, Policlinico Le Scotte, Viale Bracci 1, 53100 Siena, Italy; saracheleschi@hotmail.com (S.C.); annadepalma90@live.it (A.D.P.); pascarelli@unisi.it (N.A.P.); mauro.galeazzi@unisi.it (M.G.)

[2] Department of Medical Biotechnologies, University of Siena, Policlinico Le Scotte, Viale Bracci 1, 53100 Siena, Italy

[3] Department of Medicine, Surgery and Neurosciences, Scleroderma Unit, University of Siena, Policlinico Le Scotte, Viale Bracci 1, 53100 Siena, Italy; nicola.giordano@unisi.it

[4] Department of Medicine, Surgery and Neuroscience, Rheumatology Unit, University of Siena, Policlinico Le Scotte, Viale Bracci 1, 53100 Siena, Italy; sara_tenti@hotmail.it

* Correspondence: fioravanti7@virgilio.it.

Abstract: Oxidative stress and the overproduction of reactive oxygen species (ROS) play an important role in the pathogenesis of osteoarthritis (OA). Accumulating evidence has demonstrated the involvement of microRNAs (miRNAs) dysregulation in disease development and progression. In this study, we evaluated the effect of oxidative stress on miR-146a and miR-34a expression levels in human OA chondrocytes cultures stimulated by H_2O_2. Mitochondrial ROS production and cell apoptosis were detected by flow cytometry. The antioxidant enzymes SOD-2, CAT, GPx, the transcriptional factor NRF2 and the selected miRNAs were analyzed by qRT-PCR. The H_2O_2-induced oxidative stress was confirmed by a significant increase in superoxide anion production and of the apoptotic ratio. Furthermore, H_2O_2 significantly up-regulated the expression levels of SOD-2, CAT, GPx and NRF2, and modulated miR-146a and miR-34a gene expression. The same analyses were carried out after pre-treatment with taurine, a known antioxidant substance, which, in our experience, counteracted the H_2O_2-induced effect. In conclusion, the induction of oxidative stress affected cell apoptosis and the expression of the enzymes involved in the oxidant/antioxidant balance. Moreover, we demonstrated for the first time the modification of miR-146a and miR-34a in OA chondrocytes subjected to H_2O_2 stimulus and we confirmed the antioxidant effect of taurine.

Keywords: osteoarthritis; oxidative stress; microRNAs; chondrocytes; taurine

1. Introduction

Osteoarthritis (OA) is the most common degenerative joint disorder and it represents the main cause of pain, functional impairment and disability in adult and elderly populations [1]. OA affects the whole joint structure and it is characterized by the progressive degradation of the components of articular cartilage, thickening of subchondral bone and synovial inflammation, inducing the loss of normal joint architecture and function [2].

The release of reactive oxygen species (ROS), such as superoxide anion, hydrogen peroxide (H_2O_2) and nitric oxide, contributes to cartilage damage and a concomitant low-grade chronic inflammation [3]. ROS are free radicals containing oxygen molecules derived from cellular oxidative metabolism including enzyme activities and mitochondrial respiration, and play a pivotal role in many cellular functions. Under normal conditions, the production of endogenous ROS is balanced by the antioxidant defence system [4]. ROS have been found to be increased in several pathological conditions, including OA [5]. In OA chondrocytes, ROS overproduction causes DNA damage. Oxidative radical accumulation in the joint contributes to the inhibition of glycosaminoglycan and collagen synthesis and to the activation of metalloproteinases and aggrecanases, promoting cartilage breakdown [5]. It has also been reported that chondrocytes apoptosis related to OA can be induced by oxidative stress [6].

MicroRNAs (miRNAs) are a class of a single-stranded non-coding RNA (20–25 nucleotides in length) involved in post-transcriptional regulation of gene expression by targeting the 3′-untranslated region of the target gene messenger RNA (mRNA) and are implicated in the modulation of several cellular processes and disorders [7]. Accumulating evidence has reported differences in miRNAs expression profiles between normal and OA cartilage samples, demonstrating their role in the development and progression of OA [8–10]. Recently, some oxidative stress-responsive miRNAs were identified after the treatment of various cell types with H_2O_2 [11]; otherwise, cellular mechanisms regulating oxidative stress were fine-tuned by particular miRNAs [12]. For these reasons, miRNAs seem to be implicated in the articular damage triggered by oxidative stress in OA pathology.

Various agents, such as taurine and ascorbic acid, have been used as potent anti-oxidant substances with highly effective properties attenuating free radical toxicity [13,14]. Data from in vitro studies reported that their antioxidant property could ameliorate ROS-induced cartilage damage [15–17]. Taurine (2-aminoethane sulfonic acid) is a necessary amino acid involved in cartilage physiopathology and shows chondroprotective properties principally determining the increase in deposition of extracellular matrix components and in chondrocyte proliferation [18].

In the present study, we investigated the expression levels of miR-34a and miR-146a in human OA chondrocyte cultures stimulated with H_2O_2. To evaluate the induced oxidative stress by H_2O_2, we performed an analysis of mitochondrial ROS production and cell apoptosis; the gene expression of antioxidant enzymes and of the nuclear factor erythroid 2 like 2 (NFE2L2 or NRF2), the main transcription factor involved in redox homeostasis, which was also detected.

Furthermore, the same analyses were carried out after treatment with taurine, a known antioxidant substance.

2. Results

2.1. Mitochondrial $\bullet O_2^-$ Production

MitoSOX Red staining showed the detection of mitochondrial $\bullet O_2^-$ in chondrocyte cultures after our treatment (Figure 1). As expected, the stimulus of OA chondrocytes with H_2O_2 induced a very significant increase in $\bullet O_2^-$ production ($p < 0.001$) in comparison to basal state. No significant modification of $\bullet O_2^-$ release was observed in cells treated with the two concentrations of taurine tested alone compared with basal conditions. The effect of H_2O_2 was attenuated and significantly counteracted by the pre-treatment of the cells with taurine 100 μM and 200 μM ($p < 0.01$).

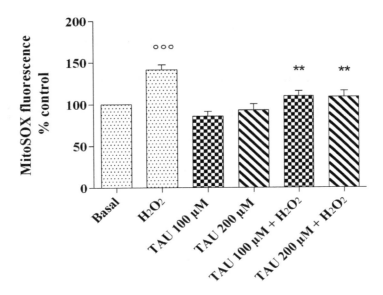

Figure 1. ROS production after 24 h of pre-treatment with taurine (TAU) at the concentrations of 100 μM and 200 μM alone and with the presence of H_2O_2 (1 μM) for 30 min. The analysis was performed by using MitoSox Red staining. Data were expressed as a percentage of mitochondrial superoxide anion ($\bullet O_2{}^-$) production in all the study conditions. The percentage was referenced to the ratio of the value of interest and basal conditions. The value of basal conditions was reported equal to 100. Data were expressed as mean $\pm$ SD of triplicate values. $^{\circ\circ\circ}$ $p < 0.001$ versus basal conditions; ** $p < 0.01$ versus H_2O_2.

2.2. Cell Viability Assay

Cell viability evaluated by (3-[4,4-dimethylthiazol-2-yl]-2,5-diphenyl-tetrazoliumbromide) MTT assay is reported in Figure 2. The chondrocytes exposed to H_2O_2 showed a significant decrease in the percentage of survival cells ($p < 0.01$); when cells were treated with taurine tested at concentrations of 100 μM and 200 μM alone or in presence of H_2O_2, no significant modifications were observed. The data was confirmed by Trypan Blue test.

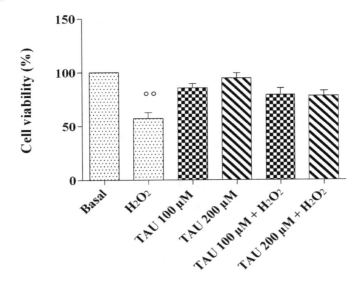

Figure 2. Evaluation of cell viability after 24 h of pre-treatment with taurine (TAU) at the concentrations of 100 μM and 200 μM alone and with the presence of H_2O_2 (1 μM) for 30 min. Data were expressed as percentage of cell viability in all the study conditions. The percentage was referenced to the ratio of the value of interest and basal conditions. The value of basal conditions was reported equal to 100. Data were expressed as mean $\pm$ SD of triplicate values. $^{\circ\circ}$ $p < 0.01$ versus basal conditions.

2.3. Apoptosis Detection

The data for chondrocyte apoptosis obtained by flow cytometry assay is reported in Figure 3. The stimulus of chondrocytes with H_2O_2 induced a significant increase in this ratio ($p < 0.05$) in comparison to the basal time. The incubation of our cultures with taurine at concentrations of 100 μM and 200 μM did not significantly modify the apoptosis ratio. The increase of apoptosis induced by H_2O_2 was significantly reduced by the treatment of the cells with taurine 200 μM ($p < 0.05$).

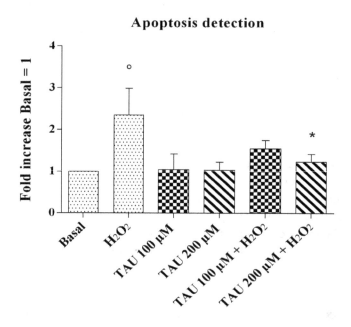

Figure 3. Apoptosis detection after 24 h of pre-treatment with taurine (TAU) at the concentrations of 100 μM and 200 μM alone and with the presence of H_2O_2 (1 μM) for 30 min. Apoptosis was measured with Alexa Fluor 488 annexin-V assay. Data were expressed as the percentage of positive cells for Annexin-V and propidium iodide (PI) in all the study conditions. The ratio of apoptosis was referenced to the ratio of the value of interest and basal conditions. The value of basal conditions was reported equal to 1. Data were expressed as mean ± SD of triplicate values. ° $p < 0.05$ versus basal conditions, * $p < 0.05$ versus H_2O_2.

2.4. Modifications of Gene Expression Levels of Factors Involved in Oxidant/Antioxidant Systems

The expression levels of catalase (*CAT*), superoxide dismutase (*SOD*)-2, glutathione peroxidase (*GPx*)4 and NRF2 are reported in Figure 4A–D. A statistically significant increase of *CAT* ($p < 0.001$, Figure 4A), *SOD-2* ($p < 0.01$, Figure 4B), and *NRF2* ($p < 0.01$, Figure 4D) expression levels was observed after stimulus with H_2O_2. No modification of gene expressions was shown after the incubation of our cultures with taurine 100 μM and 200 μM alone compared with basal conditions. The effect of H_2O_2 on *CAT* and *SOD-2* gene expression was significantly counteracted when chondrocytes were treated with taurine 100 μM and 200 μM ($p < 0.01$ for *SOD-2*, $p < 0.001$ for *CAT*). Taurine 200 μM significantly increased the expression levels of *NRF2* ($p < 0.01$) in chondrocytes stimulated with H_2O_2. No detectable changes were observed in *GPx4* expression levels after treatment (Figure 4C).

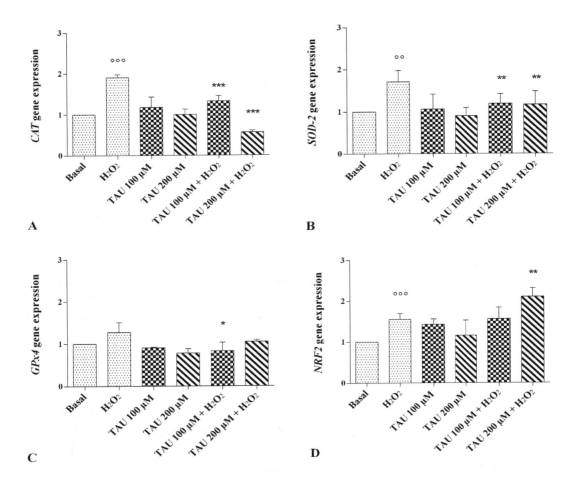

Figure 4. Evaluation of gene expression of antioxidant enzymes catalase (*CAT*) (**A**); superoxide dismutase (*SOD-2*) (**B**); glutathione peroxidase (*GPx*)-4 (**C**) and of nuclear factor erythroid 2 p45-related factor (*NRF*) 2 (**D**) by real-time PCR. This analysis was performed after 24 h of pre-treatment with taurine (TAU) at the concentrations of 100 μM and 200 μM alone and with the presence of H_2O_2 (1 μM) for 30 min. The ratio of gene expression was referenced to the ratio of the value of interest and basal conditions. The value of basal conditions was reported equal to 1. Data were expressed as mean ± SD of triplicate values. °° $p < 0.01$, °°° $p < 0.001$ versus basal conditions, * $p < 0.05$, ** $p < 0.01$, *** $p < 0.001$ versus H_2O_2.

2.5. Regulation of miR-146a and miR-34a Gene Expression

The modification of miR-146a and miR-34a expression levels is represented in Figure 5.

The stimulus of chondrocytes with H_2O_2 determined a statistically significant decrease of miR-146a ($p < 0.01$, Figure 5A) and a significant increase of miR-34a ($p < 0.001$, Figure 5B) gene expression. No modification of their expression levels was observed after the incubation of the cells with taurine at both concentrations analyzed alone, compared with the basal time. The stimulus induced by H_2O_2 was significantly reversed in chondrocytes pre-treated with taurine at both tested concentrations, inducing an increase in miR-146a and a decrease in miR-34a gene expression.

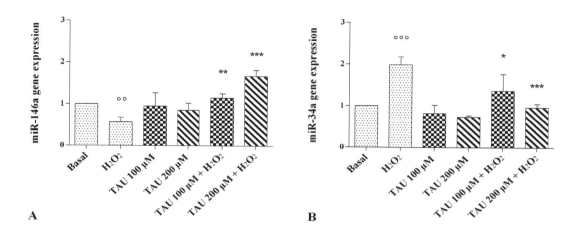

Figure 5. Evaluation of gene expression of miR-146a (**A**) and miR-34a (**B**) by real-time PCR. This analysis was performed after 24 h of pre-treatment with taurine (TAU) at the concentrations of 100 μM and 200 μM alone and with the presence of H_2O_2 (1 μM) for 30 min. The ratio of gene expression was referenced to the ratio of the value of interest and basal conditions. The value of basal conditions was reported equal to 1. Data were expressed as mean ± SD of triplicate values. °° $p < 0.01$, °°° $p < 0.001$ versus basal conditions, * $p < 0.05$, ** $p < 0.01$, *** $p < 0.001$ versus H_2O_2.

3. Discussion

ROS production is maintained at low levels in healthy articular chondrocytes where they have a role in the activation of intracellular signaling pathways fundamental to the regulation of cartilage homeostasis [4]. Otherwise, the failure in oxidant/antioxidant balance in OA cells determines an altered redox status with an excessive synthesis of ROS, as H_2O_2, $\bullet O_2^-$ and hydroxyl free radicals ($\bullet OH^-$), contributing to the pathogenesis of OA [5].

In the present study, we evaluated the effect of H_2O_2-induced oxidative stress on the expressions of some miRNAs responsive to ROS stress. To evaluate the induced stress we performed the analysis of mitochondrial $\bullet O_2^-$ production and of cell apoptosis, as well as the gene expression of antioxidant enzymes and of a transcription factor involved in redox homeostasis.

H_2O_2 is often used in in vitro models to mimic the circumstances that drive in vivo oxidative stress cartilage damage [19,20]; in fact, it can diffuse across the outer mitochondrial membrane to the cytosol regulating the activity of ROS sensors and the gene expression of antioxidant enzymes [21,22]. In addition, the exposure of articular chondrocytes to H_2O_2 induces the inhibition of proteoglycans synthesis and increases their degradation [23].

In our study, the analysis of intracellular ROS content showed upper levels of mitochondrial $\bullet O_2^-$ in chondrocytes stimulated with H_2O_2, demonstrating an increase in their production as supported by the data from the literature [24,25].

Excessive production of ROS makes chondrocytes more susceptible to oxidant-mediated apoptosis, a major factor in OA pathogenesis [26,27]. H_2O_2 influences apoptosis in chondrocytes by changing the intracellular redox state, damaging the structure and function of mitochondria and activating caspase-3 [28]. In accordance with current, state-of-the-art research, our results reported an increase in apoptotic cells after H_2O_2 exposure.

Under normal conditions, the harmful production of endogenous ROS is balanced by the antioxidant defence system including enzymes like SOD-2, CAT and GPx, to decrease oxidative stress. SOD-2 or manganese-dependent (Mn)-SOD represents one of the three SOD family members; it is located in the mitochondrial matrix of the cells and catalyzes the dismutation of superoxide anion in O_2 and H_2O_2, which is then turned in water, to CAT and GPx [29,30]. In vivo and in vitro studies have demonstrated the reduction in expression and activity of these stress-related enzymes in OA chondrocytes, leading to the loss of their ability to scavenge ROS. The resulting oxidation of intracellular and extracellular components contributes to cartilage breakdown in OA. Moreover, the products of cartilage degradation aggravate the inflammation and enhance ROS causing a vicious cycle that contributes to OA progression [31,32].

To our knowledge, no studies have been performed to evaluate the response of antioxidant enzymes to H_2O_2 in chondrocyte cultures. Conflicting results were reported from studies carried out in different cell types. Dhuna et al. [33] showed the reduced activity of SOD, CAT and GPx enzymes in C6 cells exposed to H_2O_2, whereas no modifications in their activities were observed in H_2O_2-stimulated Leydig cells in a study by Ding et al. [34]. On the other hand, an increase of SOD-2 expression levels was found in bovine normal chondrocytes cultures after stimulation with IL-1β and IL-6, which are potent proinflammatory cytokines [35].

Our results demonstrated the overexpression of *SOD-2* and *CAT* induced by H_2O_2 in OA chondrocyte cultures. The increase of these antioxidant enzymes may be interpreted as a first-line defence against a variety of ROS and proinflammatory stimuli to protect mitochondria against the deleterious effects of free radicals.

The *GPx4* mRNA levels in our cultures appeared unaffected by H_2O_2 probably due to the later responsiveness of regulatory elements of *GPx* genes to oxidative stress; analyses carried out at different time points after H_2O_2 stimulus could be useful to better investigate this aspect.

The homeostasis of the cellular redox state is also regulated by the transcription factor NRF2. In physiological conditions, NRF2 is maintained at low levels by its ubiquitination and degradation through proteasome machinery. Stress signals lead to its accumulation and its translocation into the nucleus, where it regulates the transcription of antioxidant response element-dependent genes [36]. Its target genes, including *SOD-2*, *CAT* and *GPx*, are involved in the reduction of intracellular levels of ROS, reactive nitrogen species, and electrophiles, protecting cells from oxidative damage [37]. NRF2 may exert a chondroprotective role in OA, suppressing metalloproteinases expression induced by IL-1β [38] and regulating apoptotic cell death [39]. Moreover, results from an in vivo study showed a high susceptibility to OA development in mice with NRF2 deficiency [40]. Our data demonstrated an increase of NRF2 expression levels in chondrocytes stimulated with H_2O_2, in agreement with results obtained by Bernard et al. [41] reporting its activation in OA synovial fibroblasts under abnormal ROS signaling. NRF2 response to oxidative stress may contribute to cell survival by enhancing the resistance to further damage stimuli [42].

Recently, NRF2 has been found to be regulated by miR-146a, through a miRNA-146a mimic transfection of rat hepatocytes [43]. MiRNA-146a is widely expressed in different species and tissues and plays an important role in immune and inflammatory processes [44]. Its involvement in OA pathogenesis has been investigated recently. Yamasaki et al. [45] demonstrated that miR-146a is upregulated in OA cartilage with a low grade on the Mankin scale compared with normal samples; its expression decreased with the progression of the disease. Furthermore, we reported the reduced expression of this miRNA in OA chondrocytes in comparison to healthy cells in our previous study [46]. Lately, miR-146a has been found to be modulated by H_2O_2-induced stress, highlighting its responsiveness to this kind of stimulus [11,12]. The possible modification of miR-146a caused by oxidative stress was investigated by Ji et al. [47], showing a dose-dependent upregulation of the miRNA induced by H_2O_2 in PC12 cells. Moreover, MiR-146a overexpression was also observed in HUVEC cells and is associated to the inhibition of proteins responsible for ROS generation, delaying the senescence-like phenotype [48]. On the contrary, we demonstrated the downregulation of miR-146a levels in OA chondrocytes after H_2O_2 stimulus. These controversial results could be because of the different responses by various cell types. Indeed, in PC12 cells, the use of antisense-miR-146a, reversed the *SOD-2* expression induced by H_2O_2, demonstrating the potential transcriptional regulatory role of the miRNA [47]. Our data, from OA chondrocyte cultures, seems to be in agreement with these results, confirming the interaction between miR-146a and *SOD-2* expression. So, we can speculate about the involvement of oxidative stress in the modulation of miR-146a, which, in turn, regulates the expression of *SOD-2*; these results confirm the role of oxidative stress and miR-146a in the pathogenesis of OA [11,46].

Recent evidence shows the contribution of miR-34a to the development and progression of OA; in fact, it has been found to be overexpressed in human OA chondrocytes inducing apoptosis and inhibition of cell proliferation [49]. The use of a miR-34a mimic and anti-miR-34a showed the direct targeting of the miRNA on Sirtuin (SIRT)-1/p53 signaling pathways responsible for the regulation of chondrocyte proliferation and apoptosis during OA damage [49]. Moreover, miRNA profiling analysis identified miR-34a as one of the main regulated miRNAs under oxidative stress [11]. Some studies reported the increase of miR-34a expression levels in human bronchial epithelial and hepatocellular carcinoma cell lines upon treatment with H_2O_2 in a concentration-dependent manner [11,50]. To our knowledge, no previous studies have reported the regulation of this miRNA after H_2O_2 stimulus in chondrocytes, however, our results appear to be in agreement with the available data showing an increase in miR-34a levels in chondrocyte cultures exposed to H_2O_2-induced stress.

Interestingly, Bai et al. [51] found that miR-34a regulated mitochondrial antioxidative enzymes with a concomitant modulation of intracellular ROS levels in primary mesangial cells. Our results showed the regulation of miR-34a expression levels as well as of antioxidative enzymes and intracellular ROS after H_2O_2 stimulation in OA chondrocyte cultures. These findings highlight the responsiveness of miR-34a to oxidative stress, and consequently, the possible effect of the miRNA on the oxidant/antioxidant system; on the basis of this evidence, we confirm the involvement of oxidative stress and miR-34a in the pathogenesis of OA. In the current study, we also analysed the role of taurine (100 μM and 200 μM), a well-known antioxidant substance, in counteracting the H_2O_2-induced oxidative stress. In our experimental conditions, taurine reduced the mitochondrial $\bullet O_2^-$ production and the ratio of apoptotic chondrocytes in accordance with previous results [52–55]. Recent evidence demonstrated the increase of antiapoptotic Bcl-2 protein and the reduction of proapoptotic Bax protein expression induced by taurine, confirming our data [56].

Some authors have shown the effect of taurine in preventing oxidant-induced cell damage through the reduction of ROS generation and the neutralization of the pre-existing reactive species [14,56]. Our observation showed the reduction of *SOD-2* and *CAT* expression levels in H_2O_2-stimulated chondrocyte cultures pre-treated with taurine, confirming the antioxidant property of this substance.

The chondroprotective effects of some antioxidant substances, such as gingerol and wogonin, have been demonstrated in human OA chondrocytes, through the activation of NRF2 signaling pathway [57,58]. No previous data was available regarding the effect of taurine on this transcription factor, so we showed for the first time a significant increase in *NRF2* mRNA levels in the H_2O_2-stimulated chondrocyte cultures after treatment with taurine at a concentration of 200 μM.

Furthermore, we evaluated the role of taurine in regulating the gene expression of miR-146a and miR-34a. To our knowledge, this is the first study reporting the modulation of miR-146a gene expression induced by this antioxidant substance in H_2O_2-stimulated chondrocyte cultures, underlining the responsiveness of the miRNA to oxidative stress.

On the other hand, the reduced transcription of the miR-34a by antioxidant agents such as resveratrol, statins and genistein in various cell types is well documented [59,60]. Our data appear to be in agreement with the current literature showing that the increase in miR-34a levels in chondrocytes exposed to H_2O_2 was reverted by pre-treatment of the cells with taurine, in particular at a concentration of 200 μM.

4. Materials and Methods

4.1. Cell Culture

OA human articular cartilage was obtained from the femoral heads of five patients (two men and three women) with hip OA as defined by the clinical and radiological American College of Rheumatology criteria [61], undergoing total hip replacement surgery; OA grade ranged from moderate to severe, and cartilage showed typical osteoarthritic changes, such as the presence of chondrocyte clusters, loss of metachromasia, and fibrillation (Mankin degree 3–7) [62]. OA chondrocytes originated from the area adjacent to the OA lesion. The femoral heads were provided by the Orthopaedics Surgery, University of Siena, Italy. The mean age of the patients was 70 years (range, 64–77 years). The ethics committee of the Azienda Ospedaliera Universitaria Senese/Siena University Hospital approved the use of human articular specimens (decision No. 726 of 2007), and the patients signed an informed consent.

After surgery, the cartilage was aseptically dissected and minced into small pieces. The fragments were washed in (Dulbecco's Modified Eagle Medium) DMEM with phenol red, containing 2% penicillin/streptomycin solution and 0.2% amphotericin B. The chondrocytes were isolated from the articular cartilage using sequential enzymatic digestion: 0.1% hyaluronidase for 30 min, 0.5% pronase for 1 h, and 0.2% collagenase for 1 h at 37 °C in the wash solution (DMEM + penicillin/streptomycin solution + amphotericin B). The resulting cell suspension was filtered twice using 70-μm nylon meshes, washed and centrifuged for 5 min at 700 g. The Trypan blue viability test identified a 90% to 95% cell survival. Cells were incubated for 2 weeks at 37 °C and 5% CO_2 in DMEM culture medium containing 10% fetal calf serum (FCS), 200 U/mL penicillin and 200 μg/mL streptomycin. The medium was changed three times per week. The cell morphology was examined daily with an inverted microscope (Olympus IMT-2, Tokyo, Japan) to avoid the dedifferentiation of expanded chondrocytes and to preserve their phenotypic stability.

For each single experiment, a cell culture from a unique donor was used.

4.2. Treatment of Chondrocyte Cultures

In the first passage, OA human chondrocytes were seeded in 6-well dishes at a starting density of 6×10^6 cells/well until they became confluent. Taurine (2-amino ethane sulfonic acid, purity $\geq 98\%$ high-performance liquid chromatography) was purchased from Sigma-Aldrich (Milan, Italy). It was first dissolved in dimethylsulfoxide (DMSO, Rottapharm, Monza, Italy) and then it was further diluted in culture medium immediately before treatment to achieve the final concentration required. The final concentration of DMSO during the treatment did not exceed 0.1%, and it had no effect on the growth of the cells.

The cells were pre-treated for 24 h with taurine at concentration of 100 μM and 200 μM in the presence or in the absence of H_2O_2 1 μM, added for 30 min after the incubation with the compound.

After the treatment, the media was removed, cleared through centrifugation, and stored at −80 °C and the chondrocytes were immediately processed to perform cell viability assay, real-time PCR and cytometry analysis.

4.3. Mitochondrial Superoxide Anion ($\bullet O_2^-$) Production

OA chondrocyte cultures were plated in a density of 8×10^4 cells per well in 12 multi-plates for 24 h in DMEM with 10% FCS. After that, the medium was discarded and the cells were cultured in DMEM with 2% FBS normally used for the treatment ($n = 3$ different experiments). Then, the cells were incubated in Hanks' Balanced Salt Solution (HBSS) and MitoSOX Red for 15 min at 37 °C in dark, to evaluate mitochondrial superoxide anion ($\bullet O_2^-$) production. MitoSOX was dissolved in DMSO, at a final concentration of 5 μM. Cells were then harvested by trypsin and collected into cytometry tubes and centrifuged at 1500 rpm for 10 min. Besides, cells were resuspended in saline solution before being analyzed by flow cytometry. A density of 1×10^4 cells per assay (a total of 10,000 events) were measured by flow cytometry and data were analyzed with CellQuest software (Version 4.0, Becton Dickinson, San Jose, CA, USA). Results were collected as median of fluorescence (AU) and represented the mean of three independent experiments (mean ± SD).

4.4. MTT Assay

Cell viability was evaluated immediately after the treatment by MTT assay. Chondrocytes were incubated for 3 h at 37 °C in a culture medium containing 10% of 5 mg/mL MTT (Sigma-Aldrich, Milan, Italy). After incubation, the medium was discarded and 0.2 mL of DMSO was added to each well to solubilize the formazan crystals. The absorbance was measured at 570 nm by using a microplate reader (BioTek Instruments, Inc., Winooski, VT, USA). A well without cells was used for blank measurement.

The percentage of cell survival was calculated as follows:

$$\% \text{ Survival} = (\text{Absorbance of test})/(\text{Absorbance of control}) \times 100$$

The experiments were carried out on pre-confluent cell cultures to prevent contact inhibition influencing the results. Data were expressed as OD units per 10^4 adherent cells.

4.5. Detection of Apoptotic Cells

OA chondrocyte cultures were plated in 12-well plates (8×10^4 cells per well) for 24 h in DMEM with 10% FBS. Then, the medium was discarded, and the cells were cultured in DMEM with FBS 2% normally used for the treatment. The positive control was incubated in the presence of staurosporine, at 0.2 μM (Sigma-Aldrich) for 2 h to induce apoptosis. Then, the cells were washed and harvested with trypsin, collected into cytometry tubes, and centrifuged at 1500 rpm for 10 min. The supernatant was removed, and the pellet was resuspended in 100 μL of 1× Annexin-binding buffer; 5 μL of Alexa Fluor 488 annexin-V conjugated to fluorescein (green fluorescence) and 1 μL of 100 μg/mL propidium iodide (PI) working solution were added to each 100 μL of cell suspension (ThermoFisher Scientific, Rodano, Italy). Cells were incubated at room temperature for 15 min in the dark. After incubation, 600 μL of 1 × Annexin-binding buffer was added before being analyzed by flow cytometry. A total of 1×10^4 cells per assay (10,000 events) [63] were measured on a flow cytometer. The obtained data were analyzed with CellQuest software (Becton Dickinson). The evaluation of apoptosis was performed considering staining cells simultaneously with Alexa Fluor Annexin V and PI, allowing the discrimination of intact cells (Alexa Fluor Annexin-V and PI-negative), early apoptotic state (Alexa Fluor Annexin-V-positive and PI-negative) and late apoptosis state (Alexa Fluor Annexin-V and IP positives). Results were expressed as the percentage of positive cells to each dye (total apoptosis) and represented the mean of three independent experiments (mean ± SD).

4.6. RNA Isolation and RT-qPCR

Total RNA, including miRNA, was extracted using TriPure Isolation Reagent according to the manufacturer's instructions (Roche Diagnostics GmbH, Mannheim, Germany) and was stored at −80 °C. The concentration, purity, and integrity of RNA were evaluated by measuring the OD at 260 nm and the 260/280 and 260/230 ratios by Nanodrop-1000 (Celbio, Milan, Italy). The quality of RNA was also checked through electrophoresis on agarose gel (FlashGel System, Lonza, Rockland, ME, USA). Reverse transcription for target genes was performed by using QuantiTect Reverse Transcription Kit (Qiagen, Hilden, Germany), while for miRNA, by using the cDNA miScript PCR Reverse Transcription (Qiagen, Hilden, Germany) according to the manufacturer's instructions.

Target genes and miRNAs were analyzed by real-time PCR using, respectively, QuantiFast SYBR Green PCR (Qiagen, Hilden, Germany) and miScript SYBR Green (Qiagen) kits. All qPCR reactions were performed in glass capillaries using a LightCycler 1.0 (Roche Molecular Biochemicals, Mannheim, Germany) with LightCycler Software Version 3.5. The reaction procedure for target genes amplification consisted 5 s at 95 °C, 40 cycles of 15 s at 95 °C and 30 s at 60 °C. In the last step of both protocols, the temperature was raised from 60 °C to 95 °C at 0.1 °C/step to plot the melting curve. miRNA amplification was performed at 95 °C for 15 s for HotStart polymerase activation, followed by 40 cycles of 15 s at 95 °C for denaturation, 30 s at 55 °C for annealing and 30 s at 70 °C for elongation, according to the protocol.

To further analyze the dissociation curves, we visualized the amplicons lengths in an agarose gel to confirm the correct amplification of the resulting PCR products.

For data analysis, the C_t values in each sample and the efficiencies of the primer set were calculated using LinReg Software [64] and then converted into relative quantities (RQ) and normalized according to the Pfaffl model [65].

Normalization was carried out considering the housekeeping genes as HPRT-1 for target genes and SNORD-25 for miRNAs. These genes were chosen by software geNorm [66] version 3.5.

4.7. Statistical Analysis

Three independent experiments were performed and the results were expressed as the mean $\pm$ SD of triplicate values for each experiment. Data normal distribution was confirmed by Shapiro-Wilk, D'Agostino & Pearson and Kolmogorov-Smirnov tests.

Real-time PCR were evaluated by one-way (ANOVA) with a Tukey's Post Hoc test using $2^{-\Delta\Delta C_T}$ values for each sample.

All analyses were performed using the SAS System (SAS Institute Inc., Cary, NC, USA) and GraphPad Prism 6.1. (GraphPad Software, San Diego, CA, USA). A significant effect was indicated by a p value < 0.05.

5. Conclusions

In conclusion, in this study we observed the effect of H_2O_2-induced oxidative stress on the transcriptional levels of the main factors responsible for the maintenance of homeostatic oxidant/antioxidant equilibrium in human OA chondrocyte cultures.

Furthermore, we demonstrated for the first time, the regulation of miR-146a and miR-34a gene expression in OA chondrocyte cultures stimulated with H_2O_2. On the basis of these results, we highlight the responsiveness of miR-146a and miR-34a to the variation of cellular redox state occurring during OA damage.

Moreover, we confirm the role of taurine as an antioxidant compound, able to partially counteract the oxidative stress damage induced by H_2O_2. More importantly, we provide new results about its role in modulating the expression levels of the stress-responsive miR-146a and miR-34a.

Additional studies to confirm the direct relationship between the analyzed miRNAs and oxidative changes occurring in OA chondrocytes are required. The protein levels of the analyzed antioxidant enzymes and of the transcriptional factor NRF2 should be also examined to better understand the metabolic activities induced by H_2O_2.

The identification of miR-146a and miR-34a target genes could also be interesting to deeper elucidate their mechanism of action. In fact, *SOD-2* and *NRF2* are candidate target genes of these miRNA and possible effectors of their regulative functions in oxidative stress condition.

Acknowledgments: This research did not receive any specific grant from funding agencies in the public, commercial, or not-for-profit sectors.

Author Contributions: The authors declare their participation in the drafting of this paper as specified below: Sara Cheleschi performed the cell culture, cell treatment, qRT-PCR, flow cytometry, contributed to writing the protocol and article. Anna De Palma performed the cell culture, cell treatment, qRT-PCR, flow cytometry, contributed to writing the protocol and article. Nicola Antonio Pascarelli performed data analysis. Nicola Giuseppe Giordano contributed to writing the article. Mauro Galeazzi contributed to writing the article. Sara Tenti contributed to writing the article. Antonella Fioravanti wrote the protocol and the article.

References

1. Kraus, V.B.; Blanco, F.J.; Englund, M.; Karsdal, M.A.; Lohmander, L.S. Call for standardized definitions of osteoarthritis and risk stratification for clinical trials and clinical use. *Osteoarthr. Cartil.* **2007**, *23*, 1233–1241. [CrossRef] [PubMed]
2. Glyn-Jones, S.; Palmer, A.J.; Agricola, R.; Price, A.J.; Vincent, T.L.; Weinans, H.; Carr, A.J. Osteoarthritis. *Lancet* **2017**, *386*, 376–387. [CrossRef]

3. Kapoor, M.; Martel-Pelletier, J.; Lajeunesse, D.; Pelletier, J.P.; Fahmi, H. Role of proinflammatory cytokines in the pathophysiology of osteoarthritis. *Nat. Rev. Rheumatol.* **2011**, *7*, 33–42. [CrossRef] [PubMed]

4. Henrotin, Y.E.; Bruckner, P.; Pujol, J.P. The role of reactive oxygen species in homeostasis and degradation of cartilage. *Osteoarthr. Cartil.* **2003**, *11*, 747–755. [CrossRef]

5. Lepetsos, P.; Papavassiliou, A.G. ROS/oxidative stress signaling in osteoarthritis. *Biochim. Biophys. Acta* **2016**, *1862*, 576–591. [CrossRef] [PubMed]

6. Hwang, H.S.; Kim, H.A. Chondrocyte apoptosis in the pathogenesis of osteoarthritis. *Int. J. Mol. Sci.* **2015**, *16*, 26035–26054. [CrossRef] [PubMed]

7. Nugent, M. MicroRNAs: Exploring new horizons in osteoarthritis. *Osteoarthr. Cartil.* **2016**, *24*, 573–580. [CrossRef] [PubMed]

8. Díaz-Prado, S.; Cicione, C.; Muiños-López, E.; Hermida-Gómez, T.; Oreiro, N.; Fernández-López, C.; Blanco, F.J. Characterization of microRNA expression profiles in normal and osteoarthritic human chondrocytes. *BMC Musculoskelet. Disord.* **2012**, *13*, 144. [CrossRef] [PubMed]

9. De Palma, A.; Cheleschi, S.; Pascarelli, N.A.; Tenti, S.; Galeazzi, M.; Fioravanti, A. Do microRNAs have a key epigenetic role in osteoarthritis and in mechanotransduction? *Clin. Exp. Rheumatol.* **2017**, *35*, 518–526. [PubMed]

10. Cong, L.; Zhu, Y.; Tu, G. A bioinformatic analysis of microRNAs role in osteoarthritis. *Osteoarthr. Cartil.* **2017**, *25*, 1362–1371. [CrossRef] [PubMed]

11. Wan, Y.; Cui, R.; Gu, J.; Zhang, X.; Xiang, X.; Liu, C.; Qu, K.; Lin, T. Identification of four oxidative stress-responsive microRNAs, miR-34a-5p, miR-1915-3p, miR-638, and miR-150-3p, in hepatocellular carcinoma. *Oxid. Med. Cell. Longev.* **2017**, *2017*, 5189138. [CrossRef] [PubMed]

12. Bu, H.; Wedel, S.; Cavinato, M.; Jansen-Dürr, P. MicroRNA regulation of oxidative stress-induced cellular senescence. *Oxid. Med. Cell. Longev.* **2017**, *2017*, 2398696. [CrossRef] [PubMed]

13. Padayatty, S.J.; Katz, A.; Wang, Y.; Eck, P.; Kwon, O.; Lee, J.H.; Chen, S.; Corpe, C.; Dutta, A.; Dutta, S.K.; et al. Vitamin C as an antioxidant: Evaluation of its role in disease prevention. *J. Am. Coll. Nutr.* **2003**, *22*, 18–35. [CrossRef] [PubMed]

14. Marcinkiewicz, J.; Kontny, E. Taurine and inflammatory diseases. *Amino Acids.* **2014**, *46*, 7–20. [CrossRef] [PubMed]

15. Chiu, P.R.; Hu, Y.C.; Huang, T.C.; Hsieh, B.S.; Yeh, J.P.; Cheng, H.L.; Huang, L.W.; Chang, K.L. Vitamin C protects chondrocytes against monosodium iodoacetate-induced osteoarthritis by multiple pathways. *Int. J. Mol. Sci.* **2016**, *18*, 38. [CrossRef] [PubMed]

16. Liu, C.; Cao, Y.; Yang, X.; Shan, P.; Liu, H. Tauroursodeoxycholic acid suppresses endoplasmic reticulum stress in the chondrocytes of patients with osteoarthritis. *Int. J. Mol. Med.* **2015**, *36*, 1081–1087. [CrossRef] [PubMed]

17. Chang, Z.; Huo, L.; Li, P.; Wu, Y.; Zhang, P. Ascorbic acid provides protection for human chondrocytes against oxidative stress. *Mol. Med. Rep.* **2015**, *12*, 7086–7092. [CrossRef] [PubMed]

18. Liu, Q.; Lu, Z.; Wu, H.; Zheng, L. Chondroprotective effects of taurine in primary cultures of human articular chondrocytes. *Tohoku J. Exp. Med.* **2015**, *235*, 201–213. [CrossRef] [PubMed]

19. Facchini, A.; Stanic, I.; Cetrullo, S.; Borzì, R.M.; Filardo, G.; Flamigni, F. Sulforaphane protects human chondrocytes against cell death induced by various stimuli. *J. Cell. Physiol.* **2011**, *226*, 1771–1779. [CrossRef] [PubMed]

20. D'Adamo, S.; Cetrullo, S.; Guidotti, S.; Borzì, R.M.; Flamigni, F. Hydroxytyrosol modulates the levels of microRNA-9 and its target sirtuin-1 thereby counteracting oxidative stress-induced chondrocyte death. *Osteoarthr. Cartil.* **2017**, *25*, 600–610. [CrossRef] [PubMed]

21. Khan, I.M.; Gilbert, S.J.; Caterson, B.; Sandell, L.J.; Archer, C.W. Oxidative stress induces expression of osteoarthritis markers procollagen IIA and 3B3 (−) in adult bovine articular cartilage. *Osteoarthr. Cartil.* **2008**, *16*, 698–707. [CrossRef] [PubMed]

22. Marchev, A.S.; Dimitrova, P.A.; Burns, A.J.; Kostov, R.V.; Dinkova-Kostova, A.T.; Georgiev, M.I. Oxidative stress and chronic inflammation in osteoarthritis: Can NRF2 counteract these partners in crime? *Ann. N. Y. Acad. Sci.* **2017**, *1401*, 114–135. [CrossRef] [PubMed]

23. Johnson, K.; Jung, A.; Murphy, A.; Andreyev, A.; Dykens, J.; Terkeltaub, R. Mitochondrial oxidative phosphorylation is a downstream regulator of nitric oxide effects on chondrocyte matrix synthesis and mineralization. *Arthritis Rheumatol.* **2000**, *43*, 1560–1570. [CrossRef]

24. Lim, K.M.; An, S.; Lee, O.K.; Lee, M.J.; Lee, J.P.; Lee, K.S.; Lee, G.T.; Lee, K.K.; Bae, S. Analysis of changes in microRNA expression profiles in response to the troxerutin-mediated antioxidant effect in human dermal papilla cells. *Mol. Med. Rep.* **2015**, *12*, 2650–2660. [CrossRef] [PubMed]

25. Chen, Q.; Shao, X.; Ling, P.; Liu, F.; Shao, H.; Ma, A.; Wu, J.; Zhang, W.; Liu, F.; Han, G.; et al. Low molecular weight xanthan gum suppresses oxidative stress-induced apoptosis in rabbit chondrocytes. *Carbohydr. Polym.* **2017**, *169*, 255–263. [CrossRef] [PubMed]

26. Dixon, S.J.; Stockwell, B.R. The role of iron and reactive oxygen species in cell death. *Nat. Chem. Biol.* **2014**, *10*, 9–17. [CrossRef] [PubMed]

27. Yu, S.M.; Kim, S.J. Withaferin A-caused production of intracellular reactive oxygen species modulates apoptosis via PI3K/Akt and JNKinase in rabbit articular chondrocytes. *J. Korean Med. Sci.* **2014**, *29*, 1042–1053. [CrossRef] [PubMed]

28. Dickinson, B.C.; Lin, V.S.; Chang, C.J. Preparation and use of MitoPY1 for imaging hydrogen peroxide in mitochondria of live cells. *Nat. Protoc.* **2013**, *8*, 1249–1259. [CrossRef] [PubMed]

29. Davies, C.M.; Guilak, F.; Weinberg, J.B.; Fermor, B. Reactive nitrogen and oxygen species in interleukin-1-mediated DNA damage associated with osteoarthritis. *Osteoarthr. Cartil.* **2008**, *16*, 624–630. [CrossRef] [PubMed]

30. Hosseinzadeh, A.; Jafari, D.; Kamarul, T.; Bagheri, A.; Sharifi, A.M. Evaluating the protective effects and mechanisms of diallyl disulfide on interlukin-1β-induced oxidative stress and mitochondrial apoptotic signaling pathways in cultured chondrocytes. *J. Cell. Biochem.* **2017**, *118*, 1879–1888. [CrossRef] [PubMed]

31. Aigner, T.; Fundel, K.; Saas, J.; Gebhard, P.M.; Haag, J.; Weiss, T.; Zien, A.; Obermayr, F.; Zimmer, R.; Bartnik, E. Large-scale gene expression profiling reveals major pathogenetic pathways of cartilage degeneration in osteoarthriti. *Arthritis Rheumatol.* **2006**, *54*, 3533–3544. [CrossRef] [PubMed]

32. Scott, J.L.; Gabrielides, C.; Davidson, R.K.; Swingler, T.E.; Clark, I.M.; Wallis, G.A.; Boot-Handford, R.P.; Kirkwood, T.B.; Taylor, R.W.; Young, D.A. Superoxide dismutase downregulation in osteoarthritis progression and end-stage disease. *Ann. Rheum. Dis.* **2010**, *69*, 1502–1510. [CrossRef] [PubMed]

33. Dhuna, K.; Dhuna, G.; Bhatia, V.; Singh, J.; Kamboj, S.S. Cytoprotective effect of methanolic extract of Nardostachys *jatamansi* against hydrogen peroxide induced oxidative damage in C6 glioma cells. *Acta Biochim. Pol.* **2013**, *60*, 21–31. [PubMed]

34. Ding, X.; Wang, D.; Li, L.; Ma, H. Dehydroepiandrosterone ameliorates H_2O_2-induced Leydig cells oxidation damage and apoptosis through inhibition of ROS production and activation of PI3K/Akt pathways. *Int. J. Biochem. Cell Biol.* **2016**, *70*, 126–139. [CrossRef] [PubMed]

35. Mathy-Hartert, M.; Hogge, L.; Sanchez, C.; Deby-Dupont, G.; Crielaard, J.M.; Henrotin, Y. Interleukin-1beta and interleukin-6 disturb the antioxidant enzyme system in bovine chondrocytes: A possible explanation for oxidative stress generation. *Osteoarthr. Cartil.* **2008**, *16*, 756–763. [CrossRef] [PubMed]

36. McMahon, M.; Thomas, N.; Itoh, K.; Yamamoto, M.; Hayes, J.D. Redox-regulated turnover of Nrf2 is determined by at least two separate protein domains, the redox-sensitive Neh2 degron and the redox-insensitive Neh6 degron. *J. Biol. Chem.* **2004**, *279*, 31556–31567. [CrossRef] [PubMed]

37. Kensler, T.W.; Wakabayashi, N.; Biswal, S. Cell survival responses to environmental stresses via the Keap1-Nrf2-ARE pathway. *Annu. Rev. Pharmacol. Toxicol.* **2007**, *47*, 89–116. [CrossRef] [PubMed]

38. Poulet, B.; Staines, K.A. New developments in osteoarthritis and cartilage biology. *Curr. Opin. Pharmacol.* **2016**, *28*, 8–13. [CrossRef] [PubMed]

39. Hinoi, E.; Takarada, T.; Fujimori, S.; Wang, L.; Iemata, M.; Uno, K.; Yoneda, Y. Nuclear factor E2 p45-related factor 2 negatively regulates chondrogenesis. *Bone* **2007**, *40*, 337–344. [CrossRef] [PubMed]

40. Cai, D.; Yin, S.; Yang, J.; Jiang, Q.; Cao, W. Histone deacetylase inhibition activates Nrf2 and protects against osteoarthritis. *Arthritis Res. Ther.* **2015**, *17*, 269. [CrossRef] [PubMed]

41. Bernard, K.; Logsdon, N.J.; Miguel, V.; Benavides, G.A.; Zhang, J.; Carter, A.B.; Darley-Usmar, V.M.; Thannickal, V.J. NADPH oxidase 4 (Nox4) suppresses mitochondrial biogenesis and bioenergetics in lung fibroblasts via a nuclear factor erythroid-derived 2-like 2 (Nrf2)-dependent pathway. *J. Biol. Chem.* **2017**, *292*, 3029–3038. [CrossRef] [PubMed]

42. Calabrese, V.; Cornelius, C.; Dinkova-Kostova, A.T.; Iavicoli, I.; Di Paola, R.; Koverech, A.; Cuzzocrea, S.; Rizzarelli, E.; Calabrese, E.J. Cellular stress responses, hormetic phytochemicals and vitagenes in aging and longevity. *Biochim. Biophys. Acta* **2012**, *1822*, 753–783. [CrossRef] [PubMed]

43. Smith, E.J.; Shay, K.P.; Thomas, N.O.; Butler, J.A.; Finlay, L.F.; Hagen, T.M. Age-related loss of hepatic Nrf2 protein homeostasis: Potential role for heightened expression of miR-146a. *Free Radic. Biol. Med.* **2015**, *89*, 1184–1191. [CrossRef] [PubMed]

44. Li, L.; Chen, X.P.; Li, Y.J. MicroRNA-146a and human disease. *Scand. J. Immunol.* **2010**, *71*, 227–231. [CrossRef] [PubMed]

45. Yamasaki, K.; Nakasa, T.; Miyaki, S.; Ishikawa, M.; Deie, M.; Adachi, N.; Yasunaga, Y.; Asahara, H.; Ochi, M. Expression of microRNA-146a in osteoarthritis cartilage. *Arthritis Rheumatol.* **2009**, *60*, 1035–1041. [CrossRef] [PubMed]

46. Cheleschi, S.; De Palma, A.; Pecorelli, A.; Pascarelli, N.A.; Valacchi, G.; Belmonte, G.; Carta, S.; Galeazzi, M.; Fioravanti, A. Hydrostatic Pressure Regulates MicroRNA Expression Levels in Osteoarthritic Chondrocyte Cultures via the Wnt/β-Catenin Pathway. *Int. J. Mol. Sci.* **2017**, *18*, 133. [CrossRef] [PubMed]

47. Ji, G.; Lv, K.; Chen, H.; Wang, T.; Wang, Y.; Zhao, D.; Qu, L.; Li, Y. MiR-146a regulates SOD2 expression in H_2O_2 stimulated PC12 cells. *PLoS ONE* **2013**, *8*, e69351. [CrossRef] [PubMed]

48. Vasa-Nicotera, M.; Chen, H.; Tucci, P.; Yang, A.L.; Saintigny, G.; Menghini, R.; Mahè, C.; Agostini, M.; Knight, R.A.; Melino, G.; et al. miR-146a is modulated in human endothelial cell with aging. *Atherosclerosis* **2011**, *217*, 326–330. [CrossRef] [PubMed]

49. Yan, S.; Wang, M.; Zhao, J.; Zhang, H.; Zhou, C.; Jin, L.; Zhang, Y.; Qiu, X.; Ma, B.; Fan, Q. MicroRNA-34a affects chondrocyte apoptosis and proliferation by targeting the SIRT1/p53 signaling pathway during the pathogenesis of osteoarthritis. *Int. J. Mol. Med.* **2016**, *38*, 201–209. [CrossRef] [PubMed]

50. Baker, J.R.; Vuppusetty, C.; Colley, T.; Papaioannou, A.I.; Fenwick, P.; Donnelly, L.; Ito, K.; Barne, P.J. Oxidative stress dependent microRNA-34a activation via PI3Kα reduces the expression of sirtuin-1 and sirtuin-6 in epithelial cells. *Sci. Rep.* **2016**, *6*, 35871. [CrossRef] [PubMed]

51. Bai, X.Y.; Ma, Y.; Ding, R.; Fu, B.; Shi, S.; Chen, X.M. MiR-335 and miR-34a promote renal senescence by suppressing mitochondrial antioxidative enzymes. *J. Am. Soc. Nephrol.* **2011**, *22*, 1252–1261. [CrossRef] [PubMed]

52. Liu, Q.; Wang, K.; Shao, J.; Li, C.; Li, Y.; Li, S.; Liu, X.; Han, L. Role of taurine in BDE 209-induced oxidative stress in PC12 cells. *Adv. Exp. Med. Biol.* **2017**, *975*, 897–906. [PubMed]

53. Shang, L.; Qin, J.; Chen, L.B.; Liu, B.X.; Jacques, M.; Wang, H. Effects of sodium ferulate on human osteoarthritic chondrocytes and osteoarthritis in rats. *Clin. Exp. Pharmacol. Physiol.* **2009**, *36*, 912–918. [CrossRef] [PubMed]

54. Bhatti, F.U.; Mehmood, A.; Wajid, N.; Rauf, M.; Khan, S.N.; Riazuddin, S. Vitamin E protects chondrocytes against hydrogen peroxide-induced oxidative stress in vitro. *Inflamm. Res.* **2013**, *62*, 781–789. [CrossRef] [PubMed]

55. Kim, Y.S.; Kim, E.K.; Hwang, J.W.; Kim, J.S.; Shin, W.B.; Dong, X.; Nawarathna, W.P.A.S.; Moon, S.H.; Jeon, B.T.; Park, P.J. Neuroprotective effect of Taurine-rich cuttlefish (*Sepia officinalis*) extract against hydrogen peroxide-induced oxidative stress in SH-SY5Y cells. *Adv. Exp. Med. Biol.* **2017**, *975*, 243–254. [PubMed]

56. Oliveira, M.W.; Minotto, J.B.; de Oliveira, M.R.; Zanotto-Filho, A.; Behr, G.A.; Rocha, R.F.; Moreira, J.C.; Klamt, F. Scavenging and antioxidant potential of physiological taurine concentrations against different reactive oxygen/nitrogen species. *Pharmacol. Rep.* **2010**, *62*, 185–193. [CrossRef]

57. Abusarah, J.; Benabdoune, H.; Shi, Q.; Lussier, B.; Martel-Pelletier, J.; Malo, M.; Fernandes, J.C.; de Souza, F.P.; Fahmi, H.; Benderdour, M. Elucidating the role of protandim and 6-gingerol in protection against osteoarthritis. *J. Cell. Biochem.* **2017**, *118*, 1003–1013. [CrossRef] [PubMed]

58. Khan, N.M.; Haseeb, A.; Ansari, M.Y.; Devarapalli, P.; Haynie, S.; Haqqi, T.M. Wogonin, a plant derived small molecule, exerts potent anti-inflammatory and chondroprotective effects through the activation of ROS/ERK/Nrf2 signaling pathways in human Osteoarthritis chondrocytes. *Free Radic. Biol. Med.* **2017**, *106*, 288–301. [CrossRef] [PubMed]

59. Raitoharju, E.; Lyytikäinen, L.P.; Levula, M.; Oksala, N.; Mennander, A.; Tarkka, M.; Klopp, N.; Illig, T.; Kähönen, M.; Karhunen, P.J.; et al. MiR-21, miR-210, miR-34a, and miR-146a/b are up-regulated in human atherosclerotic plaques in the Tampere Vascular Study. *Atherosclerosis* **2011**, *219*, 211–217. [CrossRef] [PubMed]

60. Zhang, H.; Zhao, Z.; Pang, X.; Yang, J.; Yu, H.; Zhang, Y.; Zhou, H.; Zhao, J. MiR-34a/sirtuin-1/foxo3a is involved in genistein protecting against ox-LDL-induced oxidative damage in HUVECs. *Toxicol. Lett.* **2017**, *277*, 115–122. [CrossRef] [PubMed]

61. Altman, R.; Alarcon, G.; Appelrouth, D.; Bloch, D.; Borestein, D.; Brandt, K.; Brown, C.; Cooke, T.D.; Daniel, W.; Feldman, D.; et al. The American College of Rheumatology criteria for the classification and reporting of osteoarthritis of the hip. *Arthritis Rheumatol.* **1991**, *34*, 505–514. [CrossRef]

62. Mankin, H.J.; Dorfman, H.; Lippiello, L.; Zarins, A. Biochemical and metabolic abnormalities in articular cartilage from osteo-arthritic human hips II. Correlation of morphology with biochemical and metabolic data. *J. Bone Jt. Surg. Am.* **1971**, *53*, 523–537. [CrossRef]

63. Fioravanti, A.; Lamboglia, A.; Pascarelli, N.A.; Cheleschi, S.; Manica, P.; Galeazzi, M.; Collodel, G. Thermal water of Vetriolo, Trentino, inhibits the negative effect of interleukin-1β; on nitric oxide production and apoptosis in human osteoarthritic chondrocyte. *J. Biol. Regul. Homeost. Agents* **2013**, *27*, 891–902. [PubMed]

64. Ramakers, C.; Ruijter, J.M.; Deprez, R.H.; Moorman, A.F. Assumption-free analysis of quantitative real-time polymerase chain reaction (PCR) data. *Neurosci. Lett.* **2003**, *339*, 62–66. [CrossRef]

65. Pfaffl, M.W. A new mathematical model for relative quantification in real RT-PCR. *Nucleic Acid Res.* **2001**, *29*, e45. [CrossRef] [PubMed]

66. Vandesompele, J.; de Preter, K.; Pattyn, F.; Poppe, B.; van Roy, N.; de Paepe, A.; Speleman, F. Accurate normalization of real-time quantitative RT-PCR data by geometric averaging of multiple internal control genes. *Genome Biol.* **2002**, *3*, research0034.1. [CrossRef] [PubMed]

2

Melatonin Protects Cholangiocytes from Oxidative Stress-Induced Proapoptotic and Proinflammatory Stimuli via miR-132 and miR-34

Ewa Ostrycharz [1,†], Urszula Wasik [1,†], Agnieszka Kempinska-Podhorodecka [1,*](ID),
Jesus M. Banales [2], Piotr Milkiewicz [3,4] and Malgorzata Milkiewicz [1](ID)

[1] Department of Medical Biology, Pomeranian Medical University, 71-111 Szczecin, Poland;
  e.ostrycharz@wp.pl (E.O.); wasikula@gmail.com (U.W.); milkiewm@pum.edu.pl (M.M.)

[2] Department of Liver and Gastrointestinal Diseases, Biodonostia Health Research Institute-Donostia
  University Hospital-Ikerbasque, CIBERehd, University of the Basque Country (UPV/EHU),
  20014 San Sebastian, Spain; jmbanales@unav.es

[3] Translational Medicine Group, Pomeranian Medical University, 71-210 Szczecin, Poland;
  p.milkiewicz@wp.pl

[4] Liver and Internal Medicine Unit, Medical University of Warsaw, 02-097 Warsaw, Poland

* Correspondence: agnieszkakempinska@interia.eu

† These authors contributed equally to this work.

Abstract: Biosynthesis of melatonin by cholangiocytes is essential for maintaining the function of biliary epithelium. However, this cytoprotective mechanism appears to be impaired in primary biliary cholangitis (PBC). MiR-132 has emerged as a mediator of inflammation in chronic liver diseases. The effect of melatonin on oxidative stress and bile acid-induced apoptosis was also examined in cholangiocyes overexpressing miR506, as a PBC-like cellular model. In PBC patients the serum levels of melatonin were found increased in comparison to healthy controls. Whereas, in cholangiocytes within cirrhotic PBC livers the melatonin biosynthetic pathway was substantially suppressed even though the expressions of melatonin rate-limiting enzyme aralkylamine N-acetyltransferase (AANAT), and CK-19 (marker of cholangiocytes) were enhanced. In cholangiocytes exposed to mitochondrial oxidative stress melatonin decreased the expression of proapoptotic stimuli (PTEN, Bax, miR-34), which was accompanied by the inhibition of a pivotal mediator of inflammatory response Nf-κB-p65 and the activation of antiapoptotic signaling (miR-132, Bcl2). Similarly, melatonin reduced bile acid-induced proapoptotic caspase 3 and Bim levels. In summary, the insufficient hepatic expression of melatonin in PBC patients may predispose cholangiocytes to oxidative stress-related damage. Melatonin, via epigenetic modulation, was able to suppress NF-κB signaling activation and protect against biliary cells apoptotic signaling.

Keywords: melatonin; primary biliary cholangitis; micro RNA; oxidative stress; apoptosis

1. Introduction

Primary biliary cholangitis (PBC) is a chronic cholestatic condition characterized by the presence of antimitochondrial autoantibodies (AMAs) and the accumulation of antigen-specific autoreactive B and T cells followed by the secretion of cytokines and biliary epithelial cell destruction [1]. The abnormal immune response during the development of PBC leads to the activation and expansion of autoreactive T and B lymphocytes followed by the production of numerous inflammatory mediators [2,3]. The ongoing inflammation in PBC patients is accompanied by a cascade of destructive events including oxidative stress followed by cholangiocyte apoptosis [4,5]. The cytoprotective mechanism against free radicals appears to be impaired in PBC. This is partially due to the decreased expression of nuclear factor-erythroid 2-related factor 2 (Nrf-2), a transcription factor that triggers an antioxidative response with the concurrent overexpression of miR-34 and miR-132, which are known to directly target the Nrf-2 gene [6].

MiR-34 and miR-132 belong to small, highly conserved, short, non-coding RNAs (microRNAs) and have pleiotropic effects on cellular homeostasis. Injured hepatocytes release reactive oxygen species (ROS) and inflammatory mediators, and miR-132 has emerged as a mediator of inflammation in chronic liver diseases. The upregulation of miR-132 has been described for the whole liver as well as in Kupffer cells isolated from mice following chronic ethanol administration [7] or bile duct ligation [8]. Similarly, miR-34 expression has been found to increase in fibrotic liver diseases in both mice and humans [6–9]. Moreover, miR-34 and miR-132 are involved in the regulation of apoptosis. MiR-132 protects against apoptosis via the downregulation of phosphatase and tensin homolog (PTEN), a well-recognized inductor of apoptosis [10,11]. In contrast, miR-34 is known to trigger apoptosis via the translational repression of either SIRT1, a deacetylase that inhibits several proapoptotic proteins including p53, or the antiapoptotic protein B-cell lymphoma-2 (Bcl-2) [12,13].

Melatonin (N-acetyl-5-methoxytryptamine) is a hormone primarily produced by the pineal gland, and its endocrine actions include regulating circadian rhythms and maintaining tissue homeostasis. Besides central nervous system melatonin is also produced in various cell types and peripheral tissues, including biliary epithelium, where it exerts paracrine and autocrine effects on cells [14–16]. Within cells, melatonin is present in different subcellular compartments, with especially high levels in the nucleus, cell membrane, and mitochondria where free radicals are generated [17]. High concentrations of melatonin also are present in bile, as it is essential for maintaining the function of the biliary epithelium via the activation of the antioxidant response and modulation of apoptosis [18]. The biosynthesis of melatonin is tightly regulated by aralkylamine N-acetyltransferase (AANAT), which is expressed in cholangiocytes and, to a lesser extent, in hepatocytes. Melatonin is known to modulate the innate immune response via the modulation of cytokine production [19], and it inhibits TNF-alpha and IL-6 expression by decreasing the expression of nuclear factor-kappa B (NF-κB) subunits, p50, and p65 [20]. Furthermore, melatonin protects against oxidative-stress damage either directly by scavenging free radicals and decreasing NO production, or indirectly, via the activation of Nrf-2 and SIRT1 signaling pathways [21,22]. Additionally, accumulated evidence suggests that melatonin is an apoptosis modulator; however, this effect depends on cellular context [23–25].

The objective of this study was to explore the expression of melatonin and its rate-limiting enzyme AANAT in the serum and livers of PBC patients. Additionally, the therapeutic effects of melatonin on oxidative stress-induced apoptosis was analyzed in vitro in human cholangiocyte (H69 cells), as well as in cholangiocytes overexpressing miR-506 (H69 miR-506), which induce PBC-like features such as an enhanced free radical generation, predisposition to bile-salt-driven apoptosis, and alterations in mitochondrial functioning.

2. Materials and Methods

2.1. Serum Samples and Liver-Tissue Specimens

This study was performed in the Liver and Internal Medicine Unit, Medical University of Warsaw between 2016 and 2018. It includes both in- and outpatients diagnosed with PBC according to the European Association for the Study of the Liver [26] criteria. Serum samples from patients with PBC ($n = 84$) and healthy individuals ($n = 58$) were collected between 8.00 and 9.00 am and stored at $-80°C$. Non-cirrhotic liver specimens ($n = 22$) were obtained through percutaneous needle liver biopsy (and immersed in RNAlater solution) from PBC patients who underwent liver biopsies for histological assessment (early-stage PBC). Samples of cirrhotic liver tissue were collected from PBC patients ($n = 24$) with histologically diagnosed cirrhosis (cirrhotic PBC) who underwent liver transplantation at Queen Elizabeth Hospital in Birmingham (historical samples from 2000 to 2001) and Hospital of Medical University of Warsaw (between 2014 and 2015). Control liver tissues ($n = 22$) with no microscopic changes indicative of liver disease as identified by a pathologist, were secured from large, margin liver resections of colorectal metastases specimens, as already described [6,27,28]. Demographic and clinical data on analyzed patients are summarized in Table 1 Written, informed consent was obtained from each patient included in the study. The study protocol was approved by the Ethics Committee of the Pomeranian Medical University (BN-001/43/06) and adhered to the ethical guidelines of the 1975 Declaration of Helsinki (6th revision, 2008).

Table 1. Demographic and laboratory features of analyzed subjects.

Parameters	Liver			Serum	
	Controls ($n = 22$)	PBC Early Stages ($n = 22$)	PBC Cirrhotic ($n = 24$)	Controls ($n = 58$)	PBC ($n = 84$)
Gender (M/F)	4/18	0/22	3/21	3/55	4/80
Age (mean ± SD)	50 ± 4	52 ± 11	57 ± 8	50 ± 5	51 ± 9
Hb (mean ± SD, NR 12–16 g/dL)	13.5 ± 0.5	12.8 ± 0.6	10.0 ± 1.5	14.6 ± 0.9	11.6 ± 1.8
Bilirubin (mean ± SD, NR mg/dL)	0.7 ± 0.2	1.3 ± 0.8	6.9 ± 6.7	0.6 ± 0.3	2.4 ± 2.7
ALP (mean ± SD, NR 40–120 IU/L)	83 ± 18	265 ± 182	560 ± 383	78 ± 20	455 ± 284
ALT (mean ± SD, NR 5–35 IU/L)	24 ± 12	81 ± 16	78 ± 15	22 ± 13	114 ± 101
AST (mean ± SD, NR 5–35 IU/L)	22 ± 11	62 ± 56	233 ± 291	21 ± 11	99 ± 83
Albumin (mean ± SD, NR 3.8–4.2 g/dL)	4.4 ± 0.6	4.0 ± 0.4	3.6 ± 0.6	4.3 ± 0.5	3.9 ± 0.6

Abbreviations: PBC—primary biliary cholangitis; Hb—hemoglobin; ALP—alkaline phosphatase; ALT—alanine aminotransferase; AST—aspartate aminotransferase; NR—normal range.

2.2. Serum Samples and Liver-Tissue Specimens Quantitative Analyses of Melatonin Protein Expression

The concentration of melatonin was estimated is sera (Cloud-Clone Corp. #CEA908G), and liver tissue (Cusabio # CSB-E08132h) according to the manufacturers' protocols.

2.3. Cell Culture

H69 (non-tumor, SV40-immortalized, human cholangiocytes), H69-miR-506 (H69 cells with experimental constitutive overexpression of miR-506) [29], and normal human cholangiocytes (NHC) [30] were grown in media containing DMEM/F12 supplemented with fetal bovine serum, penicillin/streptomycin, vitamin solution, MEM solution, CD lipid concentrate, L-glutamine, soybean trypsin inhibitor, insulin/transferrin/selenium-A, bovine pituitary extract, epidermal growth factor, 3, 3′5-triiodo-L-thyronine, dexamethasone, and forskolin. The medium for H69-miR-506 was additionally supplemented with blasticidin (Invitrogen) for miR-506 positive selection52. To induce oxidative stress, cells were incubated with either 30 μM tert-Butylhydroquinone (tBHQ) or 250 μM H_2O_2. The effect of melatonin (1 mM) on microRNA levels were assessed after 16 h of incubation. Apoptosis was induced using 100 μM sodium glycochenodeoxycholate (GCDCA), and the protective effect of melatonin (500 μM) was evaluated based on the expression of apoptotic markers, i.e.; cleaved caspase 3 (CC3) and Bim. After 24 h of incubation, proteins were isolated from the cells and the levels of CC3 and Bim was assessed via immunoblot. Experiments were repeated four times.

2.4. MicroRNA and mRNA Extraction and Quantification

Total RNA was isolated from patient livers using the RNeasy Mini kit (Qiagen, Germantown, MD, USA), according to the manufacturer's protocol. cDNA synthesis was carried out using the

SuperscriptTM II RT kit (Invitrogen, Thermo Fisher Scientific) according to the manufacturer's protocol. The transcripts of AANAT, TNF-α, PTEN, Bax, Bcl-2, p65, Nrf-2, and human 18sRNA were measured using Gene Expression Assays and the 7500 Fast Real-Time PCR System (Applied Biosystems, Thermo Fisher Scientific). Briefly, each assay comprised a 20 μL reaction mixture that contained 10 μL TaqMan®Gene Expression PCR Master Mix (Applied Biosystems), 2 μL diluted cDNA template, and 1 μL of the probe/primer assay mix.

For microRNA quantification, total RNA was isolated using the miRNeasy Mini Kit (Qiagen), and cDNA was synthesized using the TaqMan *Advanced miRNA cDNA Synthesis Kit* (Applied Biosystems) according to the manufacturer's protocol. The levels of miR-34 and miR-132 (along with miR-191-5p, which was used as a endogenous control) were measured using TaqMan®Advanced miRNA Assays (Applied Biosystems). Fluorescence data were analyzed using 7500 Software v2.0.2. (Applied Biosystems), and the expression of microRNA and target genes were calculated using the $\Delta\Delta$Ct method of relative quantification.

2.5. Protein Expression Analysis

Proteins from cirrhotic PBC liver tissues and NHC cells were extracted via homogenization in an ice-cold RIPA buffer (50 mM Tris-HCl pH = 8, 150 mM NaCl, 1% NP-40, 0.5% NaDOC, 0.1% SDS, 1 mM EDTA, 100 mM PMSF, and 100 mM NaF), which contained a protease inhibitor cocktail and PhosSTOP (Roche Diagnostics GmbH). Proteins were quantified using a bicinchoninic acid assay (Micro BCA™ Protein Assay Kit; Thermo Scientific). Protein extracts (100 μg) from each liver sample were electrophoresed on SDS polyacrylamide gels and subsequently blotted onto PVDF membranes (Thermo Scientific) under semi-dry transfer conditions. After blocking for 1 h at room temperature in TBST containing 5% (*w/v*) milk (Merck, Gernsheim, Germany), the membranes were probed with the following primary antibodies: anti-AANAT (1:500 dilution; Santa Cruz Biotechnology, Santa Cruz, CA, USA), anti-TNFα (1:1000 dilution; Santa Cruz Biotechnology), or anti-glyceraldehyde-3-phosphate dehydrogenase (GAPDH) (1:5000 dilution; Santa Cruz Biotechnology), anti-CC3 (1:500 dilution; Cell Signaling Technology, Danvers, MA, USA), and anti-Bim (1:1000 dilution; Cell Signaling Technology). For the detection of antigen-antibody complexes, a peroxidase-conjugated anti-rabbit secondary antibody (1:5000 dilution; AmershamTM, GE Healthcare, Waukesha, WI, USA) or an anti-mouse secondary antibody (1:50,000 dilution; Jackson ImmunoResearch, West Grove, PA, USA) was used. Protein expression was detected using an enhanced chemiluminescence detection system (Chemiluminescent HRP Substrate, Millipore). Bands were visualized and quantified using the MicroChemi 2.0 System and GelQuant software (DNR Bio-Imaging Systems Ltd., Neve Yamin, Israel).

2.6. Immunohistochemistry

Frozen liver sections (6 μm) derived from patients with PBC and controls were fixed in a methanol and acetone mixture (1:1) at -20 °C for 5 min. The cut order of sections was known and described. We examined the proteins of interest through immunohistochemistry analyses. Briefly, sections were treated with Avidin/Biotin Blocking Kit (#SP-2001; Vector Laboratories, USA) and then incubated in 3% H_2O_2 diluted in methanol. Sections were probed with rabbit anti-AANAT (1:500 dilution; Santa Cruz Biotechnology), anti-TNFα (1:1000 dilution; Santa Cruz Biotechnology), and anti-CK19 (1:50 dilution; Santa Cruz Biotechnology). Next, sections were incubated with either biotinylated anti-mouse/anti-rabbit IgG (Vector Laboratories) or biotinylated anti-goat IgG (Vector Laboratories). Reactions were visualized using ABC Vectastain and DAB kits (DAKO, Agilent Technologies, Denmark). The negative controls, for which the primary antibodies were omitted, were included in the analysis and uniformly demonstrated no reaction.

2.7. Statistics

Data were evaluated as mean ± standard error (SE) for continuous variables and further analyzed using (one-way/two-way) ANOVAs. A p-value < 0.05 was considered statistically significant.

Stat-View-5 Software (SAS Institute, USA) was used for all analyses. Demographic and laboratory features of analyzed subjects in Table 1 are presented as mean ± standard deviation (SD).

3. Results

3.1. Melatonin Concentration is inCreased in the Serum of Patients with PBC, But Reduced in the Liver

The serum concentration of melatonin was 7.6-fold higher in patients with PBC than in controls ($p < 0.001$); however, the expression of this hormone in the livers of PBC patients was significantly reduced (90% reduction, $p < 0.001$ vs. controls, Figure 1A). In this regard, the mRNA expression levels of the melatonin rate-limiting enzyme AANAT were not altered between cirrhotic livers from PBC patients compared to non-cirrhotic PBC or healthy controls, whereas its protein levels were found significantly increased in the liver of patients with PBC vs. healthy controls (1.4-fold; $p = 0.002$ vs. controls; Figure 1B). On the other hand, the protein levels of CK19, a marker of cholangiocytes and hepatocytes transformed into cholangiocytes in advanced fibrosis, was elevated in the liver tissue of patients with PBC when compared to healthy controls (3.5-fold, $p < 0.001$ vs. controls; Figure 1C), indicating increased ductular reaction. Similarly, the levels of the TNFα protein was substantially increased in cirrhotic PBC livers (3-fold, $p < 0.001$ vs. controls; Figure 1D).

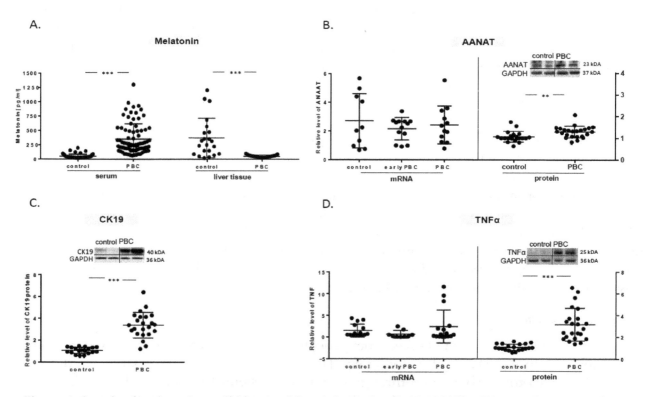

Figure 1. Levels of melatonin, aralkylamine N-acetyltransferase (AANAT), CK19, and TNFα in the livers of patients with primary biliary cholangitis (PBC). (**A**) Serum concentration of melatonin was increased in PBC patients, in contrast to liver tissue where the concentration of this hormone was decreased in comparison to controls. (**B**) The level of AANAT protein was enhanced, whereas the expression of mRNA was not change in either early-stage PBC or cirrhotic PBC. (**C**) The level of the CK19 protein was increased in the livers of PBC patients in comparison to controls. (**D**) The TNFα protein level was upregulated in PBC, while no changes in mRNA expression were observed in early-stage PBC or cirrhotic PBC. Melatonin concentration was evaluated using an ELISA assay, while AANAT, CK19, and TNFα mRNA and protein levels were evaluated using real-time PCR and Western blot, respectively. Values are shown as mean ± SE. ** $p < 0.01$ and *** $p < 0.001$.

The presence of AANAT was confirmed in the bile ducts (Figure 2A), as well as in regenerative nodules (Figure 2B). TNFα-positive cells were on the perimeter of the nodules (Figure 2E), but was barely detected in the bile ducts (Figure 2F). Antibodies against the cholangiocyte marker cytokeratin 19 (CK19) allowed for the visualization of biliary epithelial cells within bile ducts (Figure 2C) and hepatocytes transformed into cholangiocytes (Figure 2D) in cirrhotic livers.

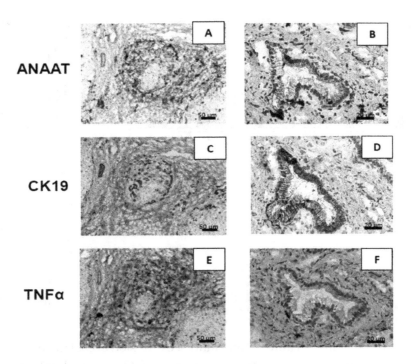

Figure 2. Hepatic expressions of aralkylamine N-acetyltransferase (AANAT), CK19, and TNFα protein in primary biliary cholangitis (PBC) patients. Positive immunohistochemical staining (dark brown) of AANAT, CK19, and TNFα in cirrhotic liver tissues were observed on the edges of regenerative nodules (**A,C,E**), and in bile ducts (**B,D,F**), respectively. (**A,B**) AANAT was ubiquitous in the liver tissue of PBC patients. (**C,D**) CK19 was localized in cholangiocytes in areas of ductular reactions. Bile ducts are marked by asterisks. (**E,F**) TNFα was localized primarily on the perimeter of nodules, but was absent in the cholangiocytes of the large bile duct.

3.2. Melatonin Modified the Expression of miR-132 and miR-34 in Cholangiocytes Subjected to Oxidative Stress

The expression of miR-132 in H69 human cholangiocytes in vitro was enhanced in response to the oxidative stress induced by both H_2O_2 (2.8-fold vs. controls; $p = 0.01$) and tBHQ (3.9-fold vs. controls, $p = 0.0004$). Yet, in H69-miR-506 cholangiocytes this response was not observed. Melatonin had a significant effect on miR-132 expression in H69 and H69 miR-506 cholangiocytes, but only when the oxidative stress was induced by tBHQ. Thus, in H69 cells cotreatment with melatonin and tBHQ further enhanced the miR-132 level (1.5-fold vs. tBHQ, $p = 0.01$, and 5.8-fold vs. controls, $p < 0.0001$). The event was different for H69-miR-506 cells, where the oxidative stress induced by tBHQ did not alter the miR-132 level. However, when the cells were coincubated with tBHQ and melatonin, the miR-132 expression was strongly upregulated in comparison to both controls and tBHQ-treated cells (3.9-fold, $p < 0.0001$, and 3.3-fold, $p < 0.001$, respectively, Figure 3A).

The miR-34 expression was only affected by H_2O_2-induced cellular oxidative stress. In H69 cells, melatonin further enhanced the H_2O_2-induced expression of miR-34 (3.3-fold vs. controls, $p = 0.02$; 6.5-fold vs. H_2O_2, $p = 0.007$; Figure 3B). Incubation of H69-miR-506 cells with H_2O_2 resulted in the induction of miR-34, which was significantly suppressed by melatonin (50% reduction, $p = 0.04$ vs. H_2O_2; Figure 3C). An analysis of the baseline expression of the microRNAs showed that in H69 cells with an experimental overexpression of miR506 the expression of miR-132 was 6-fold higher than in H69 cholangiocytes ($p = 0.04$; Figure 3B). However, there was no significant difference between those two cell lines when miR-34 expression was analyzed (Figure 3D).

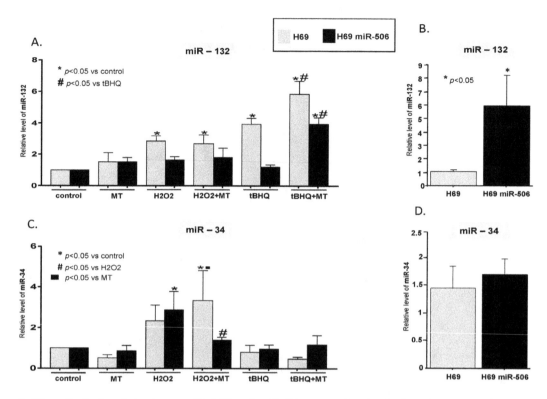

Figure 3. The effect of melatonin on miR-132 and miR-34 expression in cholangiocytes under conditions of oxidative stress. (**A**) Oxidative stress induced either by H_2O_2 or tert-Butylhydroquinone (tBHQ) enhanced the expression of miR-132 in H69 cells, whereas in H69miR-506 cells this response was only observed when the cells were treated simultaneously with tBHQ and melatonin. (**B**) The basal level of miR-132 expression was significantly increased in H69miR-506 cells in comparison to H69 cells. (**C**) H_2O_2 induced miR-34 expression in both H69 and H69miR-506 cells. Melatonin suppressed this enhancement, but only in H69miR-506 cells. (**D**) The baseline expression of miR-34 was comparable in H69miR-506 and H69 cells. Each experiment was repeated four times with similar results. MicroRNA were analyzed using real-time PCR, and the results were normalized to miR-191. Values are shown as mean ± SE.

3.3. In H69 miR-506 Cells Subjected to Oxidative Stress, Melatonin Modulated Apoptosis and the Expression of Transcription Factors Regulating Inflammation

In looking for the mechanism related to the protective role of melatonin against oxidative stress in cholangiocytes, we analyzed the interplay between melatonin and the expression of anti- and proapoptotic genes. Melatonin suppressed PTEN expression in tBHQ-treated H69-miR-506 cells ($p = 0.02$ vs. tBHQ; Figure 4A). Moreover, coincubation with melatonin and tBHQ prevented the induction of proapoptotic Bax expression (3.6-fold reduction vs. tBHQ-treated cells, $p = 0.002$, and a 20% reduction vs. controls $p = 0.03$; Figure 4B) in these cells. In contrast, melatonin had the opposite effect on the antiapoptotic protein Bcl-2 and upregulated the Bcl-2 expression in cells exposed to oxidative stress induced by tBHQ (2.9-fold increase vs. tBHQ, $p = 0.03$, and 2.7-fold increase vs. controls, $p = 0.03$, Figure 4C). Furthermore, melatonin affected the level of the Bcl-2 protein when oxidative stress was induced by H_2O_2 (2.5-fold induction vs. controls, $p = 0.04$, Figure 4D). We also observed that incubation with tBHQ enhanced the expression of p65, the subunit of NF-κB nuclear factor (2-fold vs. nontreated cells, $p = 0.008$), and Nrf-2, the marker of oxidative stress (3-fold vs. nontreated cells, $p = 0.002$), and that this upregulation was restrained when the cells were additionally treated with melatonin (80% reduction vs. tBHQ, $p = 0.02$; and 60% reduction vs. tBHQ, $p = 0.02$; respectively, Figure 4E,F).

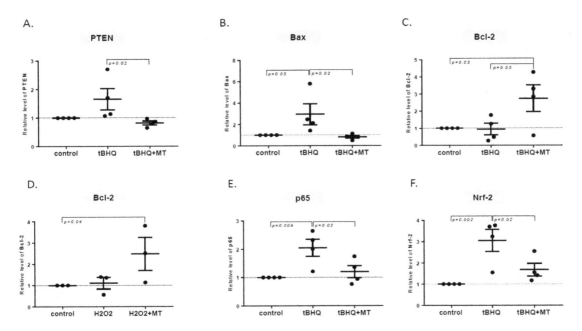

Figure 4. The effect of melatonin on apoptosis- and inflammation-related factors in H69-miR-506 cells subjected to oxidative stress. Melatonin inhibited the tert-Butylhydroquinone (tBHQ)-induced upregulation of proapoptotic factors (**A**) phosphatase and tensin homolog and (**B**) Bax, and induced expression of antiapoptotic factor B-cell lymphoma-2 in cells treated with (**C**) tBHQ and (**D**) H_2O_2. In cells subjected to mitochondrial oxidative stress, melatonin suppressed the expression of (**E**) p65, a subunit of nuclear factor-kappa B associated with inflammation, and (**F**) nuclear factor-erythroid 2-related factor 2, a marker of oxidative stress. Each experiment was repeated four times with similar results. All factors were estimated using real-time PCR. The results were normalized to 18S RNA. Values are shown as mean ± SE.

3.4. Melatonin Restrained Bile Salt-Induced Apoptosis of Cholangiocytes

Melatonin suppressed the protein level of CC3 in NHC cells incubated with GCDCA. GCDCA induced CC3 expression (1.6-fold, $p = 0.04$ vs. control; Figure 5A), which was inhibited by melatonin (2.7-fold reduction vs. GCDCA, $p = 0.007$). Similarly, the enhanced expression of the proapoptotic Bim protein was induced by GCDCA (1.76-fold, $p = 0.01$ vs. control), and melatonin cotreatment downregulated the Bim level in those cells (2.2-fold reduction vs. GCDCA, $p = 0.003$; Figure 5B).

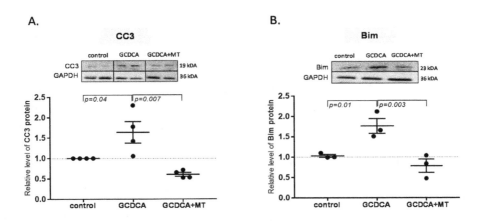

Figure 5. Melatonin protected against toxic bile acid-induced apoptosis in normal human cholangiocytes (NHCs). Melatonin suppressed the expression of sodium glycochenodeoxycholate (GCDCA)-induced (**A**) active form of caspase-3, cleaved caspase-3 (CC3), and (**B**) proapoptotic protein Bim. Protein levels were evaluated using Western blot and quantified relative to glyceraldehyde-3-phosphate dehydrogenase. Values are shown as mean ± SE.

4. Discussion

While the beneficial effects of melatonin using models of liver damage have been previously described, there has not, thus far, been experimental studies investigating the effect of melatonin within the context of cholangiopathy characterized by cholestasis, ductopenia, and biliary inflammation in primary biliary cholangitis.

The elevated serum concentration of melatonin in PBC patients suggests that the overproduction of this hormone in pineal sites is a compensatory mechanism to alleviate an exacerbated immune response and oxidative stress during the development of cholangitis. In contrast, the hepatic concentration of melatonin is substantially reduced in cirrhotic PBC. This implies that neither pineal-gland secreted melatonin is capable of providing sufficient protection of liver cells, nor is melatonin biosynthesis by cholangiocytes impaired. Consequently, melatonin does not elicit its endocrine and paracrine effects on cells, which are constantly subjected to inflammatory insults and free-radical generation. Moreover, in this study we showed that a low hepatic concentration of melatonin is not the result of an insufficient expression of the rate-limiting enzyme AANAT, as the protein level of this enzyme was significantly enhanced in PBC livers in comparison to controls. Furthermore, the presence of AANAT in pathologically altered liver cells, including cholangiocytes, was confirmed by immunohistochemical analysis. The function of AANAT is spatially and temporally regulated, and its activity strongly depends on post-translational modifications, notably phosphorylation [31,32]. The balance between phosphorylation and dephosphorylation of AANAT is maintained by cAMP and calcium ions, and both molecules are important in cholangiocyte biology. For instance, in PBC, cAMP in large cholangiocytes and calcium in small cholangiocytes are involved in cholangiocyte proliferation [33]. The homeostasis of both molecules may be altered due to the overexpression of miR-506, which has been reported in PBC [34]. Furthermore, miR-506 targets the type III inositol 1, 4, 5-trisphosphate receptor, and alterations of its level leads to the deterioration of calcium homeostasis [35]. Further studies are needed to investigate whether changes in the intracellular conditions in PBC cholangiocytes result in inappropriate post-translational modifications of AANAT and the inadequate biosynthesis of melatonin.

Melatonin has been demonstrated to ameliorate liver damage by decreasing oxidative stress, inflammatory responses, and biliary senescence [36,37]. In human cholangiopathies such as PBC and PSC, an initial balance between cholangiocyte apoptosis and compensatory cholangiocyte proliferation is followed by a failure in the proliferative capacity of cholangiocytes, and enhanced apoptosis favors evolution toward ductopenia. The downregulation of melatonin in the livers of PBC patients may significantly decrease the ability of cholangiocytes to counteract disease-related insults. Cholangiocytes that undergo chronic injury are more prone to apoptosis. We have previously shown that PBC is characterized by the enhanced expression of two oxidative-stress related microRNAs, namely miR-132 and miR-34, which are implicated in the modulation of apoptosis [6]. MicroRNA-132 is engaged in maintaining cellular and tissue homeostasis via the regulation of apoptosis, autophagy, and cell proliferation [11,38]. In H69 cells, the expression of miR-132 was induced following treatment with both H_2O_2 and tBHQ, and melatonin further enhanced its expression under tBHQ-induced oxidative stress. In contrast, in H69 miR-506 cells, neither H_2O_2 nor tBHQ induced oxidative stress was able to elevate the level of miR132. This may be due to the fact that the baseline expression of miR-132 was significantly higher in H69 miR-506 cells in comparison to H69 cells. It is worth mentioning that H69 miR-506 cells (a cellular model of PBC) [35] are characterized by chronic stress due to a stable overexpression of miR-506.

Intriguingly, melatonin-dependent, enhanced expression of miR-132 was observed only when cells were subjected to tBHQ, but not H_2O_2 incubation. The action of tBHQ is related to the mitochondria, an important site of a free-radical generation, and a key cellular compartment involved in apoptosis [39]. Melatonin acts as a free radical scavenger, and may directly inhibit the stress-mediated release of cytochrome c and, thus, block caspase activation and apoptosis by the activation of the MT1 receptor located in the outer mitochondrial membrane [40]. We undertook additional studies to demonstrate a causal link between melatonin and the development of apoptosis in response to oxidative stress. Melatonin substantially enhanced the expression of miR-132 in cholangiocytes exposed to tBHQ, which was accompanied by the downregulation of both PTEN and the proapoptotic Bax protein, along with the upregulation of the antiapoptotic protein Bcl-2. These observations indicate there is a role for melatonin in protecting against apoptosis induced by mitochondrial oxidative stress. Our study is in accordance with observations suggesting a positive effect of melatonin treatment by which the upregulation of miR-132 inhibited the PTEN-dependent proapoptotic pathway in primary cortical neurons [38]. Similarly, following ischemic neuronal injury, melatonin reverses the proapoptotic phenotype by blocking Bax activity and inducing the Bcl-2 protein [41]. The induction of miR-132-5p has been reported to upregulate the antiapoptotic protein Bcl-2 [42]. We also found that melatonin suppressed miR-34 in H_2O_2-treated cholangiocytes and was associated with the upregulation of antiapoptotic Bcl2. This beneficial effect of melatonin in suppressing miR-34 is important as this microRNA promotes apoptosis and senescence, inhibits proliferation, and leads to marked alterations in Bcl-2 and p53 expressions [43,44].

Our research also provides data showing the effects of melatonin treatment in protecting against GCDCA-induced apoptosis in human cholangiocytes. High levels of BA, including GCDCA at pathophysiological levels, have been shown to injure mitochondria by changing their membrane potential and releasing cytochrome c [45,46]. Here, we showed that melatonin, via reduced levels of cleaved caspase-3 and the proapoptotic protein Bim, may protect against the development of apoptosis.

The anti-inflammatory properties of melatonin have been extensively studied in animal and cellular models [20,47]; however, its role in cholestatic liver disease needs to be further elucidated. Experimental and clinical data have suggested that melatonin reduces chronic and acute inflammation by modulating serum inflammatory factors and through the inhibition of prostanoids, leukotrienes, and nitric oxide production along with the inhibition of nuclear factor-kappa B [48–50]. Our data demonstrate that melatonin suppressed NF-κB signaling activation by reducing p65 expression in tBHQ-oxidative stress induced in cholangiocytes with PBC-like phenotypes. Furthermore, in this cellular model of PBC, the marker of oxidative stress, Nrf-2, was downregulated following cotreatment with melatonin. Consistent with our observations are reports showing that the inhibition of NF-κB signaling contributes to melatonin alleviating inflammation, downregulating mtROS production [51], and preventing invasiveness in HepG2 liver cancer cells [52].

In conclusion, the results of this study suggest that decreased melatonin secretion in the livers of PBC patients is not due to an insufficient hepatic expression of the AANAT enzyme and interestingly, the biosynthesis of this enzyme was even increased. The inadequate expression of melatonin predisposes liver cells to immune- and oxidative stress-related damage. Our findings demonstrate the beneficial effects of melatonin supplementation in cholagiocytes with PBC-like phenotypes. Melatonin, via epigenetic modulation, was able to suppress NF-κB signaling activation and protect against apoptotic signaling induced by either oxidative stress or high concentrations of bile salt. In view of significantly higher serum level of melatonin in patients with PBC as compared to controls, simple oral supplementation of melatonin does not seem to address this problem at the tissue level. Delivery of melatonin to the liver using a carefully selected carrier such as bile acids could be a possible venue for further investigations.

Author Contributions: Conceptualization, M.M. and U.W.; methodology, U.W.; E.O.; A.K.-P.; software, E.O. and U.W.; validation, M.M.; and A.K.-P.; formal analysis, U.W.; E.O. and A.K.-P.; investigation, M.M. and U.W.; data curation, M.M.; writing—original draft preparation, E.O.; and U.W.; writing—review and editing, M.M.; visualization, E.O.; U.W. and A.K.-P.; supervision, P.M.; M.M.; J.M.B.; project administration, M.M.; funding acquisition, M.M. All authors have read and agreed to the published version of the manuscript.

References

1.   Hirschfield, G.M.; Gershwin, M.E. The immunobiology and pathophysiology of primary biliary cirrhosis. *Annu. Rev. Pathol.* **2013**, *8*, 303–330. [CrossRef]

2. Yang, C.Y.; Ma, X.; Tsuneyama, K.; Huang, S.; Takahashi, T.; Chalasani, N.P.; Bowlus, C.L.; Yang, G.X.; Leung, P.S.; Ansari, A.A.; et al. IL-12/Th1 and IL-23/Th17 biliary microenvironment in primary biliary cirrhosis: Implications for therapy. *Hepatology* **2014**, *59*, 1944–1953. [CrossRef]

3. Shimoda, S.; Tsuneyama, K.; Kikuchi, K.; Harada, K.; Nakanuma, Y.; Nakamura, M.; Ishibashi, H.; Hisamoto, S.; Niiro, H.; Leung, P.S.; et al. The role of natural killer (NK) and NK T cells in the loss of tolerance in murine primary biliary cirrhosis. *Clin. Exp. Immunol.* **2012**, *168*, 279–284. [CrossRef]

4. Bell, L.N.; Wulff, J.; Comerford, M.; Vuppalanchi, R.; Chalasani, N. Serum metabolic signatures of primary biliary cirrhosis and primary sclerosing cholangitis. *Liver Int.* **2015**, *35*, 263–274. [CrossRef]

5. Salunga, T.L.; Cui, Z.G.; Shimoda, S.; Zheng, H.C.; Nomoto, K.; Kondo, T.; Takano, Y.; Selmi, C.; Alpini, G.; Gershwin, M.E.; et al. Oxidative stress-induced apoptosis of bile duct cells in primary biliary cirrhosis. *J. Autoimmun.* **2007**, *29*, 78–86. [CrossRef]

6. Wasik, U.; Milkiewicz, M.; Kempinska-Podhorodecka, A.; Milkiewicz, P. Protection against oxidative stress mediated by the Nrf2/Keap1 axis is impaired in Primary Biliary Cholangitis. *Sci. Rep.* **2017**, *7*, 44769. [CrossRef]

7. Szabo, G.; Bala, S. MicroRNAs in liver disease. *Nat. Rev. Gastroenterol. Hepatol.* **2013**, *10*, 542–552. [CrossRef]

8. Mann, J.; Chu, D.C.; Maxwell, A.; Oakley, F.; Zhu, N.L.; Tsukamoto, H.; Mann, D.A. MeCP2 controls an epigenetic pathway that promotes myofibroblast transdifferentiation and fibrosis. *Gastroenterology* **2010**, *138*, 705–714. [CrossRef]

9. Pogribny, I.P.; Starlard-Davenport, A.; Tryndyak, V.P.; Han, T.; Ross, S.A.; Rusyn, I.; Beland, F.A. Difference in expression of hepatic microRNAs miR-29c, miR-34a, miR-155, and miR-200b is associated with strain-specific susceptibility to dietary nonalcoholic steatohepatitis in mice. *Lab Invest.* **2010**, *90*, 1437–1446. [CrossRef]

10. Tan, D.X.; Manchester, L.C.; Reiter, R.J.; Qi, W.B.; Zhang, M.; Weintraub, S.T.; Cabrera, J.; Sainz, R.M.; Mayo, J.C. Identification of highly elevated levels of melatonin in bone marrow: Its origin and significance. *Biochim. Biophys. Acta.* **1999**, *1472*, 206–214. [CrossRef]

11. Mziaut, H.; Henniger, G.; Ganss, K.; Hempel, S.; Wolk, S.; McChord, J.; Chowdhury, K.; Ravassard, P.; Knoch, K.P.; Krautz, C.; et al. MiR-132 controls pancreatic beta cell proliferation and survival through Pten/Akt/Foxo3 signaling. *Mol Metab.* **2020**, *31*, 150–162. [CrossRef]

12. Yamakuchi, M.; Lowenstein, C.J. MiR-34, SIRT1 and p53: The feedback loop. *Cell Cycle.* **2009**, *8*, 712–715. [CrossRef]

13. Hermeking, H. The miR-34 family in cancer and apoptosis. *Cell Death Differ.* **2010**, *17*, 193–199. [CrossRef]

14. Renzi, A.; DeMorrow, S.; Onori, P.; Carpino, G.; Mancinelli, R.; Meng, F.; Venter, J.; White, M.; Franchitto, A.; Francis, H.; et al. Modulation of the biliary expression of aralkylamine N-acetyltransferase alters the autocrine proliferative responses of cholangiocytes in rats. *Hepatology* **2013**, *57*, 1130–1141. [CrossRef]

15. Luchetti, F.; Canonico, B.; Betti, M.; Arcangeletti, M.; Pilolli, F.; Piroddi, M.; Canesi, L.; Papa, S.; Galli, F. Melatonin signaling and cell protection function. *FASEB J.* **2010**, *24*, 3603–3624. [CrossRef]

16. Han, Y.; Onori, P.; Meng, F.; DeMorrow, S.; Venter, J.; Francis, H.; Franchitto, A.; Ray, D.; Kennedy, L.; Greene, J.; et al. Prolonged exposure of cholestatic rats to complete dark inhibits biliary hyperplasia and liver fibrosis. *Am. J. Physiol Gastrointest. Liver Physiol.* **2014**, *307*, G894–G904. [CrossRef]

17. Venegas, C.; Garcia, J.A.; Escames, G.; Ortiz, F.; Lopez, A.; Doerrier, C.; Garcia-Corzo, L.; Lopez, L.C.; Reiter, R.J.; Acuna-Castroviejo, D. Extrapineal melatonin: Analysis of its subcellular distribution and daily fluctuations. *J. Pineal Res.* **2012**, *52*, 217–227. [CrossRef]

18. Tan, D.; Manchester, L.C.; Reiter, R.J.; Qi, W.; Hanes, M.A.; Farley, N.J. High physiological levels of melatonin in the bile of mammals. *Life Sci.* **1999**, *65*, 2523–2529. [CrossRef]

19. Carrillo-Vico, A.; Lardone, P.J.; Alvarez-Sanchez, N.; Rodriguez-Rodriguez, A.; Guerrero, J.M. Melatonin: Buffering the immune system. *Int. J. Mol Sci.* **2013**, *14*, 8638–8683. [CrossRef]

20. Cuesta, S.; Kireev, R.; Forman, K.; Garcia, C.; Escames, G.; Ariznavarreta, C.; Vara, E.; Tresguerres, J.A. Melatonin improves inflammation processes in liver of senescence-accelerated prone male mice (SAMP8). *Exp. Gerontol.* **2010**, *45*, 950–956. [CrossRef]

21. Singh, A.K.; Haldar, C. Melatonin modulates glucocorticoid receptor mediated inhibition of antioxidant response and apoptosis in peripheral blood mononuclear cells. *Mol. Cell Endocrinol.* **2016**, *436*, 59–67. [CrossRef]

22. Shi, S.; Lei, S.; Tang, C.; Wang, K.; Xia, Z. Melatonin attenuates acute kidney ischemia/reperfusion injury in diabetic rats by activation of the SIRT1/Nrf2/HO-1 signaling pathway. *Biosci. Rep.* **2019**, *39*. [CrossRef]

23. Martin-Renedo, J.; Mauriz, J.L.; Jorquera, F.; Ruiz-Andres, O.; Gonzalez, P.; Gonzalez-Gallego, J.

Melatonin induces cell cycle arrest and apoptosis in hepatocarcinoma HepG2 cell line. *J. Pineal Res.* **2008**, *45*, 532–540. [CrossRef]

24. Laothong, U.; Hiraku, Y.; Oikawa, S.; Intuyod, K.; Murata, M.; Pinlaor, S. Melatonin induces apoptosis in cholangiocarcinoma cell lines by activating the reactive oxygen species-mediated mitochondrial pathway. *Oncol. Rep.* **2015**, *33*, 1443–1449. [CrossRef]

25. Xu, G.; Zhao, J.; Liu, H.; Wang, J.; Lu, W. Melatonin Inhibits Apoptosis and Oxidative Stress of Mouse Leydig Cells via a SIRT1-Dependent Mechanism. *Molecules* **2019**, *24*, 3084. [CrossRef]

26. European Association for the Study of the Liver. EASL Clinical Practice Guidelines: The diagnosis and management of patients with primary biliary cholangitis. *J. Hepatol.* **2017**, *67*, 145–172. [CrossRef]

27. Kempinska-Podhorodecka, A.; Milkiewicz, M.; Wasik, U.; Ligocka, J.; Zawadzki, M.; Krawczyk, M.; Milkiewicz, P. Decreased Expression of Vitamin D Receptor Affects an Immune Response in Primary Biliary Cholangitis via the VDR-miRNA155-SOCS1 Pathway. *Int. J. Mol Sci.* **2017**, *18*, 289. [CrossRef]

28. Kopycinska, J.; Kempinska-Podhorodecka, A.; Haas, T.; Elias, E.; DePinho, R.A.; Paik, J.; Milkiewicz, P.; Milkiewicz, M. Activation of FoxO3a/Bim axis in patients with Primary Biliary Cirrhosis. *Liver Int.* **2013**, *33*, 231–238. [CrossRef]

29. Ananthanarayanan, M.; Banales, J.M.; Guerra, M.T.; Spirli, C.; Munoz-Garrido, P.; Mitchell-Richards, K.; Tafur, D.; Saez, E.; Nathanson, M.H. Post-translational regulation of the type III inositol 1, 4, 5-trisphosphate receptor by miRNA-506. *J. Biol. Chem.* **2015**, *290*, 184–196. [CrossRef]

30. Loarca, L.; De Assuncao, T.M.; Jalan-Sakrikar, N.; Bronk, S.; Krishnan, A.; Huang, B.; Morton, L.; Trussoni, C.; Bonilla, L.M.; Krueger, E.; et al. Development and characterization of cholangioids from normal and diseased human cholangiocytes as an in vitro model to study primary sclerosing cholangitis. *Lab Invest.* **2017**, *97*, 1385–1396. [CrossRef]

31. Choi, B.H.; Chae, H.D.; Park, T.J.; Oh, J.; Lim, J.; Kang, S.S.; Ha, H.; Kim, K.T. Protein kinase C regulates the activity and stability of serotonin N-acetyltransferase. *J. Neurochem.* **2004**, *90*, 442–454. [CrossRef]

32. Schomerus, C.; Korf, H.W. Mechanisms regulating melatonin synthesis in the mammalian pineal organ. *Ann. N. Y. Acad. Sci.* **2005**, *1057*, 372–383. [CrossRef]

33. Hall, C.; Sato, K.; Wu, N.; Zhou, T.; Kyritsi, K.; Meng, F.; Glaser, S.; Alpini, G. Regulators of Cholangiocyte Proliferation. *Gene Expr.* **2017**, *17*, 155–171. [CrossRef]

34. Banales, J.M.; Saez, E.; Uriz, M.; Sarvide, S.; Urribarri, A.D.; Splinter, P.; Tietz Bogert, P.S.; Bujanda, L.; Prieto, J.; Medina, J.F.; et al. Up-regulation of microRNA 506 leads to decreased Cl^-/HCO_3^- anion exchanger 2 expression in biliary epithelium of patients with primary biliary cirrhosis. *Hepatology* **2012**, *56*, 687–697. [CrossRef]

35. Erice, O.; Munoz-Garrido, P.; Vaquero, J.; Perugorria, M.J.; Fernandez-Barrena, M.G.; Saez, E.; Santos-Laso, A.; Arbelaiz, A.; Jimenez-Aguero, R.; Fernandez-Irigoyen, J.; et al. MicroRNA-506 promotes primary biliary cholangitis-like features in cholangiocytes and immune activation. *Hepatology* **2018**, *67*, 1420–1440. [CrossRef]

36. Hu, S.; Yin, S.; Jiang, X.; Huang, D.; Shen, G. Melatonin protects against alcoholic liver injury by attenuating oxidative stress, inflammatory response, and apoptosis. *Eur. J. Pharmacol.* **2009**, *616*, 287–292. [CrossRef]

37. Wu, N.; Meng, F.; Zhou, T.; Han, Y.; Kennedy, L.; Venter, J.; Francis, H.; DeMorrow, S.; Onori, P.; Invernizzi, P.; et al. Prolonged darkness reduces liver fibrosis in a mouse model of primary sclerosing cholangitis by miR-200b down-regulation. *FASEB J.* **2017**, *31*, 4305–4324. [CrossRef]

38. Zhao, Y.; Zhao, R.; Wu, J.; Wang, Q.; Pang, K.; Shi, Q.; Gao, Q.; Hu, Y.; Dong, X.; Zhang, J.; et al. Melatonin protects against Abeta-induced neurotoxicity in primary neurons via miR-132/PTEN/AKT/FOXO3a pathway. *Biofactors* **2018**, *44*, 609–618. [CrossRef]

39. Imhoff, B.R.; Hansen, J.M. Tert-butylhydroquinone induces mitochondrial oxidative stress causing Nrf2 activation. *Cell Biol. Toxicol.* **2010**, *26*, 541–551. [CrossRef]

40. Suofu, Y.; Li, W.; Jean-Alphonse, F.G.; Jia, J.; Khattar, N.K.; Li, J.; Baranov, S.V.; Leronni, D.; Mihalik, A.C.; He, Y.; et al. Dual role of mitochondria in producing melatonin and driving GPCR signaling to block cytochrome c release. *Proc. Natl. Acad. Sci. USA* **2017**, *114*, E7997–E8006. [CrossRef]

41. Sun, F.Y.; Lin, X.; Mao, L.Z.; Ge, W.H.; Zhang, L.M.; Huang, Y.L.; Gu, J. Neuroprotection by melatonin against ischemic neuronal injury associated with modulation of DNA damage and repair in the rat following a transient cerebral ischemia. *J. Pineal Res.* **2002**, *33*, 48–56. [CrossRef]

42. Shrestha, T.; Takahashi, T.; Li, C.; Matsumoto, M.; Maruyama, H. Nicotine-induced upregulation of

miR-132-5p enhances cell survival in PC12 cells by targeting the anti-apoptotic protein Bcl-2. *Neurol. Res.* **2020**, *42*, 405–414. [CrossRef]

43. Chen, F.; Hu, S.J. Effect of microRNA-34a in cell cycle, differentiation, and apoptosis: A review. *J. Biochem. Mol. Toxicol.* **2012**, *26*, 79–86. [CrossRef]

44. Lin, Q.; Mao, Y.; Song, Y.; Huang, D. MicroRNA34a induces apoptosis in PC12 cells by reducing Bcell lymphoma 2 and sirtuin1 expression. *Mol. Med. Rep.* **2015**, *12*, 5709–5714. [CrossRef]

45. Cai, S.Y.; Ouyang, X.; Chen, Y.; Soroka, C.J.; Wang, J.; Mennone, A.; Wang, Y.; Mehal, W.Z.; Jain, D.; Boyer, J.L. Bile acids initiate cholestatic liver injury by triggering a hepatocyte-specific inflammatory response. *JCI. Insight.* **2017**, *2*, e90780. [CrossRef]

46. Botla, R.; Spivey, J.R.; Aguilar, H.; Bronk, S.F.; Gores, G.J. Ursodeoxycholate (UDCA) inhibits the mitochondrial membrane permeability transition induced by glycochenodeoxycholate: A mechanism of UDCA cytoprotection. *J. Pharmacol. Exp. Ther.* **1995**, *272*, 930–938.

47. Liu, Z.; Gan, L.; Xu, Y.; Luo, D.; Ren, Q.; Wu, S.; Sun, C. Melatonin alleviates inflammasome-induced pyroptosis through inhibiting NF-kappaB/GSDMD signal in mice adipose tissue. *J. Pineal Res.* **2017**, *63*, e12414. [CrossRef]

48. Garcia, J.A.; Volt, H.; Venegas, C.; Doerrier, C.; Escames, G.; Lopez, L.C.; Acuna-Castroviejo, D. Disruption of the NF-kappaB/NLRP3 connection by melatonin requires retinoid-related orphan receptor-alpha and blocks the septic response in mice. *FASEB J.* **2015**, *29*, 3863–3875. [CrossRef]

49. Rahim, I.; Djerdjouri, B.; Sayed, R.K.; Fernandez-Ortiz, M.; Fernandez-Gil, B.; Hidalgo-Gutierrez, A.; Lopez, L.C.; Escames, G.; Reiter, R.J.; Acuna-Castroviejo, D. Melatonin administration to wild-type mice and nontreated NLRP3 mutant mice share similar inhibition of the inflammatory response during sepsis. *J. Pineal Res.* **2017**, *63*, e12410. [CrossRef]

50. Mauriz, J.L.; Collado, P.S.; Veneroso, C.; Reiter, R.J.; Gonzalez-Gallego, J. A review of the molecular aspects of melatonin's anti-inflammatory actions: Recent insights and new perspectives. *J. Pineal Res.* **2013**, *54*, 1–14. [CrossRef]

51. Chen, F.; Jiang, G.; Liu, H.; Li, Z.; Pei, Y.; Wang, H.; Pan, H.; Cui, H.; Long, J.; Wang, J.; et al. Melatonin alleviates intervertebral disc degeneration by disrupting the IL-1beta/NF-kappaB-NLRP3 inflammasome positive feedback loop. *Bone Res.* **2020**, *8*, 10. [CrossRef]

52. Ordonez, R.; Carbajo-Pescador, S.; Prieto-Dominguez, N.; Garcia-Palomo, A.; Gonzalez-Gallego, J.; Mauriz, J.L. Inhibition of matrix metalloproteinase-9 and nuclear factor kappa B contribute to melatonin prevention of motility and invasiveness in HepG2 liver cancer cells. *J. Pineal Res.* **2014**, *56*, 20–30. [CrossRef]

3

The Role of MicroRNAs in Diabetes-Related Oxidative Stress

Mirza Muhammad Fahd Qadir [1,2,†], Dagmar Klein [1,†], Silvia Álvarez-Cubela [1], Juan Domínguez-Bendala [1,2,3,*] and Ricardo Luis Pastori [1,4,*]

[1]   Diabetes Research Institute, University of Miami Miller School of Medicine, Miami, FL 33136, USA; fahd.qadir@med.miami.edu (M.M.F.Q.); dklein@med.miami.edu (D.K.); salvarez@med.miami.edu (S.Á.-C.)

[2]   Department of Cell Biology and Anatomy, University of Miami Miller School of Medicine, Miami, FL 33136, USA

[3]   Department of Surgery, University of Miami Miller School of Medicine, Miami, FL 33136, USA

[4]   Department of Medicine, Division of Metabolism, Endocrinology and Diabetes, University of Miami Miller School of Medicine, Miami, FL 33136, USA

*   Correspondence: jdominguez2@med.miami.edu (J.D.-B.); rpastori@med.miami.edu (R.L.P.)

†   These authors contributed equally to this work.

Abstract: Cellular stress, combined with dysfunctional, inadequate mitochondrial phosphorylation, produces an excessive amount of reactive oxygen species (ROS) and an increased level of ROS in cells, which leads to oxidation and subsequent cellular damage. Because of its cell damaging action, an association between anomalous ROS production and disease such as Type 1 (T1D) and Type 2 (T2D) diabetes, as well as their complications, has been well established. However, there is a lack of understanding about genome-driven responses to ROS-mediated cellular stress. Over the last decade, multiple studies have suggested a link between oxidative stress and microRNAs (miRNAs). The miRNAs are small non-coding RNAs that mostly suppress expression of the target gene by interaction with its 3'untranslated region (3'UTR). In this paper, we review the recent progress in the field, focusing on the association between miRNAs and oxidative stress during the progression of diabetes.

Keywords: diabetes; beta cells; oxidative stress; microRNAs

---

1. Introduction

Diabetes, which affects approximately 422 million people worldwide, is a disease characterized by the loss of glycemic control, which causes side effects such as polyuria, glycosuria, weight loss, neuropathies, retinopathy, and renal plus vascular diseases. Because diabetes results in the loss of glucose homeostasis, it is associated with high morbidity and mortality [1]. The most prevalent forms of this disease are Type 1 (T1D) and Type 2 diabetes (T2D). Both types are characterized by hyperglycemia due to either insufficient insulin production (T1D) or loss of cellular sensitivity to insulin, known as insulin resistance (T2D). Insulin-producing beta cells reside in the pancreas within clusters of endocrine cells called "Islets of Langerhans". Islets are dispersed throughout the pancreas, representing around 2% of the overall pancreatic tissue [2]. Beta cells are essential for blood glucose homeostasis. Their dysregulation is linked to both forms of diabetes. In T1D, the primary targets of autoimmunity are beta cells [3]. In T2D, insulin resistance (i.e., the inability of cells to respond to insulin to take up glucose) leads to excessive insulin production by beta cells, resulting in their exhaustion and eventual death [4]. Strong evidence indicates that T2D is associated with a deficit in beta cell mass [5], which leads to long lasting inefficient glycemic control leading to toxic amount of glucose.

Hyperglycemia is responsible for the development of severe complications such as microvascular, neuropathic, and macrovascular problems, which affect the quality and expectancy of life [6,7].

Since beta cells have notoriously low proliferating rates in adults, replenishing beta cell mass remains one of the greatest challenges of modern biology [8,9]. Even a partial restoration of insulin production in the pancreas could be therapeutically sufficient, judging by the fact that even after 80% loss of beta cell mass, T1D patients remain asymptomatic [10]. Although each of the two diabetes types has a different etiology, they are both greatly affected by cellular oxidative stress. On the one hand, oxidative stress in T1D originates from T cell-mediated autoimmunity targeting beta cells through the generation of proinflammatory cytokines. In addition, low tissue expression of antioxidative enzymes and antioxidative agents make affected individuals vulnerable to damage induced by reactive oxygen species (ROS) and reactive nitrogen species (RNS) originating from hypoxia or cytokine-mediated oxidative stress. A well-balanced equilibrium between oxidative molecules and antioxidative defenses is critical for physiological cell functions. On the other hand, type 2 diabetes is a metabolic syndrome where a group of conditions such as hypertension, glucose intolerance, insulin resistance, obesity, and dyslipidemia result in cellular oxidative stress across tissues [11,12]. Specifically, abdominal obesity has been shown to be a source of proinflammatory cytokines and, consequently, leads to insulin resistance.

Numerous studies have recently reported a strong link between oxidative stress and microRNAs (miRNAs). MiRNAs are post-transcriptional regulators, approximately 18 to 23 nucleotides long, that suppress gene expression by specific interaction with target genes [13]. The miRNAs have a role in controlling cellular redox homeostasis between highly reactive oxidative and antioxidative species. Current reports show that changes in miRNA levels contribute to persistent cellular oxidative stress, eventually leading to the development of diseases. Publications over the last few years increasingly support the link between miRNAs and oxidative stress in diabetes. A better understanding of the molecular mechanisms influencing the relationship between miRNAs and oxidative stress in diabetes could be useful to the development of therapeutic approaches that improve beta cell survival under metabolic stress. In this paper, we review the progress made in this field, describing mechanistic miRNA-driven gene regulation during oxidative stress and diabetes progression.

2. Overview of MicroRNA Biology: MiRNA Regulation and Their Role in Islets and Diabetes

The discovery of microRNA (miRNA) over twenty-five years ago revolutionized the field of cell biology and molecular biology. The first well-characterized small RNAs were lin-4 and let-7 [14–16] both of which have been found to be involved in control of early development, while let-7 has been found highly conserved across animal species [17]. According to a conservative analysis from ENCODE (Encyclopedia of DNA Elements) [18], an international consortium funded by the National Human Genome Research Institute (NHGRI) to study the human genome, 62% of the genome bases are transcribed into RNA of more than 200 bases long, of which only 5% corresponds to exons. Therefore, most of the transcribed RNA does not code for proteins and is designated as non-protein coding RNA (ncRNA). MiRNAs, a subset of ncRNAs, are small single stranded gene products of 18 to 23 nts, with an important role in post-transcriptional regulation of gene expression [13,19]. Almost half of the human miRNA genes are located in intergenic regions of the genome. Most of the other half are located in intronic regions of protein-coding genes, whereas some are found within exons [20]. The most common miRNA biogenesis pathway is known as the canonical pathway, although some miRNAs take alternative biogenesis routes [21,22]. In the canonical pathway, miRNA genes are transcribed by RNA polymerase II (Pol-II) to primary miRNAs (pri-miRNAs), which are processed in the nucleus by a microprocessor complex composed of human ribonuclease III (Drosha) and the DGCR8 (DiGeorge syndrome critical region 8) to a pre-miR stem loop precursor of approximately 60 to 70 nt [13,23

The pre-miRNA stem loop is actively transported to cytoplasm by exportin 5, where it is cleaved by Dicer, another member of the ribonuclease III protein family, into approximately 18 to 23 nucleotide double-stranded mature miRNA [13]. One strand arises from the 5′ end of the stem-loop and the other strand from the 3′ end, termed -5p and -3p, respectively. The miRNA is then incorporated into a ribonucleoprotein complex known as RISC (RNA-induced silencing complex) containing the essential silencing protein Argonaute 2 (Argo2) [24]. Argonautes belong to a highly conserved protein family. Together with small RNAs, such as miRNAs, they form ribonucleoprotein complexes (RNPs) that regulate post-transcriptional gene pathways. If the complementarity with the target mRNA is extensive, as is the case for the homeobox HOXB8 mRNA and miR-196, the Argonaute protein cleaves the mRNA [25]. However, in eukaryotes, the most frequent forms of silencing are by inhibition of translation or mRNA destabilization by polyA shortening [26].

Only the active mature RNA strand, known as a guide strand, is preserved and loaded on RISC, while the other complementary strand, designated as * strand, and known as a passenger strand, is degraded [24]. Many miRNAs retain both 5′ and 3′ strands, which are then incorporated into RISC complexes, generating miR-5p, as well as miR-3p. The choice of miR-5p or -3p as active mature miRNAs depends mostly on cell type [27]. It appears that the decision to select the guide strand from the miRNA duplex generated by Dicer is partly due to thermodynamics considerations. The strand with the weakest binding at its 5′ end is more likely to become the guide strand. In many human miRNAs, the guide strand is U-biased at the 5′ end with an excess of purines, while the passenger strand is C-biased with an excess of pyrimidines. Proteins such as Dicer, Argo2, and others participate in this decision as well. However, the mechanism is basically unknown [28]. The miRNA leads the RISC to a target mRNA. The single strand miRNA-RISC-Argo2 complex principally functions to inhibit target gene expression through recognition of partially complementary sequences in messenger RNA (mRNA), thus regulating mRNA translation by inhibiting gene expression and protein translation. The recognition sequence on the target mRNA is usually found at the 3′ UTR and is recognized by the "seed" sequence, two to eight nucleotides long, located at the 5′ domain of the miRNA. The MiRNAs target specific genes, which in turn may be targeted by many different miRNAs, hence regulating entire critical cellular expression networks (Figure 1).

It has been estimated that over 60% of human protein-coding genes are targets of miRNAs [29].

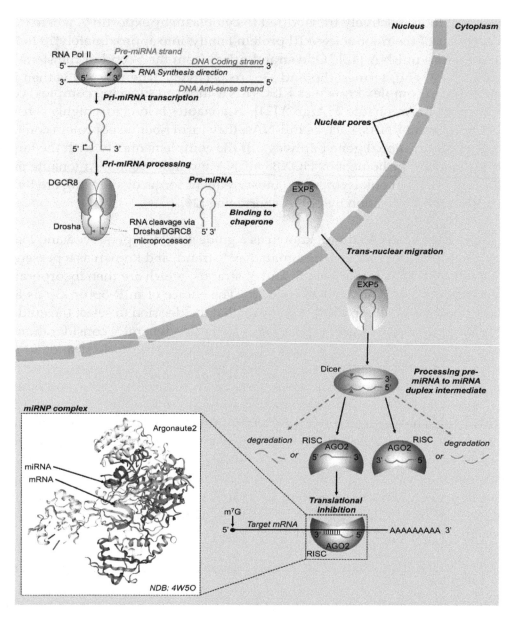

Figure 1. Canonical microRNA biogenesis and RNA targeting. In vertebrates, RNA polymerase-II transcribes primary miRNA genes (pri-miRNAs), which contain a hairpin-loop along with 5′ and 3′ flanking regions. DGCR8 (DiGeorge critical region 8) and a Drosha molecule combine to form the microprocessor complex which binds with pri-miRNA and cleaves it at specific sites (red arrowheads). The resulting precursor miRNA (pre-miRNA) contains a phosphate on its 5′ end and a hydroxyl group on its 3′ end along with a 2 to 3 nucleotide over-hang. Subsequently, the nuclear chaperone Exportin 5 (EXP5) binds to pre-miRNA molecules and transports pre-miRNA molecules to the cytoplasm via transnuclear migration, where Dicer, another RNAse III enzyme, binds to pre-miRNA molecules, cleaves them at specific regions, and releases a miRNA duplex intermediate. Argonaute 2 (AGO2) and other proteins assemble with miRNA molecules released from the miRNA duplex intermediate, together forming the RNA induced silencing complex (RISC). The 3′ or 5′ miRNA containing RISCs may bind to target regions and either result in translational repression, mRNA degradation, or in some cases translational activation. Inset shows a crystal structure of human Argonaute 2 bound to a guide and target RNA [30].

To date, the human genome contains 1917 annotated hairpin precursors, and 2654 mature sequences which are annotated in the Wellcome Trust Sanger Institute miRNA database [31] (http://www.mirbase.org/cgi-bin/mirna_summary.pl?org=hsa). miRNAs play a fundamental role in regulation of gene expression in key biological events such as cell proliferation, differentiation, death, and malignant transformation [13,32–35]. Consequently, impairment of miRNA expression is the underlying cause of many diseases. The miRNAs are mostly intracellular, but they are also found circulating in the body fluids, such as plasma or urine. They are extremely stable in human fluids, and therefore are well suited as clinical biomarkers [36]. They are protected from nucleases either by forming ribonucleoprotein particles (RNPs) with RNA-interacting proteins such as the RISC protein Ago2 or enclosed in extracellular vehicles (EVs) such as exosomes, present in and released by the majority of cell types [37]. The exosome-mediated transfer of mRNAs and miRNAs is a mechanism of cellular communication and genetic exchange among cells. The biogenesis, mode of action and suitability of circulating miRNAs as biomarkers for several diseases, is a hot research topic in biomedicine. Numerous studies suggest that miRNAs have an active role in pancreas organogenesis and in islet function [38–42]. An important study regarding miRNAs and their role in islet development is a report on the deletion of *Dicer1* in pancreatic progenitors. *Dicer1* is an enzyme involved in miRNA maturation, and its loss results in a marked reduction of endocrine cells [40]. Likewise, deletion of *Dicer1* in embryonic beta cells results in fewer beta cells, and impaired glucose tolerance [43,44]. There is evidence that miRNAs are involved in the pathogenesis of diabetes. Comprehensive reviews describing miRNAs in the context of T1D, T2D, and other diabetes models have recently become available. Furthermore, the role of miRNAs in tissues targeted by insulin, and in healthy or stressed islets, have been reported [45–48]. We have previously identified a subset of miRNAs differentially expressed in developing human islets, in human developing pancreas, and in alpha and beta cells of adult human islets [49–52]. These observations set the stage for studies to specifically assess the role of miRNAs and their target molecules in endocrine differentiation. In fact, many studies, including ours, identified individual miRNAs enriching endocrine tissue such as, miR-375 and miR-7, with the role in beta cell differentiation and function [53–57]. The same miRNAs have an important role in in vitro human stem cell differentiation into beta cells [58–61]. On the basis of the information presented above, it can be implied that oxidative stress affecting deregulation of miRNA networks, which is important for acquisition and maintenance of beta cell identity or proper cellular function and metabolism, contributes to the development of diabetes [62].

3. Overview of Oxidative Stress in Glucose Metabolism

The term oxidative stress refers to an imbalance between cellular oxidants and antioxidants [63,64]. Oxidative stress can be classified into the following two major groups: Endogenous (mitochondrial, peroxisomes, lipoxygenases, NADPH oxidase (NOX), and cytochrome P450) and exogenous (UV and ionizing radiation, chemotherapeutics, inflammatory cytokines, and environmental toxins). Oxidative stress is an accumulation of reactive oxygen species (ROS) above physiological levels, where ROS molecules oxidize cellular components stochastically, leading to progressive cellular damage. Under physiological conditions, the utmost ROS generation occurs in mitochondria, accounting for the transformation of 1% to 2% of oxygen molecules into superoxide anions [65]. Adenosine 5'-triphosphate (ATP) molecules are the major cellular energy currency. Generation of ATP in mitochondria, results in the production of ROS which occurs on two occasions with electron transport chain, at complex-I (NADH dehydrogenase) and at complex-III (ubiquinone-cytochrome c reductase). ATPs are first generated in the breakdown of glucose molecules during glycolysis. Glycolysis of one glucose molecule yields two pyruvate molecules with a net gain of only two ATP molecules. The greatest contributor to ATP production is the subsequent metabolism of pyruvate in the mitochondria through the tricarboxylic acid cycle, followed by oxidation of its energy mediators, NADH and FADH2, in the electron transport chain. In this process, known as oxidative phosphorylation, electrons are transferred from electron donors to electron acceptors via redox reactions. Oxidative phosphorylation, hypothetically, generates a maximum of 36 ATP molecules per glucose molecule. Oxygen is the final electron acceptor, generating H_2O. Incomplete transfer of electrons to oxygen results in the production of reactive oxygen species (ROS) such as superoxide or peroxide anions. Superoxide is rapidly converted [66] into peroxide (H_2O_2) by the enzyme superoxide dismutase (SOD). Hydrogen peroxide, in turn, is either neutralized to H_2O and O_2 by glutathione peroxidase (Gpx, in the mitochondria), or detoxified by catalase in peroxisomes. Increased levels of Cu (copper) and Fe (iron) and significantly decreased levels of Zn (zinc) in the serum of T2D patients and their first degree relatives (FDR) could be either triggering factors for the development of diabetes or a consequence of the illness [67]. H_2O_2 can be converted into highly reactive radical hydroxyl (HO·), the neutral form of the hydroxide ion, via the Fenton reaction. Hydroxyl radicals target the DNA base deoxyguanosine with great efficiency [65,68].

A discrete amount of ROS is necessary for efficient cellular physiological function. For example, ROS are one of the metabolic signals for insulin secretion [69] and play an essential role as promoter of natural defenses [70,71]. If the production of ROS during mitochondrial oxidative phosphorylation is not well balanced by antioxidative activity, ROS become toxic [66]. Even though oxidative phosphorylation is a significant contributor to the formation of ROS, recent studies have identified other cellular sources of ROS, such as peroxisomes, endoplasmic reticulum, and plasma membrane, which could contribute to tissue oxidative damage [72]. ROS are free radicals and, because they have unpaired valence electrons, they are extremely reactive with many electron donor molecules such as membrane lipids, proteins, and DNA, leading to potential toxicity. Overproduction of ROS causes oxidative stress associated with numerous diseases and aging.

The interaction of ROS with the cell membrane's polyunsaturated fatty acids generates a lipid peroxidation chain reaction with the production of toxic and highly reactive aldehyde metabolites such as malondialdehyde (MDA) [73,74]. MDA causes a reduction of cell membrane fluidity and function [75]. ROS cause oxidative damage of proteins by direct interaction either on amino acid residues or cofactors or by indirect oxidation via lipid peroxidation end products [76,77]. Likewise, ROS target pyrimidine and purine bases, as well as the deoxyribose moiety of genomic and mitochondrial DNA, causing cellular damage such as strand breakage, nucleotide removal, and DNA-protein binding. Extensive damage that cannot be corrected by cellular DNA repair could result in permanent impairment followed by apoptosis [78].

As far as islet beta cells are concerned, they are highly susceptible to ROS-mediated damage because of insufficient amounts of antioxidative compounds such as glutathione, and the naturally low expression of antioxidative enzymes such as the mitochondrial SOD (Mn-SOD), cytoplasmic Cu/Zn SOD, glutathione peroxidase (GPx), and catalase [79]. Several examples also illustrate the critical role of antioxidative defenses in the vascular system in diabetes. For example, cardiomyocytes in diabetes overexpress SOD or catalase, protecting cardiac mitochondria from extensive oxidative damage. SOD also prevents morphological abnormalities in diabetic hearts, correcting the aberrant contractility [80,81]. Two emerging crucial regulators of antioxidative stress responses are the uncoupling protein 2 (UCP2) and the transcription factor NRF2 (NFE2L2). UCP2, originally thought to function in adaptive thermogenesis similar to UCP1, is now considered to be primarily a regulator of ROS generation in mitochondria. UCP2 is a proton channel protein localized on the inner mitochondrial membrane that reduces the electrochemical gradient on both sides of the membrane, decreases ROS production, and protects against oxidative damage in mitochondria [82]. UCP2 has a critical role in the regulation of glucose homeostasis and in oxidative stress-mediated vascular diseases [83,84]. As for NRF2, it controls the transcription of key components of many antioxidative responses by binding to antioxidant response (ARE) elements in the promoter regions of target genes such as members of the glutathione and thioredoxin antioxidant systems and NAPDH (nicotinamide adenine dinucleotide phosphate) regeneration [85]. NRF2-mediated antioxidative responses are dysfunctional in diabetes [86] and dysregulation of the NRF2 redox pathway affects healing of diabetic wounds [87].

4. Oxidative Stress Generated by T Cell-Mediated Recognition of Beta Cells

T1D is an autoimmune disease characterized by T cell-mediated recognition and destruction of insulin-producing beta cells [88]. The beta cells are destroyed during the inflammatory phase known as insulitis. Insulitis is a significant component of T1D pathology and is characterized by infiltration of islets by immune and inflammatory cells. The leucocytic infiltration in insulitis is relatively subtle and transient, and therefore is detected mostly in cases with recent onset of the disease (less than one year [89]. There is limited knowledge about autoreactive T cells and autoantigens involved in the development of T1D. A primary autoantigen that activates autoreactive T cells is insulin [90]. Current views on T1D onset suggest that autoimmune destruction by insulitis is secondary to primary invasion of macrophages and dendritic cells activated by intercellular ROS from resident pancreatic phagocytes. Stimulated macrophages and dendritic cells will induce inflammatory genes and carry beta cell antigens specifically to lymph nodes, where T cells are activated. The activated T cells will specifically destroy beta cells through proinflammatory cytokine insults and more intracellular ROS formation [91]. So far, there is no cure for autoimmune T1D. Treatment is mostly focused on intensive insulin therapy aiming at tight glycemic control, which can significantly reduce debilitating long-term complications. There is a genetic predisposition for T1D. The strongest associations point at HLA class II, specifically haplotypes DRB1and DQB1 [92]. Although the autoreactive antigens and self-reactive T cells involved in autoimmune attack in T1D are well documented, the mechanism is not yet completely understood, however, the contribution of ROS and proinflammatory cytokines in beta cell death is fully substantiated [93]. The immune-mediated recognition of beta cells by autoreactive T cells and cytotoxic CD8T cells generates ROS and proinflammatory cytokines, inducing beta cell destruction and enhancing the effector response of islet-specific self-reactive CD4 T cells and cytotoxic CD8 T cells [94]. The proinflammatory milieu includes cytokines such as INFg, TNFa, IL-6, IL-12p70 and IL-1b, and ROS [95]. The destructive effect of ROS is amplified by the generation of reactive nitrogen species (RNS), which are extremely toxic free radicals such as free radical nitric oxide (NO) produced by IL-1b in beta cells. The IL-1b activates the enzyme nitric oxide synthase (iNOS), catalyzing production of nitric oxide and ultimately the superoxide ROS [96],. NO interacts with superoxide to generate the highly destructive molecule peroxynitrite. Both NO derived RNS and ROS cause beta cell damage using different pathways [97]. It is important to emphasize that an unbalanced ratio of oxidative to antioxidative events is what causes free radical toxicity. This has been illustrated by a recent study showing the dual role, protective or toxic, of NO in beta cells [98]. As stated above, insulitis and beta cell destruction are the crucial components of T1D pathology, but these are observed only in a limited proportion of islets at any given time, even at the time of diagnosis. Other factors, such as intercellular oxidative stress, precede insulitis [99]. This raises the possibility that in addition to the immune-mediated damaging effect of insulitis, a high level of dysfunction of beta cell contributes to T1D pathology as well. Interestingly, the lipid peroxidation, and oxidative stress detected by the presence of malondialdehyde in plasma of nondiabetic first degree relatives of the patients with T1D [100] supports the observation that oxidative stress can be clinically detected before the onset of diabetes.

5. Oxidative Stress and Metabolic Syndrome and Insulin Resistance in T2 Diabetes

T2D is currently considered a metabolic and inflammatory disease closely associated with metabolic syndrome, a group of conditions such as high blood pressure, glucose intolerance, insulin resistance, obesity, and dyslipidemia [101]. In many cases, a pre-T2D condition known as pre-diabetes is the prelude to the development of the disease. Pre-diabetes is characterized by impaired glucose tolerance and a state of mild hyperglycemia, not high enough to be diagnosed as diabetes, but leading to glucose intolerance. In addition, the main features of pre-diabetes are metabolic abnormalities similar to T2D, with essential roles of proinflammatory cytokines and free fatty acids (FFA), which are elevated in obesity and T2D as well. These factors initiate oxidative stress-mediated pathways, eventually resulting in beta cell dysfunction, impaired insulin secretion, and insulin resistance of peripheral tissue. Many studies indicate that oxidative stress originates before hyperglycemia, which in turn significantly contributes to the later complications of T2D (similar to those of T1D), such as vascular damage, retinopathy, nephropathy, and neuropathy [102]. In vitro and in vivo studies have indicated that the major oxidative stress-mediated pathways activated by hyperglycemia and ROS are JNK/SAPK, p38 MAPK, NF-kB, and the hexosamine biosynthetic pathway [103]. The first two, JNK/SAPK and p38

MAPK, contribute to the development of insulin resistance via direct and indirect phosphorylation of serine and threonine residues of insulin receptors [104,105]. Numerous studies link transcription factor NF-kB with regulation of gene-associated complications of diabetes [106]. In addition, hyperglycemia and oxidative stress mediate their actions through other signaling pathways such as advanced glycation end products (AGEs). AGEs refer to a group of heterogeneous compounds formed by the Maillard reaction process that involves the non-enzymatic glycation of proteins, lipids, and nucleic acids by reducing sugars and aldehydes. AGEs function through the multiligand immunoglobulin superfamily receptor for advanced glycation end products (RAGEs). The AGE compounds directly affect proteins of the mitochondrial respiratory chain to generate reactive oxygen species (ROS) [107]. AGE and RAGE are involved in diabetes vascular pathologies as well [108]. They also activate production of the second messenger signaling lipid diacylglycerol leading to activation of several isoforms of the protein kinase C (PKC). Isoforms of PKC are implicated in generating insulin resistance [109–111]. Last, but not least, AGE increases utilization of the polyol pathway that will decrease the cofactor NAPDH, and therefore directly affects the production of antioxidative glutathione [112,113]. As described above, multiple signaling pathways contribute to oxidative stress-mediated damage leading to T2D. Therefore, dysregulation of miRNAs controlling these pathways can certainly contribute to development and persistence of diabetes.

6. MicroRNAs in Diabetic Oxidative Stress

We reviewed research articles in PubMed, primarily focusing on studies describing changes in the expression of miRNAs due to oxidative stress in the context of diabetes and their target components controlling mechanism of oxidative stress homeostasis.

This review does not include studies dealing with miRNAs induced by proinflammatory cytokines generated by T1D autoimmune attack on beta cells. Thorough reviews have been written on this topic [46,114–116]. Table 1 lists the miRNAs reported as having an effect on oxidative stress in diabetes, the source of oxidative stress and the observed effect, target tissue or organ, and target genes. A few miRNAs, with known target tissue but unknown gene targets are included as well. Ten miRNAs identified in Table 1, overlap with a previous in silico analysis of miRNAs in human cells regulated in vitro by oxidative stress [117]. These are let-7f, miR-9, miR-16, miR-21, miR-22, miR-29b, miR-99a, miR-141, miR-144, and miR-200c. In order to make this overview of miRNAs and their targets in oxidative stress and diabetes easy to follow, we organized the miRNAs by their function in the affected tissues and organs.

Table 1. Selected PubMed articles describing miRNAs in diabetic oxidative stress.

Source of Oxidative Stress	Differentially Expressed miRNAs	Target Tissue/Organ	Target Gene	Reference
T2D	miR-203↓	Cardiac tissue	PIK3CA	[118]
T2D	miR-30e-5p↓	Kidney and vasculature	UCP2, MUC17, UBE2I	[119]
Diabetic retinopathy, hyperglycemia	miR-455-5p↓	Retinal epithelial cells	SOCS3	[120]
Diabetic nephropathy, hyperglycemia	miR-214↓	Kidney tissue	-	[121]
Insulin synthesis	miR-15a↑	Beta cells	UCP2	[122]
Kidney fibrosis	miR-30e↓	Tubular epithelial cells	UCP2	[123]
DCM	miR-30c↓	Cardiac tissue	PGC-1β	[124]
T2D	miR-233↓	Hepatic tissue	KEAP1	[125]
T1D, Diabetic nephropathy	miR-146a↓	Neural tissue, kidney tissue	-	[126,127]
DCM	miR-503↑	Cardiac tissue	NRF2	[128,129]
Diabetic Retinopathy	miR-365↓	Retinal tissue	TIMP3	[130]
Gestational Diabetes	miR-129-2↑	Murine neural tube	PGC-1α	[131]
Hyperglycemia	miR-106b↑	Pancreatic islets	SIRT1	[132]
Diabetic nephropathy	miR-106a↓	Murine neural tissue	ALOX15	[133]
Diabetic retinopathy	miR-7-5p↑	Retinal tissue	EPAC1	[134]
Diabetic neurotoxicity	miR-302↓	Neural tissue	PTEN	[135]
T2D	miR-17↓	Skeletal muscle	GLUT4	[136]

Table 1. *Cont.*

Source of Oxidative Stress	Differentially Expressed miRNAs	Target Tissue/Organ	Target Gene	Reference
Diabetic retinopathy, hyperglycemia	miR-145↓	Retinal epithelial cells	TLR4	[137]
Diabetic nephropathy, hyperglycemia	miR-25↓	Neural tissue, kidney tissue	PTEN, CDC42	[138–140]
TXNIP overexpression	miR-200b↑	Beta cells	ZEB1	[141]
Diabetic mice	miR-200c↑	Vasculature	ZEB1	[142]
Diabetic Mice	miR-200a/b↓	Vasculature	OGT	[143]
DCM	miR-92a↑	Vasculature	HMOX1	[144,145]
T2D	miR-200b/c↑ and miR-429↑	Vasculature	ZEB1	[146]
T2D, T1D	miR-200c↑	Murine arteries	SIRT1, FOXO1, eNOS	[147]
Long-term diabetes	miR-126↑	Vasculature, skeletal muscles	SIRT1, SOD	[148]
T2D	miR-133a↓	Murine gastric smooth muscle cells	RhoA/Rho kinase	[149]
Hyperglycemia, T2D, T1D	miR-21↑	Vasculature, β-cells, Cardiac tissue	KIRT1, FOXO1, NRF2, SOD2, PPARA	[150–152]
T1D model	miR-200b↑	Murine retinal cells	OXR1	[153]
T2D	miR-15a↑	Plasma	AKT3	[154]
Diabetic embryopathy	miR-27a↑	Murine embryos, kidney tissue	NRF2	[129,155]
STZ-diabetic mice	miR-34a↑	β-cells, vasculature	SIRT1	[156]
Endothelial cells, vascular stress	miR-204↑	Vascular wall /endothelium in vivo	SIRT1	[157]
Cardiomyocytes apoptosis	miR-675↓	Vasculature	VDAC1	[158]
T1D, Diabetic retinopathy	miR-195↑	Cardiac tissue, β-cells	CASP3, MFN2	[159,160]
Gestational diabetes, hyperglycemia	miR-322↓	Murine Embryos, Neurons	TRAF3	[161]
T2D	miR-126↓	Vasculature	VEGFR2	[162]
T2D	miR-27b↓	Vasculature, wounds	SHC1, SEMA6A, TSP-1, TSP-2	[163]
Hyperglycemia, Polyol pathway	miR-200a-3p↑, miR-141-3p↑	Kidney tissue	KEAP1, TGFβ1/2	[164]
STZ mice	miR-1↓, miR-499↓, miR-133a/b↓ and miR-21↑	Cardiac tissue	ASPH	[165]
Persistent UPR IRE1α deficiency	miR-200↑, miR-466h-5p↑	Vasculature, wounds	ANGPT1	[166]
T2D, DCM	miR-9-5p↑	Retinal tissue	ELAVL1	[167]
T2D	miR-99a↑	Vasculature	IGF1R, MTOR	[168]
Hyperlipidemia	miR-155-5p↑	β-cells	MAFB	[169]
T1D NOD islets	miR-29c↑	β-cells	MCL1	[170]
T2D, glucose and lipid oxidation	miR-29↑	Skeletal muscle	-	[171]
Diabetic nephropathy	miR-29↑	Regulation of inflammatory cytokines	TTP	[172]
Diabetic heart T2D	miR-29↑	Cardio-metabolic disorders	Lypla 1	[173]
Gestational diabetes	Circular RNAs: circ-5824↓, circ-3636↓, circ-0395↓	Human placenta	(In silico analysis) AGE- and RAGE-related genes	[174]

6.1. Vascular Endothelial Cells, Diabetic Cardiomyopathy, and Muscle

MiR-21 is a miRNA related to diabetes. The expression of miR-21 is increased in the plasma of patients with impaired glucose tolerance and with T2D [150]. It has been proposed that circulating extracellular vesicles carrying miR-21 could be used as a marker of developing type 1 diabetes [175]. It has been found that miR-21 increases susceptibility to oxidative stress induced by fluctuating glucose levels in primary pooled human umbilical vein endothelial cells (HUVECs), by targeting genes regulating homeostasis of intracellular ROS, such as KRIT1, NRF2, and SOD2 [151]. A reduced expression of miR-21 protects against cardiac remodeling in diabetic cardiomyopathy (DCM). An in vivo experiment in mice confirmed, that suppression of miR-21 stimulates the nuclear hormone receptor PPAR (peroxisome proliferator activated receptor), known to regulate homeostasis in response to glucose and lipid levels. The PPAR initiates nuclear translocation of NRF2, and thus the antioxidative response of NRF2 protects from DCM [152]. MiR-21 also regulates the signaling pathway of the intracellular AGE–RAGE interaction and targets TIMP3, an inhibitor of extracellular matrix degradation in diabetic neuropathy [176].

Similarly, in a rat model of DCM, the expression of miR-503 is increased in myocardial cells and has a deleterious role by targeting NRF2 and antioxidant response element (ARE) signaling pathway as well [128]. The cluster of miR-200 is an important player in oxidative response in diabetes [177]. It is formed by the following five evolutionary conserved miRNAs: miR-200a, miR-200b, miR-200c, miR-141, and miR-429. These miRNAs can be grouped according to their seed sequences into subgroup I, miR-200a and miR-141 (AACACUG), and subgroup II composed of miR-200b, miR-200c, and miR-429 (AAUACUG), suggesting that miRNAs in each subgroup will target different genes. Several reports indicate that the miR-200 family has a role in the development of endothelial inflammation present in diabetic vascular complications and cardiovascular diseases. In many instances, the action of miR-200 is via targeting the (zinc finger E-box-binding homeobox) ZEB1. ZEB1 has a role in epithelial–mesenchymal transition (EMT) [141] and is associated with the inhibition of apoptosis. The thioredoxin-interacting protein, TXNIP, is induced in vivo by hyperglycemia and it inhibits the antioxidative function of thioredoxin resulting in accumulation of reactive oxygen species, cellular stress, and induction of the miR-200 family which induces apoptosis through inhibition of ZEB1. Likewise, inhibition of miR-200c restores endothelial function in diabetic mice through upregulation of ZEB1 [177], and in HUVEC under oxidative conditions miR-200 expression is increased which suppress ZEB1 causing apoptosis. Overexpression of ZEB1 in the cells reversed the effect [178]. Downregulation of ZEB1, by miR-200a/b/c and miR-429, contributes to activation of proinflammatory genes in vascular smooth muscle cells of diabetic mice [146]. Furthermore, the miR-200 family negatively regulates beta cell survival in type 2 diabetes in vivo. Overexpression of miR-200, in mice, causes beta cell death and is sufficient to render T2D lethal [179].

In addition, the family of miRNA-200 has been reported to exhibit a protective effect in diabetic oxidative stress by targeting high glucose-induced O-linked N-acetylglucosamine transferase (OGT), whose enzymatic activity is associated with diabetic complications, and endothelial inflammation in mice with diabetes. Experiments with human aortic endothelial cells (HAEC) confirmed miR-200 silencing OGT by direct binding to 3'UTR of mRNA [143].

Another important antioxidative gene that is regulated by the family of miR-200 is Sirtuin 1 (SIRT1) [177]. SIRT1 is NAD+-dependent deacetylase that controls histone chromatin proteins as well as non-histone proteins, many of them are transcription factors such as fork-head box O1 (FOXO)1. To date, seven sirtuins have been identified. They are associated with several cellular processes, such as energy balance, stress resistance, and insulin resistance. Some are located in the cytoplasm and others are located in the nucleus or mitochondria [180]. SIRT1, -2, -3, and -6 have a function in oxidative stress. By targeting SIRT1, endothelial nitric oxide synthase (eNOS) and FOXO1 miR-200 impairs their regulatory circuit and promotes ROS production and endothelial dysfunction [147]. It has been shown that miR-200 targets these three genes in vitro in HUVEC cells. The in vitro results were validated in three in vivo models of oxidative stress, human skin fibroblasts from old donors, femoral arteries from old mice, and a murine model of hindlimb ischemia [147].

In endothelial cells, SIRT1 is targeted by other miRNAs, increasing diabetes-related oxidative stress. Examples include the following: miR-34 induces endothelial inflammation by downregulating SIRT1 [156] and targeting SIRT1; miR-204 promotes vascular endoplasmic reticulum (ER) stress, inflammation, and dysfunction in mice; downregulation of miR-204 activates protection against ER stress through an increase of SIRT1 expression [157]; miR-106b targets SIRT1 in mouse insulinoma cell line NIT-1, rendering them vulnerable to hyperglycemia induced by 30mM glucose; and in vivo suppression of miR-106b increases expression of SIRT1 and reduces cardiovascular damage in diabetic mice [132].

Furthermore, it has been shown, in a mouse model of peripheral arterial disease, that the more abundant circulating form of unacylated ghrelin (UnAG) exerts its protective effect from ROS imbalance in endothelial cells via induction of miR-126, a known endothelial miRNA. By targeting vascular cell adhesion molecule 1 (VCAM1), miR-126 indirectly activates SIRT1 and SOD to induce resistance to oxidative stress [148].

MiR-9 plays a positive role in oxidative stress-mediated cardiomyopathy in T2D. In vitro experiments with immortalized cardiomyocyte culture and samples of failing heart tissue collected at the time of transplantation confirmed that downregulation of miR-9 in human cardiomyocytes results in higher expression of its target ELAV-like protein 1 (ELAVL1), a ubiquitously expressed RNA binding protein that stabilizes inflammatory mRNAs by binding to ARE domains and thus leading to cardiomyocyte death [167]. Another miRNA with a protective role in diabetic cardiomyopathy is miR-30c. MiR-30c targets PGC-1β, one of important coactivators of PPAR alpha and mitochondrial key regulator. Knockdown of PGC1 beta reduces excessive ROS and myocardial lipid accumulation which decreases cardiac dysfunction in diabetes [124].

Numerous studies report miR-29 family participation in oxidative stress-mediated inflammatory response in diabetes. The miR-29 family consists of three members divided into two clusters that are transcribed polycistronically; the miR-29a/b-1 cluster is localized on human chromosome 7 and the miR-29c/b-2 cluster on chromosome 1 [181]. The miR-29s are known to be regulated in multiple tissues. Hyperinsulinemia dramatically reduces their expression, while hyperglycemia induces it. Experiments with MIN6 insulinoma beta cell line determined that miR-29 targets a member of the BCL2 family, an antiapoptotic protein, the MCL1 (myeloid cell leukemia 1) (MCL-1) gene. Interestingly, in humans, repression of MCL1 is related to diabetes mellitus-associated cardiomyocyte disorganization [182]. Since circulating miR-29 has been reported in newly diagnosed T2D patients and, furthermore, upregulation of miR-29 expression contributes to development of the first stage of type 1 diabetes mellitus in the T1D model of NOD mice [170], there is the possibility that miR-29 regulates MCL1 at different stages of the disease.

There are instances that indicate the miR-29 cluster family has a protective role against oxidative stress conditions. Its elevated expression has been associated with a compensatory mechanism for heart hypertrophy and fibrosis due to age increased oxidative stress, modulating targets such as DNA methylases and collagens [183]. A protective role in endothelial dysfunction in cardiometabolic disorders found in T2D has been reported. MiR-29 is upregulated in T2D arterioles to compensate for endothelial dysfunction. Specifically, miR-29 targets Lypla 1 (lysophospholipase I), a gene that negatively regulates production of NO, required for vasodilation. Lypla 1 depalmitoylates eNOS (nitric oxide synthase), reducing NO in endothelial cells [173].

The expression of miR-29a and miR-29c in skeletal muscle of patients with type 2 diabetes are upregulated which suppresses glucose and lipid metabolism possibly by targeting insulin receptor substrate 1 (IRS1) and phosphoinositide 3 kinase (PI3K). Both genes are involved in glucose insulin regulation, moreover they control lipid oxidation by targeting peroxisome activated receptor gamma coactivator1alpha (PGC1alpha). In vivo overexpression of miR-29 in mouse tibias anterior muscle resulted in a decrease of glucose uptake and glycogen content. MiR-29 acts as an important regulator of insulin stimulated glucose metabolism [171].

6.2. Retina Cells

Oxidative stress and hypoxia cause retinopathy by induction of miR-7 that negatively regulates the RAPGEF3/EPAC-1 (rap guanine nucleotide exchange factor 3). EPAC-1 is an accessory protein for cAMP activation and stimulation for survival and growth in response to extracellular signals [134]. MiR-7-mediated decrease of EPAC1 expression results in endothelial hyperpermeability and loss of (endothelial nitric oxide synthase) eNOS activity in murine experimental retinopathy. EPAC-1 is associated with cAMP-induced vascular relaxation in endothelial cells via eNOS and amelioration of endothelial hyperpermeability induced by inflammatory mediators [134]. Development of retinopathy in T2D is associated with miR-15 as well. This miRNA is mostly found in the pancreas, where it plays an important role in beta cell insulin secretion. Interestingly, miR-15 has been detected in the plasma of T2D patients, where its amount corelated with the severity of the disease. Experiments with the rat beta cell line INS1 showed that the concentration of miR-15 in the cells increases when cultured in high glucose media. Coculture of INS1 insulinoma cells with Muller cells (retinal glial cells) showed a clear transfer of miR-15 into Muller cells, and the transfer was achieved by exosomes. The deleterious effect of miR-15 in the retina is via targeting AKT3, an isoform of the AKT gene (serine/threonine kinase 1). Loss of AKT3 in the tissue increases intracellular content of ROS, leading to cellular apoptosis. These results also prove that under pathological conditions some miRNAs can travel from tissue to tissue through exosome transfer [154]. Incidentally, persistent exposure to high glucose causes intracellular accumulation of insulin in beta cells mediated by suppression of the UPC2 gene by miR-15a. High glucose treatment for a short time induces miR-15a, while longer exposure suppresses the expression. It has been found that inhibition of UPC2 by miR-15a increases O_2 consumption beta cell function and insulin synthesis [122].

Oxidative stress in retinal glial Muller cells induces upregulation of miR-365 causing damage by targeting TIMP3, the protein that inhibits matrix metalloproteinases and has antioxidative properties [130]. MiR-455-5p may have a positive role in diabetic retinopathy. Upregulation of miR-455-5p attenuates high glucose-triggered oxidative stress injury by targeting SOCS3 (suppressor of cytokine signaling 3) mRNA. SOCS3 downregulation decreases production of intracellular ROS, malondialdehyde (MDA) content, and NADPH oxidase 4 expression, while enhancing superoxide dismutase, catalase, and GPX activities [120].

6.3. Diabetic Wound

Moreover, the miR-200 family has an effect on the pathology of diabetic skin ulcers by targeting the angiogenic factor angiopoietin 1 (ANGPT1), resulting in disrupted angiogenesis. In diabetic wound healing, hyperglycemia-mediated oxidative stress produces an unmodulated, persistent unfolded protein response (UPR), generating deficiency in inositol-requiring enzyme 1 (IRE1α), a primary UPR transducer that modulates expression of mRNAs and miRNAs. This deficiency leads to the upregulation of the miR-200 family and miR-466, both targeting ANGPT1. Angiogenesis may be rescued by upregulation of IRE-1a, which attenuates maturation of both miRNAs [166].

6.4. Kidney Tissues and Functions

Another miRNA that interferes with ROS homeostasis in diabetes via targeting NRF2 is miR-27a. The adipokine omentin 1 restores renal function of type 2 diabetic db/db mice through suppression of miR-27a, which upregulates NRF2 and decreases oxidative stress [155]. NRF2/KEAP1 is a master antioxidant pathway regulating redox under nonstressed and stressed conditions. Under nonstressed conditions, NRF2 is anchored by a repressor KEAP1 in cytoplasm. A stressed situation releases KEAP1 and the stabilized NRF2 relocates to nucleus, where it binds to the antioxidant response element (ARE) activating transcription of antioxidant proteins [184]. In experiments with mice rendered diabetic with streptozotocin, hyperglycemia activates the polyol pathway in renal mesangial cells. The polyol pathway is involved in microvascular damage to retina in diabetes. On the one hand, activation of the polyol pathway increases the activity of aldose reductase which in turn decreases expression of miR-200a and miR-141. These miRNAs are regulators of KEAP-1. Their low expression enhances suppressive activity of KEAP-1 on NRF2. The suppressed transcription factor, NRF2, cannot activate transcription of antioxidant genes resulting in an increase of ROS and oxidative stress. On the other hand, aldose reductase deficiency in the renal cortex upregulates miR-200 and miR-141, which releases the KEAP-1 suppression of NRF2 and ameliorates the oxidative stress and downregulates TGF-beta, preventing kidney fibrosis [164]. The NRF2/KEAP1 pathway is also regulated in other organs under oxidative stress damage, such as in the pathological process of liver injury in T2DM. In this case, miR-233 targets KEAP1 allowing the released NFR2 to migrate to the nucleus and activate synthesis of antioxidative mRNAs and proteins such as SOD and HO-1 [125].

Endothelial dysfunction in cardiovascular disease is also affected by CKD (chronic kidney disease). CKD is caused by the accumulation of uremic toxin which upregulates miR-92a. The miRNA can be detected in the patient's serum, which could be useful for diagnostic purposes. Uremic toxins generated oxidative stress results in downregulation of endothelial protective factors such as SIRT1 and eNOS [144]. At this time, it is not known if this is through direct or indirect regulation. Additionally, miR-92a is upregulated in diabetic aortic endothelium of C57BL-db/db mice and in renal arteries from human diabetic subjects. MiR-92a downregulates expression of heme oxygenase 1 (HO-1), an endothelial protective enzyme synthesized through NRF2 binding to the ARE sequence in the nucleus. The resulting oxidative stress impairs endothelium dependent relaxation. The suppression of miR-92 restores the endothelial function and the expression of HO-1 [145]. The expression of miR-25 in diabetic mouse kidneys and in human peripheral blood of patients with diabetes is much lower than in non-diabetic subjects. MiR-25 has a protective role in ROS-mediated diabetic kidney disease, by direct regulation of the Ras-related gene CDC42. The CDC42 gene belongs to the family of Rho small GTPases which are central regulators of actin reorganization and have a role in nephrotic pathogenesis. An increase of miR-25 expression represses glomerular fibrosis [139]. Some of the intracellular effects of ROS are mediated by regulation of the PTEN/PI3K/AKT pathway [185]. Blood samples and kidney tissue from diabetic subjects show downregulation of miR-25. Gain and loss of function performed with the human kidney cell line HK2 confirmed the crucial role of miR-25 protection against dysfunction and apoptosis of renal tubular epithelial cells. MiR-25 inhibits the apoptotic effect of hyperglycemia-mediated ROS in renal tubular epithelial cells by targeting PTEN. Knockout of PTEN activates the PI3K/AKT. PTEN is a dual protein and lipid phosphatase whose main substrate is phosphatidyl-inositol,3,4,5 triphosphate (PIP3). PTEN catalysis dephosphorization of PIP3 to PIP2 which represses the antiapoptotic signaling pathway of PI3k/AKT. Knockout of PTEN by miR-25 activates the AKT pathway ameliorating ROS and apoptosis [140]. Some miRNAs exert their antioxidative role by regulating the expression of UCP2 (uncoupling protein 2) which attenuates ROS activity in mitochondria. In HK2 (kidney cortex and proximal tubule cell line), it has been shown that miR-214 suppresses oxidative stress in diabetic nephropathy via the ROS/Akt/mTOR signaling pathway and enhancing UCP2 expression [121]. On the other hand, an experiment in a diabetic mouse model showed that miR-30e targets directly UCP2 in kidney cells, thus mediating the TGF-β1-induced epithelial-mesenchymal transition and kidney fibrosis [123]. In diabetic nephropathy, miRNA-29c contributes to the progression of the disease by regulating proinflammatory cytokines via targeting tristetraprolin (TTP) mRNA [172]. Experiments were performed in kidney tissues from DN patients and controls. TTP has anti-inflammatory effects by enhancing the decay of mRNAs bearing the adenosine/uridine-rich element (ARE) present in the 3'UTR of cytokine transcripts such as Il-6 and TNF alpha. Additional experiments with cultured podocytes confirmed the findings. Finally, miR-21, a diabetes-related miRNA, described above, has a role in diabetic nephropathy by regulating TIMP3, an inhibitor of extracellular matrix degradation [176], involved in mesangial expansion characteristic of diabetic nephropathy.

6.5. Diabetic Neuropathy

In the case of diabetic peripheral neuropathy, PKC activity is linked to a protective role of miR-25. MiR-25 downregulates production of AGE and RAGE, reduces activation of PKC, and reduces NAPDH oxidase activity probably via regulation of NOX4, an isoform of the NOX family. NOX4 protects vasculature against inflammatory stress. Experiments to clarify the protective role of miR-25 in diabetic neuropathy were done with sciatic nerve from db/db diabetic mouse model and BALB/c healthy counterparts. The conclusions were confirmed with cultured Schwann cells [138]. Modulation of the PTEN/AKT pathway is also critical to attenuate the oxidative stress mediated by extracellular amyloid-β (Aβ) peptides in diabetic neurotoxicity. Activation of the AKT pathway through direct targeting of PTEN by miR-302 attenuates amyloid beta induced toxicity in neurons and activated AKT signaling, which subsequently stabilizes NRF2 and synthesis of cytoprotective protein HO-1 [135].

Finally, as stated above, we have not included in this review the miRNAs involved in the oxidative stress caused by the effect of proinflammatory cytokines in beta cells. However, beta cells are also the target of other oxidative sources such as oxidized LDL (low density of lipoproteins). Oxidative stress induced the generation of oxidized LDL in hyperlipidemia conditions. Oxidized LDL enhances the activity of LPS (lysophosphatidylcholine) increasing the expression of miR-155-5p in murine pancreatic beta cells. MiR-155 targets MAFB (v-maf musculoaponeurotic fibrosarcoma oncogene family, protein B), enhancing the transcription of IL-6 that stimulates the production of GLP-1 in alpha cells, which suppresses glucagon secretion from alpha cells and stimulates insulin secretion from beta cells in a glucose-dependent manner. Through this mechanism, miR-155-5p improves the adaptation of beta cells to insulin resistance and protection of islets from stress [169].

6.6. Gestational Diabetes

As discussed previously, the miR-29 family is regulated in multiple tissues. Although in most cases it has a deleterious and proinflammatory effect, in some organs the effect of miR-29 alleviates symptoms. In rats, miR-29b has a positive effect on gestational diabetes mellitus by targeting PI3K/Akt signal. Administration of miR-29 mimics reduced markers indicating oxidative stress, increased super oxide dismutase (SOD), catalase [165], and decreased malondialdehyde (MDA) in liver tissues of GDM rats [186]. Maternal diabetes and hyperglycemia dysregulate mitochondrial function through activation of protein kinase C (PKC) isoforms that have a role in the diabetic embryopathy. One of the isoforms of PKCα upregulates expression of miR-129-2, which targets the PGC-1α, the ligand of PPAR alpha (peroxisome proliferator activated receptor alpha). PGC1 alpha is a positive regulator of mitochondrial function and its downregulation by miR-129-2 mediates teratogenicity of hyperglycemia leading to NTDs (Neural tube defects in embryos) [131]. On the other hand, in the case of oxidative stress induced in embryo by maternal diabetes, inhibition of miR-27a increases NRF2 expression, which restores the homeostasis [129].

More recently, specific circular RNAs (circRNAs) interacting with miRNAs were identified in placentas from women with gestation diabetes mellitus that may regulate the AGE–RAGE interaction [174]. The circRNAs have their 5′ end and 3′ end covalently bond and are generated by a process known as back splicing, in which an upstream splice acceptor is joined to a downstream splice donor. They are expressed in various types of cells and tissues and, although little is known about their biological role, some act as gene regulators. In particular, several circRNAs have been described as acting as miRNA silencers or "sponges" by containing miRNA target sequences, in different type of cells including beta cells [187,188]. The differentially expressed circRNAs have been analyzed by Kyoto Encyclopedia of Genes and Genomes (KEGG) enrichment and circRNA–miRNA interaction, according to the sponge molecular interaction. The KEGG analysis predicted that circRNAs are likely to be involved in advanced glycation end products receptor for advanced glycation end products, AGE-RAGE, signaling pathways in diabetic complications. The expression of three circRNAs, circ-5824, circ-3636, and circ-0395, are downregulated in placentas of GDM. The circRNA–miRNA interaction analysis showed that miR-1273g-3p activated by acute glucose fluctuation is also involved in the progression of several complications caused by diabetes and it could be a potential gene of interest in GDM [174].

Figure 2 shows a scheme depicting the group of selected miRNAs described above and in Table with their role in regulation of oxidative stress in diabetes

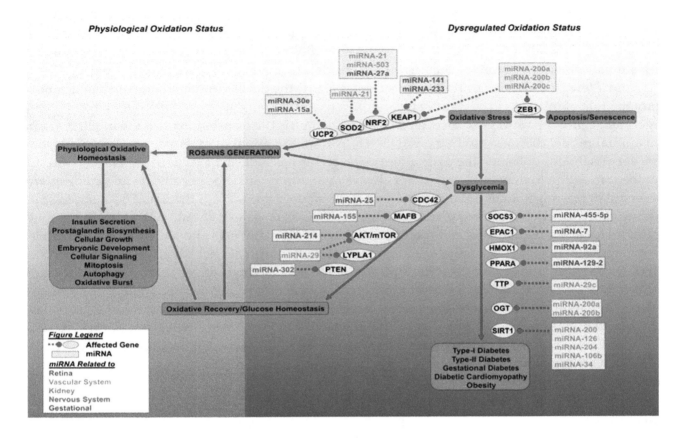

Figure 2. Dysregulated oxidative stress and microRNAs result in loss of glucose homeostasis. This figure outlines the effect of aberrant accumulation of cellular reactive oxygen species (ROS) and reactive nitrogen species (RNS). Cellular oxidative status is maintained by SOD2, NRF2, and UCP2, which allows for a spectrum of physiological functions carried out by the cell. Excessive ROS and RNS generation led to dysglycemia or cellular senescence. The miRNA molecules can target NRF2 (miRNA-21, miRNA-27a, miRNA-503, miRNA-233), SOD2 (miRNA-21), and UCP2 (miR-30e and miR-15a), leading to loss of oxidative regulation and the initiation of oxidative stress. Cellular oxidative stress can lead to either dysglycemia or cellular senescence. Cellular senescence is mediated by the inhibition of zinc finger E-box binding homeobox 1 (ZEB1) by miR-200 family miRNAs. Dysglycemia develops when O-linked β-N-acetylglucosamine transferase (OGT) and NAD-dependent deacetylase sirtuin-1 (SIRT1) are targeted by specific miRNAs. Oxidative stress driven dysglycemia rapidly initiates the expression of miRNA molecules which target suppressor of cytokine signaling 3 (SOCS3), exchange factor directly activated by cAMP 1 (EPAC1), and heme oxygenase (decycling) 1 (HMOX1), Peroxisome proliferator-activated receptor alpha (PPARA), mitochondrial uncoupling protein 2 (UCP2), and tristetraprolin (TTP), leading to decreased expression of these genes and the advance of diabetes. Alternatively, recovery can occur by miRNA directed targeting of genes involved in dysglycemia, they include: Cell division control protein 42 homolog (CDC42), V-maf musculoaponeurotic fibrosarcoma oncogene homolog B (MAFB), protein kinase B and mammalian target of rapamycin (AKT/mTOR), acyl-protein thioesterase 1 (LYPLA1) and phosphatase and tensin homolog (PTEN). Recovery of glucose homeostasis results in oxidative normalization and cellular homeostasis. Different colors of miRNA denote affected organ.

7. Conclusions

In diabetes, hyperglycemia induces intense oxidative stress that can no longer be modulated by the cellular antioxidative response, thus leading to accumulation of ROS. Overall, this process causes pancreatic beta cell dysfunction and unpaired glucose tolerance response, both of which have a deleterious effect on many types of cells and tissues. miRNAs have a critical role in the molecular mechanism involved in this process. Many of the studies reviewed here were performed in in vitro with animal cell lines or primary cells, in animal models (some in combination with tissues), some *in silico*, and a few cases in human tissues. It is expected that the development of new transgenic mice to study the role of miRNAs in oxidative stress will be useful to confirm or even discover novel potential targets and cellular pathways. However, the real challenge is the translation of all the in vitro, *in silico*, and animal model discovery to human diabetes. Although animal models, especially rodents, have been very useful for obtaining the basic information on the mechanism of several diseases, it is also true that the translation to human disease is not always straightforward. Specifically, many strategies were successful in treating autoimmune diabetes in rodent models, but none of them had been successful in treating human T1D. Furthermore, human basic and clinical research should aim at developing new strategies focusing on miRNAs and their target genes to cure diabetes and its complications. One of the emerging strategies is the use of a combination of human primary cells derived from human stem cell differentiation and organoid cultures plus genome editing alternatives to investigate the causes and role of miRNAs in oxidative stress in diabetes, as well as to screen for potential drugs to treat or alleviate its effects. However, it is important to remember that, currently, therapeutic approaches based on manipulation of miRNA expression are more effective in vitro than in vivo because of difficulties with specific delivery. As we have presented in this review, miRNAs are of variable nature, depending very much on the external and internal triggers. Therefore, it is of utmost importance to determine their specific targets and approach the treatment from that direction.

Author Contributions: M.M.F.Q., conceptualization, writing and design of figures; D.K., writing and critical reading; S.Á.-C., writing and critical reading; J.D.-B., writing and critical reading; R.L.P., conceptualization, writing, and revision.

References

1. World Health Organization. *Global Report on Diabetes*; World Health Organization: Geneva, Switzerland, 2018.
2. Saito, K.; Iwama, N.; Takahashi, T. Morphometrical Analysis on Topographical Difference in Size Distribution, Number and Volume of Islets in the Human Pancreas. *Tohoku J. Exp. Med.* **1978**, *124*, 177–186. [CrossRef] [PubMed]
3. Burrack, A.L.; Martinov, T.; Fife, B.T. T Cell-Mediated Beta Cell Destruction: Autoimmunity and Alloimmunity in the Context of Type 1 Diabetes. *Front. Endocrinol.* **2017**, *8*, 343. [CrossRef] [PubMed]
4. Cantley, J.; Ashcroft, F.M. Q&A: Insulin secretion and type 2 diabetes: Why do beta-cells fail? *BMC Biol.* **2015**, *13*, 33.
5. Meier, J.J.; Bonadonna, R.C. Role of reduced beta-cell mass versus impaired beta-cell function in the pathogenesis of type 2 diabetes. *Diabetes Care* **2013**, *36*, S113–S119. [CrossRef] [PubMed]
6. Kaufman, F.R. Type 1 diabetes mellitus. *Pediatr. Rev.* **2003**, *24*, 291–300. [CrossRef]
7. Soltész, G. Diabetes in the young: A paediatric and epidemiological perspective. *Diabetologia* **2003**, *46*, 447–454. [CrossRef]
8. Butler, P.C. The replication of beta cells in normal physiology, in disease and for therapy. *Nat. Clin. Pract. Endocrinol. Metab.* **2007**, *3*, 758–768. [CrossRef]
9. Gregg, B.E. Formation of a human beta-cell population within pancreatic islets is set early in life. *J. Clin. Endocrinol. Metab.* **2012**, *97*, 3197–3206. [CrossRef]
10. Chen, C.; Cohrs, C.M.; Stertmann, J.; Bozsak, R.; Speier, S. Human beta cell mass and function in diabetes: Recent advances in knowledge and technologies to understand disease pathogenesis. *Mol. Metab.* **2017**, *6*, 943–957. [CrossRef]
11. Ginsberg, H.N.; Maccallum, P.R. The obesity, metabolic syndrome, and type 2 diabetes mellitus pandemic:

Part I. Increased cardiovascular disease risk and the importance of atherogenic dyslipidemia in persons with the metabolic syndrome and type 2 diabetes mellitus. *J. CardioMetabolic Syndr.* **2009**, *4*, 113–119. [CrossRef]

12. Shin, J.-A.; Lee, J.-H.; Lim, S.-Y.; Ha, H.-S.; Kwon, H.-S.; Park, Y.; Lee, W.-C.; Kang, M.-I.; Yim, H.-W.; Yoon, K.-H.; et al. Metabolic syndrome as a predictor of type 2 diabetes, and its clinical interpretations and usefulness. *J. Diabetes Investig.* **2013**, *4*, 334–343. [CrossRef] [PubMed]

13. Bartel, D.P. MicroRNAs: Genomics, biogenesis, mechanism, and function. *Cell* **2004**, *116*, 281–297. [CrossRef]

14. Reinhart, B.J.; Slack, F.J.; Basson, M.; Pasquinelli, A.E.; Bettinger, J.C.; Rougvie, A.E.; Horvitz, H.R.; Ruvkun, G. The 21-nucleotide let-7 RNA regulates developmental timing in Caenorhabditis elegans. *Nature* **2000**, *403*, 901–906. [CrossRef] [PubMed]

15. Wightman, B.; Ha, I.; Ruvkun, G. Posttranscriptional regulation of the heterochronic gene lin-14 by lin-4 mediates temporal pattern formation in C. elegans. *Cell* **1993**, *75*, 855–862. [CrossRef]

16. Lee, R.; Gowrishankar, T.; Basch, R.; Patel, P.; Golan, D. Cell shape-dependent rectification of surface receptor transport in a sinusoidal electric field. *Biophys. J.* **1993**, *64*, 44–57. [CrossRef]

17. Pasquinelli, A.E.; Reinhart, B.J.; Slack, F.; Martindale, M.Q.; Kuroda, M.I.; Maller, B.; Hayward, D.C.; Ball, E.E.; Degnan, B.; Müller, P.; et al. Conservation of the sequence and temporal expression of let-7 heterochronic regulatory RNA. *Nature* **2000**, *408*, 86–89. [CrossRef]

18. Davis, C.A. The Encyclopedia of DNA elements (ENCODE): Data portal update. *Nucleic Acids Res.* **2018**, *46*, D794–D801. [CrossRef]

19. Ambros, V. microRNAs: Tiny regulators with great potential. *Cell* **2001**, *107*, 823–826. [CrossRef]

20. Hinske, L.C.G.; Galante, P.A.; Kuo, W.P.; Ohno-Machado, L. A potential role for intragenic miRNAs on their hosts' interactome. *BMC Genom.* **2010**, *11*, 533. [CrossRef]

21. Cifuentes, D.; Xue, H.; Taylor, D.W.; Patnode, H.; Mishima, Y.; Cheloufi, S.; Ma, E.; Mane, S.; Hannon, G.J.; Lawson, N.D.; et al. A novel miRNA processing pathway independent of Dicer requires Argonaute2 catalytic activity. *Science* **2010**, *328*, 1694–1698. [CrossRef]

22. Liu, Y.P.; Karg, M.; Harwig, A.; Herrera-Carrillo, E.; Jongejan, A.; Van Kampen, A.; Berkhout, B. Mechanistic insights on the Dicer-independent AGO2-mediated processing of AgoshRNAs. *RNA Boil.* **2015**, *12*, 92–100. [CrossRef] [PubMed]

23. Barad, O.; Mann, M.; Chapnik, E.; Shenoy, A.; Blelloch, R.; Barkai, N.; Hornstein, E. Efficiency and specificity in microRNA biogenesis. *Nat. Struct. Mol. Boil.* **2012**, *19*, 650–652. [CrossRef] [PubMed]

24. Hammond, S.M. An overview of microRNAs. *Adv. Drug Deliv. Rev.* **2015**, *87*, 3–14. [CrossRef] [PubMed]

25. Yekta, S.; Shih, I.-H.; Bartel, D.P. MicroRNA-Directed Cleavage of HOXB8 mRNA. *Science* **2004**, *304*, 594–596. [CrossRef]

26. Sarshad, A.A.; Juan, A.H.; Muler, A.I.C.; Anastasakis, D.G.; Wang, X.; Genzor, P.; Feng, X.; Tsai, P.-F.; Sun, H.-W.; Haase, A.D.; et al. Argonaute-miRNA Complexes Silence Target mRNAs in the Nucleus of Mammalian Stem Cells. *Mol. Cell* **2018**, *71*, 1040–1050. [CrossRef]

27. Ohanian, M.; Humphreys, D.T.; Anderson, E.; Preiss, T.; Fatkin, D. A heterozygous variant in the human cardiac miR-133 gene, MIR133A2, alters miRNA duplex processing and strand abundance. *BMC Genet.* **2013**, *14*, 18. [CrossRef]

28. Meijer, H.A.; Smith, E.M.; Bushell, M. Regulation of miRNA strand selection: Follow the leader? *Biochem. Soc. Trans.* **2014**, *42*, 1135–1140. [CrossRef]

29. Friedman, R.C. Most mammalian mRNAs are conserved targets of microRNAs. *Genome Res.* **2009**, *19*, 92–105. [CrossRef]

30. Schirle, N.T.; Sheu-Gruttadauria, J.; Macrae, I.J. Structural basis for microRNA targeting. *Science* **2014**, *346*, 608–613. [CrossRef]

31. Kozomara, A.; Birgaoanu, M.; Griffiths-Jones, S. miRBase: From microRNA sequences to function. *Nucleic Acids Res.* **2019**, *47*, D155–D162. [CrossRef]

32. Ambros, V. The functions of animal microRNAs. *Nature* **2004**, *431*, 350–355. [CrossRef] [PubMed]

33. Bernstein, E.; Kim, S.Y.; Carmell, M.A.; Murchison, E.P.; Alcorn, H.; Li, M.Z.; Mills, A.A.; Elledge, S.J.; Anderson, K.V.; Hannon, G.J. Dicer is essential for mouse development. *Nat. Genet.* **2003**, *35*, 215–217. [CrossRef] [PubMed]

34. Sun, X.; Zhang, Z.; McManus, M.T.; Harfe, B.D.; Harris, K.S. Dicer function is essential for lung epithelium morphogenesis. *Dev. Boil.* **2006**, *295*, 460. [CrossRef]

35. Kosik, K.S. The neuronal microRNA system. *Nat. Rev. Neurosci.* **2006**, *7*, 911–920. [CrossRef] [PubMed]

36. Guay, C.; Regazzi, R. Circulating microRNAs as novel biomarkers for diabetes mellitus. *Nat. Rev. Endocrinol.* **2013**, *9*, 513–521. [CrossRef] [PubMed]

37. Guay, C.; Regazzi, R. Exosomes as new players in metabolic organ cross-talk. *Diabetes Obes. Metab.* **2017**, *19*, 137–146. [CrossRef] [PubMed]

38. Baroukh, N. MicroRNA-124a regulates Foxa2 expression and intracellular signaling in pancreatic beta-cell lines. *J. Biol. Chem.* **2007**, *282*, 19575–19588. [CrossRef]

39. Joglekar, M.V.; Joglekar, V.M.; Hardikar, A.A. Expression of islet-specific microRNAs during human pancreatic development. *Gene Expr. Patterns* **2009**, *9*, 109–113. [CrossRef]

40. Lynn, F.C.; Skewes-Cox, P.; Kosaka, Y.; McManus, M.; Harfe, B.D.; German, M.S. MicroRNA Expression Is Required for Pancreatic Islet Cell Genesis in the Mouse. *Diabetes* **2007**, *56*, 2938–2945. [CrossRef]

41. Poy, M.N.; Eliasson, L.; Krutzfeldt, J.; Kuwajima, S.; Ma, X.; Macdonald, P.E.; Pfeffer, S.; Tuschl, T.; Rajewsky, N.; Rorsman, P.; et al. A pancreatic islet-specific microRNA regulates insulin secretion. *Nature* **2004**, *432*, 226–230. [CrossRef]

42. Kaspi, H.; Pasvolsky, R.; Hornstein, E. Could microRNAs contribute to the maintenance of beta cell identity? *Trends Endocrinol. Metab.* **2014**, *25*, 285–292. [CrossRef] [PubMed]

43. Kālis, M.; Bolmeson, C.; Esguerra, J.L.S.; Gupta, S.; Edlund, A.; Tormo-Badia, N.; Speidel, D.; Holmberg, D.; Mayans, S.; Khoo, N.K.S.; et al. Beta-Cell Specific Deletion of Dicer1 Leads to Defective Insulin Secretion and Diabetes Mellitus. *PLoS ONE* **2011**, *6*, e29166. [CrossRef] [PubMed]

44. Mandelbaum, A.D.; Melkman-Zehavi, T.; Oren, R.; Kredo-Russo, S.; Nir, T.; Dor, Y.; Hornstein, E. Dysregulation of Dicer1 in Beta Cells Impairs Islet Architecture and Glucose Metabolism. *Exp. Diabetes Res.* **2012**, *2012*, 1–8. [CrossRef] [PubMed]

45. Berry, C.; Lal, M.; Binukumar, B.K. Crosstalk Between the Unfolded Protein Response, MicroRNAs, and Insulin Signaling Pathways: In Search of Biomarkers for the Diagnosis and Treatment of Type 2 Diabetes. *Front. Endocrinol.* **2018**, *9*, 210. [CrossRef] [PubMed]

46. Dotta, F. MicroRNAs: Markers of beta-cell stress and autoimmunity. *Curr. Opin. Endocrinol. Diabetes Obes.* **2018**, *25*, 237–245. [CrossRef]

47. Feng, J.; Xing, W.; Xie, L. Regulatory Roles of MicroRNAs in Diabetes. *Int. J. Mol. Sci.* **2016**, *17*, 1729. [CrossRef]

48. Lapierre, M.P.; Stoffel, M. MicroRNAs as stress regulators in pancreatic beta cells and diabetes. *Mol. Metab.* **2017**, *6*, 1010–1023. [CrossRef]

49. Bravo-Egana, V. Quantitative differential expression analysis reveals miR-7 as major islet microRNA. *Biochem. Biophys. Res. Commun.* **2008**, *366*, 922–926. [CrossRef]

50. Correa-Medina, M.; Bravo-Egana, V.; Rosero, S.; Ricordi, C.; Edlund, H.; Diez, J.; Pastori, R.L. MicroRNA miR-7 is preferentially expressed in endocrine cells of the developing and adult human pancreas. *Gene Expr. Patterns* **2009**, *9*, 193–199. [CrossRef]

51. Rosero, S.; Bravo-Egana, V.; Jiang, Z.; Khuri, S.; Tsinoremas, N.; Klein, D.; Sabates, E.; Correa-Medina, M.; Ricordi, C.; Domínguez-Bendala, J.; et al. MicroRNA signature of the human developing pancreas. *BMC Genom.* **2010**, *11*, 509. [CrossRef]

52. Klein, D.; Misawa, R.; Bravo-Egana, V.; Vargas, N.; Rosero, S.; Piroso, J.; Ichii, H.; Umland, O.; Zhijie, J.; Tsinoremas, N.; et al. MicroRNA Expression in Alpha and Beta Cells of Human Pancreatic Islets. *PLoS ONE* **2013**, *8*, e55064. [CrossRef] [PubMed]

53. Kloosterman, W.P.; Lagendijk, A.K.; Ketting, R.F.; Moulton, J.D.; Plasterk, R.H.A. Targeted inhibition of miRNA maturation with morpholinos reveals a role for miR-375 in pancreatic islet development. *PLoS Boil.* **2007**, *5*, e203. [CrossRef] [PubMed]

54. Kredo-Russo, S.; Ness, A.N.; Mandelbaum, A.D.; Walker, M.D.; Hornstein, E. Regulation of Pancreatic microRNA-7 Expression. *Exp. Diabetes Res.* **2012**, *2012*, 1–7. [CrossRef] [PubMed]

55. Latreille, M. MicroRNA-7a regulates pancreatic beta cell function. *J. Clin. Investig.* **2014**, *124*, 2722–2735. [CrossRef] [PubMed]

56. Nieto, M.; Hevia, P.; Garcia, E.; Klein, D.; Álvarez-Cubela, S.; Bravo-Egana, V.; Rosero, S.; Molano, R.D.; Vargas, N.; Ricordi, C.; et al. Antisense miR-7 Impairs Insulin Expression in Developing Pancreas and in Cultured Pancreatic Buds. *Cell Transplant.* **2012**, *21*, 1761–1774. [CrossRef] [PubMed]

57. Poy, M.N.; Hausser, J.; Trajkovski, M.; Braun, M.; Collins, S.; Rorsman, P.; Zavolan, M.; Stoffel, M. miR-375 maintains normal pancreatic alpha- and beta-cell mass. *Proc. Natl. Acad. Sci. USA* **2009**, *106*, 5813–5818. [CrossRef]

58. López-Beas, J.; Capilla-Gonzalez, V.; Aguilera, Y.; Mellado, N.; Lachaud, C.C.; Martín, F.; Smani, T.; Soria, B.; Hmadcha, A. miR-7 Modulates hESC Differentiation into Insulin-Producing Beta-like Cells and Contributes to Cell Maturation. *Mol. Ther. Nucleic Acids* **2018**, *12*, 463–477. [CrossRef]

59. Nathan, G. MiR-375 promotes redifferentiation of adult human beta cells expanded in vitro. *PLoS ONE* **2015**, *10*, e0122108. [CrossRef]

60. Shaer, A.; Azarpira, N.; Karimi, M.H. Differentiation of Human Induced Pluripotent Stem Cells into Insulin-Like Cell Clusters with miR-186 and miR-375 by using chemical transfection. *Appl. Biochem. Biotechnol.* **2014**, *174*, 242–258. [CrossRef]

61. Wei, R.; Yang, J.; Liu, G.-Q.; Gao, M.-J.; Hou, W.-F.; Zhang, L.; Gao, H.-W.; Liu, Y.; Chen, G.-A.; Hong, T.-P. Dynamic expression of microRNAs during the differentiation of human embryonic stem cells into insulin-producing cells. *Gene* **2013**, *518*, 246–255. [CrossRef]

62. Martinez-Sanchez, A.; Rutter, G.A.; Latreille, M. MiRNAs in beta-Cell Development, Identity, and Disease. *Front. Genet.* **2016**, *7*, 226. [PubMed]

63. Sies, H. Oxidative stress: Oxidants and antioxidants. *Exp. Physiol.* **1997**, *82*, 291–295. [CrossRef] [PubMed]

64. Sies, H.; Cadenas, E. Oxidative stress: Damage to intact cells and organs. *Philos Trans. R Soc. Lond. B Biol. Sci.* **1985**, *311*, 617–631. [CrossRef] [PubMed]

65. Cadenas, E.; Davies, K.J. Mitochondrial free radical generation, oxidative stress, and aging. *Free. Radic. Boil. Med.* **2000**, *29*, 222–230. [CrossRef]

66. Cheifetz, S.; Li, I.W.; McCulloch, C.A.; Sampath, K.; Sodek, J. Influence of osteogenic protein-1 (OP-1;BMP-7) and transforming growth factor-beta 1 on bone formation in vitro. *Connect. Tissue Res.* **1996**, *35*, 71–78. [CrossRef] [PubMed]

67. Atari-Hajipirloo, S.; Valizadeh, N.; Khadem-Ansari, M.-H.; Rasmi, Y.; Kheradmand, F. Altered Concentrations of Copper, Zinc, and Iron are Associated With Increased Levels of Glycated Hemoglobin in Patients With Type 2 Diabetes Mellitus and Their First-Degree Relatives. *Int. J. Endocrinol. Metab.* **2016**, *14*, 33273. [CrossRef]

68. Giulivi, C.; Boveris, A.; Cadenas, E. Hydroxyl Radical Generation during Mitochondrial Electron-Transfer and the Formation of 8-Hydroxydesoxyguanosine in Mitochondrial-DNA. *Arch. Biochem. Biophys.* **1995**, *316*, 909–916. [CrossRef]

69. Pi, J.; Bai, Y.; Zhang, Q.; Wong, V.; Floering, L.M.; Daniel, K.; Reece, J.M.; Deeney, J.T.; Andersen, M.E.; Corkey, B.E.; et al. Reactive Oxygen Species as a Signal in Glucose-Stimulated Insulin Secretion. *Diabetes* **2007**, *56*, 1783–1791. [CrossRef]

70. Roy, J.; Galano, J.-M.; Durand, T.; Le Guennec, J.-Y.; Lee, J.C.-Y. Physiological role of reactive oxygen species as promoters of natural defenses. *FASEB J.* **2017**, *31*, 3729–3745. [CrossRef]

71. Pizzino, G.; Irrera, N.; Cucinotta, M.; Pallio, G.; Mannino, F.; Arcoraci, V.; Squadrito, F.; Altavilla, D.; Bitto, A. Oxidative Stress: Harms and Benefits for Human Health. *Oxidative Med. Cell. Longev.* **2017**, *2017*, 1–13. [CrossRef]

72. Di Meo, S.; Reed, T.T.; Venditti, P.; Victor, V.M. Role of ROS and RNS Sources in Physiological and Pathological Conditions. *Oxidative Med. Cell. Longev.* **2016**, *2016*, 1–44. [CrossRef] [PubMed]

73. Barrera, G.; Pizzimenti, S.; Daga, M.; Dianzani, C.; Arcaro, A.; Cetrangolo, G.P.; Giordano, G.; Cucci, M.A.; Graf, M.; Gentile, F. Lipid Peroxidation-Derived Aldehydes, 4-Hydroxynonenal and Malondialdehyde in Aging-Related Disorders. *Antioxidants* **2018**, *7*, 102. [CrossRef] [PubMed]

74. Vaca, C.; Wilhelm, J.; Harms-Ringdahl, M. Interaction of lipid peroxidation products with DNA. A review. *Mutat. Res. Genet. Toxicol.* **1988**, *195*, 137–149. [CrossRef]

75. Ayala, A.; Muñoz, M.F.; Argüelles, S. Lipid Peroxidation: Production, Metabolism, and Signaling Mechanisms of Malondialdehyde and 4-Hydroxy-2-Nonenal. *Oxidative Med. Cell. Longev.* **2014**, *2014*, 1–31. [CrossRef] [PubMed]

76. Schieber, M.; Chandel, N.S. ROS function in redox signaling and oxidative stress. *Curr. Boil.* **2014**, *24*, R453–R462. [CrossRef]

77. Gaschler, M.M.; Stockwell, B.R. Lipid peroxidation in cell death. *Biochem. Biophys. Res. Commun.* **2017**, *482*, 419–425. [CrossRef]

78. Cadet, J.; Wagner, J.R. DNA Base Damage by Reactive Oxygen Species, Oxidizing Agents, and UV Radiation. *Cold Spring Harb. Perspect. Boil.* **2013**, *5*, a012559. [CrossRef]

79. Lenzen, S. Chemistry and biology of reactive species with special reference to the antioxidative defence status in pancreatic beta-cells. *Biochim. Biophys. Acta Gen. Subj.* **2017**, *1861*, 1929–1942. [CrossRef]

80. Shen, X.; Zheng, S.; Metreveli, N.S.; Epstein, P.N. Protection of cardiac mitochondria by overexpression of MnSOD reduces diabetic cardiomyopathy. *Diabetes* **2006**, *55*, 798–805. [CrossRef]

81. Ye, G.; Metreveli, N.S.; Donthi, R.V.; Xia, S.; Xu, M.; Carlson, E.C.; Epstein, P.N. Catalase protects cardiomyocyte function in models of type 1 and type 2 diabetes. *Diabetes* **2004**, *53*, 1336–1343. [CrossRef]

82. Brand, M.D.; Esteves, T.C. Physiological functions of the mitochondrial uncoupling proteins UCP2 and UCP3. *Cell Metab.* **2005**, *2*, 85–93. [CrossRef] [PubMed]

83. Pierelli, G.; Stanzione, R.; Forte, M.; Migliarino, S.; Perelli, M.; Volpe, M.; Rubattu, S. Uncoupling Protein 2: A Key Player and a Potential Therapeutic Target in Vascular Diseases. *Oxidative Med. Cell. Longev.* **2017**, *2017*, 1–11. [CrossRef] [PubMed]

84. Giralt, M.; Villarroya, F. Mitochondrial Uncoupling and the Regulation of Glucose Homeostasis. *Curr. Diabetes Rev.* **2017**, *13*, 386–394. [CrossRef] [PubMed]

85. Tonelli, C.; Chio, I.I.C.; Tuveson, D.A. Transcriptional Regulation by Nrf2. *Antioxid. Redox Signal* **2018**, *29*, 1727–1745. [CrossRef] [PubMed]

86. Assar, M.E.; Angulo, J.; Rodriguez-Manas, L. Diabetes and ageing-induced vascular inflammation. *J. Physiol.* **2016**, *594*, 2125–2146. [CrossRef]

87. Rabbani, P.S. Dysregulation of Nrf2/Keap1 Redox Pathway in Diabetes Affects Multipotency of Stromal Cells. *Diabetes* **2019**, *68*, 141–155. [CrossRef] [PubMed]

88. Simmons, K.M.; Michels, A.W. Type 1 diabetes: A predictable disease. *World J. Diabetes* **2015**, *6*, 380–390. [CrossRef]

89. Veld, P.I. Insulitis in human type 1 diabetes: A comparison between patients and animal models. *Semin. Immunopathol.* **2014**, *36*, 569–579. [CrossRef]

90. Nakayama, M. Insulin as a key autoantigen in the development of type 1 diabetes. *Diabetes Metab. Res. Rev.* **2011**, *27*, 773–777. [CrossRef]

91. Delmastro, M.M.; Piganelli, J.D. Oxidative Stress and Redox Modulation Potential in Type 1 Diabetes. *Clin. Dev. Immunol.* **2011**, *2011*, 1–15. [CrossRef]

92. Atkinson, M.A.; Eisenbarth, G.S. Type 1 diabetes: New perspectives on disease pathogenesis and treatment. *Lancet* **2001**, *358*, 221–229. [CrossRef]

93. Padgett, L.E.; Broniowska, K.A.; Hansen, P.A.; Corbett, J.A.; Tse, H.M. The role of reactive oxygen species and proinflammatory cytokines in type 1 diabetes pathogenesis. *Ann. N. Y. Acad. Sci.* **2013**, *1281*, 16–35. [CrossRef] [PubMed]

94. Feduska, J.M.; Tse, H.M. The proinflammatory effects of macrophage-derived NADPH oxidase function in autoimmune diabetes. *Free. Radic. Boil. Med.* **2018**, *125*, 81–89. [CrossRef] [PubMed]

95. Ali, M.K. Diabetes: An Update on the Pandemic and Potential Solutions. In *Cardiovascular, Respiratory, and Related Disorders*; World Bank and Oxford University Press: Washington, DC, USA, 2017.

96. Heitmeier, M.R. Pancreatic beta-cell damage mediated by beta-cell production of interleukin-1. A novel mechanism for virus-induced diabetes. *J. Biol. Chem.* **2001**, *276*, 11151–11158. [CrossRef]

97. Meares, G.P. Differential responses of pancreatic beta-cells to ROS and RNS. *Am. J. Physiol. Endocrinol. Metab.* **2013**, *304*, E614–E622. [CrossRef] [PubMed]

98. Oleson, B.J. Nitric Oxide Suppresses beta-Cell Apoptosis by Inhibiting the DNA Damage Response. *Mol. Cell Biol.* **2016**, *36*, 2067–2077. [CrossRef] [PubMed]

99. Pugliese, A. Insulitis in the pathogenesis of type 1 diabetes. *Pediatr. Diabetes* **2016**, *17*, 31–36. [CrossRef]

100. Matteucci, E.; Cinapri, V.; Quilici, S.; Forotti, G.; Giampietro, O. Oxidative stress in families of type 1 diabetic patients. *Diabetes Res. Clin. Pr.* **2000**, *50*, 308. [CrossRef]

101. Saklayen, M.G. The Global Epidemic of the Metabolic Syndrome. *Curr. Hypertens. Rep.* **2018**, *20*, 12. [CrossRef]

102. Newsholme, P. Oxidative stress pathways in pancreatic beta cells and insulin sensitive cells and tissues-importance to cell metabolism, function and dysfunction. *Am. J. Physiol. Cell Physiol.* **2019**. [CrossRef]

103. Evans, J.L.; Goldfine, I.D.; Maddux, B.A.; Grodsky, G.M. Are oxidative stress-activated signaling pathways mediators of insulin resistance and beta-cell dysfunction? *Diabetes* **2003**, *52*, 1–8. [CrossRef] [PubMed]

104. Gao, D.; Nong, S.; Huang, X.; Lu, Y.; Zhao, H.; Lin, Y.; Man, Y.; Wang, S.; Yang, J.; Li, J. The Effects of Palmitate on Hepatic Insulin Resistance Are Mediated by NADPH Oxidase 3-derived Reactive Oxygen Species through JNK and p38MAPK Pathways*. *J. Boil. Chem.* **2010**, *285*, 29965–29973. [CrossRef] [PubMed]

105. Guo, X.-X.; An, S.; Yang, Y.; Liu, Y.; Hao, Q.; Tang, T.; Xu, T.-R. Emerging role of the Jun N-terminal kinase interactome in human health. *Cell Boil. Int.* **2018**, *42*, 756–768. [CrossRef] [PubMed]

106. Patel, S.; Santani, D. Role of NF-kappa B in the pathogenesis of diabetes and its associated complications. *Pharmacol. Rep.* **2009**, *61*, 595–603. [CrossRef]

107. Chilelli, N.C.; Burlina, S.; Lapolla, A. AGEs, rather than hyperglycemia, are responsible for microvascular complications in diabetes: A "glycoxidation-centric" point of view. *Nutr. Metab. Cardiovasc. Dis.* **2013**, *23*, 913–919. [CrossRef] [PubMed]

108. Jud, P.; Sourij, H.; Philipp, J.; Harald, S. Therapeutic options to reduce advanced glycation end products in patients with diabetes mellitus: A review. *Diabetes Res. Clin. Pr.* **2019**, *148*, 54–63. [CrossRef] [PubMed]

109. Brandon, A.E.; Liao, B.M.; Diakanastasis, B.; Parker, B.L.; Raddatz, K.; McManus, S.A.; O'Reilly, L.; Kimber, E.; Van Der Kraan, A.G.; Hancock, D.; et al. Protein Kinase C Epsilon Deletion in Adipose Tissue, but Not in Liver, Improves Glucose Tolerance. *Cell Metab.* **2019**, *29*, 183–191. [CrossRef]

110. Fleming, A.K.; Storz, P. Protein kinase C isoforms in the normal pancreas and in pancreatic disease. *Cell. Signal.* **2017**, *40*, 1–9. [CrossRef]

111. Moser, B.; Schumacher, C.; Von Tscharner, V.; Clark-Lewis, I.; Baggiolini, M. Neutrophil-activating peptide 2 and gro/melanoma growth-stimulatory activity interact with neutrophil-activating peptide 1/interleukin 8 receptors on human neutrophils. *J. Boil. Chem.* **1991**, *266*, 10666–10671.

112. Giacco, F.; Brownlee, M. Oxidative stress and diabetic complications. *Circ. Res.* **2010**, *107*, 1058–1070. [CrossRef]

113. Yan, L.-J. Redox imbalance stress in diabetes mellitus: Role of the polyol pathway. *Anim. Model. Exp. Med.* **2018**, *1*, 7–13. [CrossRef] [PubMed]

114. Isaacs, S.R. MicroRNAs in Type 1 Diabetes: Complex Interregulation of the Immune System, beta Cell Function and Viral Infections. *Curr. Diabetes Rep.* **2016**, *16*, 133. [CrossRef] [PubMed]

115. Pileggi, A.; Klein, D.; Fotino, C.; Bravo-Egana, V.; Rosero, S.; Doni, M.; Podetta, M.; Ricordi, C.; Molano, R.D.; Pastori, R.L. MicroRNAs in islet immunobiology and transplantation. *Immunol. Res.* **2013**, *57*, 185–196. [CrossRef] [PubMed]

116. Zheng, Y.; Wang, Z.; Zhou, Z. miRNAs: Novel regulators of autoimmunity-mediated pancreatic beta-cell destruction in type 1 diabetes. *Cell Mol. Immunol.* **2017**, *14*, 488–496. [CrossRef]

117. Engedal, N.; Žerovnik, E.; Rudov, A.; Galli, F.; Olivieri, F.; Procopio, A.D.; Rippo, M.R.; Monsurrò, V.; Betti, M.; Albertini, M.C. From Oxidative Stress Damage to Pathways, Networks, and Autophagy via MicroRNAs. *Oxidative Med. Cell. Longev.* **2018**, *2018*, 4968321. [CrossRef]

118. Yang, X.; Li, X.; Lin, Q.; Xu, Q. Up-regulation of microRNA-203 inhibits myocardial fibrosis and oxidative stress in mice with diabetic cardiomyopathy through the inhibition of PI3K/Akt signaling pathway via PIK3CA. *Gene* **2019**, *715*, 143995. [CrossRef]

119. Dieter, C.; Assmann, T.S.; Costa, A.R.; Canani, L.H.; De Souza, B.M.; Bauer, A.C.; Crispim, D. MiR-30e-5p and MiR-15a-5p Expressions in Plasma and Urine of Type 1 Diabetic Patients With Diabetic Kidney Disease. *Front. Genet.* **2019**, *10*, 563. [CrossRef]

120. Chen, P.; Miao, Y.; Yan, P.; Wang, X.J.; Jiang, C.; Lei, Y. MiR-455-5p ameliorates HG-induced apoptosis, oxidative stress and inflammatory via targeting SOCS3 in retinal pigment epithelial cells. *J. Cell. Physiol.* **2019**, *234*, 21915–21924. [CrossRef]

121. Yang, S.; Fei, X.; Lu, Y.; Xu, B.; Ma, Y.; Wan, H. miRNA-214 suppresses oxidative stress in diabetic nephropathy via the ROS/Akt/mTOR signaling pathway and uncoupling protein 2. *Exp. Ther. Med.* **2019**, *17*, 3530–3538. [CrossRef]

122. Sun, L.-L.; Jiang, B.-G.; Li, W.-T.; Zou, J.-J.; Shi, Y.-Q.; Liu, Z.-M. MicroRNA-15a positively regulates insulin synthesis by inhibiting uncoupling protein-2 expression. *Diabetes Res. Clin. Pr.* **2011**, *91*, 94–100. [CrossRef]

123. Jiang, L. A microRNA-30e/mitochondrial uncoupling protein 2 axis mediates TGF-beta1-induced tubular epithelial cell extracellular matrix production and kidney fibrosis. *Kidney Int.* **2013**, *84*, 285–296. [CrossRef] [PubMed]

124. Yin, Z. MiR-30c/PGC-1beta protects against diabetic cardiomyopathy via PPARalpha. *Cardiovasc. Diabetol.* **2019**, *18*, 7. [CrossRef] [PubMed]

125. Ding, X.; Jian, T.; Wu, Y.; Zuo, Y.; Li, J.; Lv, H.; Ma, L.; Ren, B.; Zhao, L.; Li, W.; et al. Ellagic acid ameliorates oxidative stress and insulin resistance in high glucose-treated HepG2 cells via miR-223/keap1-Nrf2 pathway. *Biomed. Pharmacother.* **2019**, *110*, 85–94. [CrossRef]

126. Kubota, K.; Nakano, M.; Kobayashi, E.; Mizue, Y.; Chikenji, T.; Otani, M.; Nagaishi, K.; Fujimiya, M. An enriched environment prevents diabetes-induced cognitive impairment in rats by enhancing exosomal miR-146a secretion from endogenous bone marrow-derived mesenchymal stem cells. *PLoS ONE* **2018**, *13*, e0204252. [CrossRef] [PubMed]

127. Wan, R.J.; Li, Y.H. MicroRNA-146a/NAPDH oxidase4 decreases reactive oxygen species generation and inflammation in a diabetic nephropathy model. *Mol. Med. Rep.* **2018**, *17*, 4759–4766. [CrossRef] [PubMed]

128. Miao, Y.; Wan, Q.; Liu, X.; Wang, Y.; Luo, Y.; Liu, D.; Lin, N.; Zhou, H.; Zhong, J. miR-503 Is Involved in the Protective Effect of Phase II Enzyme Inducer (CPDT) in Diabetic Cardiomyopathy via Nrf2/ARE Signaling Pathway. *BioMed Res. Int.* **2017**, *2017*, 1–10. [CrossRef] [PubMed]

129. Zhao, Y. Oxidative stress-induced miR-27a targets the redox gene nuclear factor erythroid 2-related factor 2 in diabetic embryopathy. *Am. J. Obstet. Gynecol.* **2018**, *218*, 136 e1–136 e10. [CrossRef]

130. Wang, J.; Zhang, J.; Chen, X.; Yang, Y.; Wang, F.; Li, W.; Awuti, M.; Sun, Y.; Lian, C.; Li, Z.; et al. miR-365 promotes diabetic retinopathy through inhibiting Timp3 and increasing oxidative stress. *Exp. Eye Res.* **2018**, *168*, 89–99. [CrossRef]

131. Wang, F.; Xu, C.; Reece, E.A.; Li, X.; Wu, Y.; Harman, C.; Yu, J.; Dong, D.; Wang, C.; Yang, P.; et al. Protein kinase C-alpha suppresses autophagy and induces neural tube defects via miR-129-2 in diabetic pregnancy. *Nat. Commun.* **2017**, *8*, 15182. [CrossRef]

132. Chen, D.-L.; Yang, K.-Y.; Chen, D.; Yang, K. Berberine Alleviates Oxidative Stress in Islets of Diabetic Mice by Inhibiting miR-106b Expression and Up-Regulating SIRT1. *J. Cell. Biochem.* **2017**, *118*, 4349–4357. [CrossRef]

133. Wu, Y. MiR-106a Associated with Diabetic Peripheral Neuropathy through the Regulation of 12/15-LOX-meidiated Oxidative/Nitrative Stress. *Curr. Neurovasc. Res.* **2017**, *14*, 117–124. [CrossRef] [PubMed]

134. Garcia-Morales, V.; Friedrich, J.; Jorna, L.M.; Campos-Toimil, M.; Hammes, H.-P.; Schmidt, M.; Krenning, G. The microRNA-7-mediated reduction in EPAC-1 contributes to vascular endothelial permeability and eNOS uncoupling in murine experimental retinopathy. *Acta Diabetol.* **2017**, *54*, 581–591. [CrossRef] [PubMed]

135. Li, H.H. miR-302 Attenuates Amyloid-beta-Induced Neurotoxicity through Activation of Akt Signaling. *J. Alzheimers Dis.* **2016**, *50*, 1083–1098. [CrossRef] [PubMed]

136. Xiao, D.; Zhou, T.; Fu, Y.; Wang, R.; Zhang, H.; Li, M.; Lin, Y.; Li, Z.; Xu, C.; Yang, B.; et al. MicroRNA-17 impairs glucose metabolism in insulin-resistant skeletal muscle via repressing glucose transporter 4 expression. *Eur. J. Pharmacol.* **2018**, *838*, 170–176. [CrossRef] [PubMed]

137. Hui, Y.; Yin, Y. MicroRNA-145 attenuates high glucose-induced oxidative stress and inflammation in retinal endothelial cells through regulating TLR4/NF-kappaB signaling. *Life Sci.* **2018**, *207*, 212–218. [CrossRef] [PubMed]

138. Zhang, Y.; Song, C.; Liu, J.; Bi, Y.; Li, H. Inhibition of miR-25 aggravates diabetic peripheral neuropathy. *NeuroReport* **2018**, *29*, 945–953. [CrossRef] [PubMed]

139. Liu, Y.; Li, H.; Liu, J.; Han, P.; Li, X.; Bai, H.; Zhang, C.; Sun, X.; Teng, Y.; Zhang, Y.; et al. Variations in MicroRNA-25 Expression Influence the Severity of Diabetic Kidney Disease. *J. Am. Soc. Nephrol.* **2017**, *28*, 3627–3638. [CrossRef]

140. Li, H.; Zhu, X.; Zhang, J.; Shi, J. MicroRNA-25 inhibits high glucose-induced apoptosis in renal tubular epithelial cells via PTEN/AKT pathway. *Biomed. Pharmacother.* **2017**, *96*, 471–479. [CrossRef]

141. Filios, S.R.; Xu, G.; Chen, J.; Hong, K.; Jing, G.; Shalev, A. MicroRNA-200 Is Induced by Thioredoxin-interacting Protein and Regulates Zeb1 Protein Signaling and Beta Cell Apoptosis*. *J. Boil. Chem.* **2014**, *289*, 36275–36283. [CrossRef]

142. Zhang, H.; Liu, J.; Qu, D.; Wang, L.; Luo, J.-Y.; Lau, C.W.; Gao, Z.; Tipoe, G.L.; Lee, H.K.; Ng, C.F.; et al. Inhibition of miR-200c restores endothelial function in diabetic mice through suppression of COX-2. *Diabetes* **2016**, *65*, 1196–1207. [CrossRef]

143. Lo, W.-Y.; Yang, W.-K.; Peng, C.-T.; Pai, W.-Y.; Wang, H.-J. MicroRNA-200a/200b Modulate High Glucose-Induced Endothelial Inflammation by Targeting O-linked N-Acetylglucosamine Transferase Expression. *Front. Physiol.* **2018**, *9*, 355. [CrossRef] [PubMed]

144. Shang, F.; Wang, S.-C.; Hsu, C.-Y.; Miao, Y.; Martin, M.; Yin, Y.; Wu, C.-C.; Wang, Y.-T.; Wu, G.; Chien, S.; et al. MicroRNA-92a Mediates Endothelial Dysfunction in CKD. *J. Am. Soc. Nephrol.* **2017**, *28*, 3251–3261. [CrossRef] [PubMed]

145. Gou, L.; Zhao, L.; Song, W.; Wang, L.; Liu, J.; Zhang, H.; Huang, Y.; Lau, C.W.; Yao, X.; Tian, X.Y.; et al. Inhibition of miR-92a Suppresses Oxidative Stress and Improves Endothelial Function by Upregulating Heme Oxygenase-1 in db/db Mice. *Antioxid. Redox Signal.* **2018**, *28*, 358–370. [CrossRef] [PubMed]

146. Reddy, M.A.; Jin, W.; Villeneuve, L.; Wang, M.; Lanting, L.; Todorov, I.; Kato, M.; Natarajan, R. Pro-inflammatory role of microrna-200 in vascular smooth muscle cells from diabetic mice. *Arter. Thromb. Vasc. Boil.* **2012**, *32*, 721–729. [CrossRef]

147. Carlomosti, F.; D'Agostino, M.; Beji, S.; Torcinaro, A.; Rizzi, R.; Zaccagnini, G.; Maimone, B.; Di Stefano, V.; De Santa, F.; Cordisco, S.; et al. Oxidative Stress-Induced miR-200c Disrupts the Regulatory Loop Among SIRT1, FOXO1, and eNOS. *Antioxid. Redox Signal.* **2017**, *27*, 328–344. [CrossRef]

148. Togliatto, G. Unacylated ghrelin induces oxidative stress resistance in a glucose intolerance and peripheral artery disease mouse model by restoring endothelial cell miR-126 expression. *Diabetes* **2015**, *64*, 1370–1382. [CrossRef]

149. Mahavadi, S.; Sriwai, W.; Manion, O.; Grider, J.R.; Murthy, K.S. Diabetes-induced oxidative stress mediates upregulation of RhoA/Rho kinase pathway and hypercontractility of gastric smooth muscle. *PLoS ONE* **2017**, *12*, 0178574. [CrossRef]

150. La Sala, L.; Mrakic-Sposta, S.; Tagliabue, E.; Prattichizzo, F.; Micheloni, S.; Sangalli, E.; Specchia, C.; Uccellatore, A.C.; Lupini, S.; Spinetti, G.; et al. Circulating microRNA-21 is an early predictor of ROS-mediated damage in subjects with high risk of developing diabetes and in drug-naïve T2D. *Cardiovasc. Diabetol.* **2019**, *18*, 18. [CrossRef]

151. La Sala, L.; Mrakic-Sposta, S.; Micheloni, S.; Prattichizzo, F.; Ceriello, A. Glucose-sensing microRNA-21 disrupts ROS homeostasis and impairs antioxidant responses in cellular glucose variability. *Cardiovasc. Diabetol.* **2018**, *17*, 105. [CrossRef]

152. Gao, L.; Liu, Y.; Guo, S.; Xiao, L.; Wu, L.; Wang, Z.; Liang, C.; Yao, R.; Zhang, Y. LAZ3 protects cardiac remodeling in diabetic cardiomyopathy via regulating miR-21/PPARa signaling. *Biochim. Biophys. Acta BBA Mol. Basis Dis.* **2018**, *1864*, 3322–3338. [CrossRef]

153. Murray, A.R.; Chen, Q.; Takahashi, Y.; Zhou, K.K.; Park, K.; Ma, J.-X. MicroRNA-200b Downregulates Oxidation Resistance 1 (Oxr1) Expression in the Retina of Type 1 Diabetes Model. *Investig. Opthalmol. Vis. Sci.* **2013**, *54*, 1689–1697. [CrossRef] [PubMed]

154. Kamalden, T.A.; Macgregor-Das, A.M.; Kannan, S.M.; Dunkerly-Eyring, B.; Khaliddin, N.; Xu, Z.; Fusco, A.P.; Abu Yazib, S.; Chow, R.C.; Duh, E.J.; et al. Exosomal MicroRNA-15a Transfer from the Pancreas Augments Diabetic Complications by Inducing Oxidative Stress. *Antioxid. Redox Signal.* **2017**, *27*, 913–930. [CrossRef] [PubMed]

155. Song, J.; Zhang, H.; Sun, Y.; Guo, R.; Zhong, D.; Xu, R.; Song, M. Omentin-1 protects renal function of mice with type 2 diabetic nephropathy via regulating miR-27a-Nrf2/Keap1 axis. *Biomed. Pharmacother.* **2018**, *107*, 440–446. [CrossRef] [PubMed]

156. Li, Q.; Kim, Y.-R.; Vikram, A.; Kumar, S.; Kassan, M.; Gabani, M.; Lee, S.K.; Jacobs, J.S.; Irani, K. P66Shc-Induced MicroRNA-34a Causes Diabetic Endothelial Dysfunction by Downregulating Sirtuin1. *Arter. Thromb. Vasc. Boil.* **2016**, *36*, 2394–2403. [CrossRef]

157. Kassan, M.; Vikram, A.; Li, Q.; Kim, Y.-R.; Kumar, S.; Gabani, M.; Liu, J.; Jacobs, J.S.; Irani, K. MicroRNA-204 promotes vascular endoplasmic reticulum stress and endothelial dysfunction by targeting Sirtuin1. *Sci. Rep.* **2017**, *7*, 9308. [CrossRef]

158. Li, X.; Wang, H.; Yao, B.; Xu, W.; Chen, J.; Zhou, X. lncRNA H19/miR-675 axis regulates cardiomyocyte apoptosis by targeting VDAC1 in diabetic cardiomyopathy. *Sci. Rep.* **2016**, *6*, 36340. [CrossRef]

159. Zheng, D.; Ma, J.; Yu, Y.; Li, M.; Ni, R.; Wang, G.; Chen, R.; Li, J.; Fan, G.-C.; Lacefield, J.C.; et al. Silencing of miR-195 reduces diabetic cardiomyopathy in C57BL/6 mice. *Diabetologia* **2015**, *58*, 1949–1958. [CrossRef]

160. Zhang, R.; Garrett, Q.; Zhou, H.; Wu, X.; Mao, Y.; Cui, X.; Xie, B.; Liu, Z.; Cui, D.; Jiang, L.; et al. Upregulation of miR-195 accelerates oxidative stress-induced retinal endothelial cell injury by targeting mitofusin 2 in diabetic rats. *Mol. Cell. Endocrinol.* **2017**, *452*, 33–43. [CrossRef]

161. Gu, H. The miR-322-TRAF3 circuit mediates the pro-apoptotic effect of high glucose on neural stem cells. *Toxicol. Sci.* **2015**, *144*, 186–196. [CrossRef]

162. Wu, K.; Yang, Y.; Zhong, Y.; Ammar, H.M.; Zhang, P.; Guo, R.; Liu, H.; Cheng, C.; Koroscil, T.M.; Chen, Y.; et al. The effects of microvesicles on endothelial progenitor cells are compromised in type 2 diabetic patients via downregulation of the miR-126/VEGFR2 pathway. *Am. J. Physiol. Metab.* **2016**, *310*, E828–E837. [CrossRef]

163. Wang, J.M. MicroRNA miR-27b rescues bone marrow-derived angiogenic cell function and accelerates wound healing in type 2 diabetes mellitus. *Arter. Thromb. Vasc. Biol.* **2014**, *34*, 99–109. [CrossRef] [PubMed]

164. Wei, J. Aldose reductase regulates miR-200a-3p/141-3p to coordinate Keap1-Nrf2, Tgfbeta1/2, and Zeb1/2 signaling in renal mesangial cells and the renal cortex of diabetic mice. *Free Radic. Biol. Med.* **2014**, *67*, 91–102. [CrossRef] [PubMed]

165. Yildirim, S.S.; Akman, D.; Catalucci, D.; Turan, B. Relationship Between Downregulation of miRNAs and Increase of Oxidative Stress in the Development of Diabetic Cardiac Dysfunction: Junctin as a Target Protein of miR-1. *Cell Biophys.* **2013**, *67*, 1397–1408. [CrossRef] [PubMed]

166. Wang, J.M. Inositol-Requiring Enzyme 1 Facilitates Diabetic Wound Healing Through Modulating MicroRNAs. *Diabetes* **2017**, *66*, 177–192. [CrossRef] [PubMed]

167. Jeyabal, P.; Thandavarayan, R.A.; Joladarashi, D.; Babu, S.S.; Krishnamurthy, S.; Bhimaraj, A.; Youker, K.A.; Kishore, R.; Krishnamurthy, P. MicroRNA-9 inhibits hyperglycemia-induced pyroptosis in human ventricular cardiomyocytes by targeting ELAVL1. *Biochem. Biophys. Res. Commun.* **2016**, *471*, 423–429. [CrossRef]

168. Zhang, Z.-W.; Guo, R.-W.; Lv, J.-L.; Wang, X.-M.; Ye, J.-S.; Lu, N.-H.; Liang, X.; Yang, L.-X. MicroRNA-99a inhibits insulin-induced proliferation, migration, dedifferentiation, and rapamycin resistance of vascular smooth muscle cells by inhibiting insulin-like growth factor-1 receptor and mammalian target of rapamycin. *Biochem. Biophys. Res. Commun.* **2017**, *486*, 414–422. [CrossRef]

169. Zhu, M. Hyperlipidemia-Induced MicroRNA-155-5p Improves beta-Cell Function by Targeting Mafb. *Diabetes* **2017**, *66*, 3072–3084. [CrossRef]

170. Roggli, E. Changes in microRNA expression contribute to pancreatic beta-cell dysfunction in prediabetic NOD mice. *Diabetes* **2012**, *61*, 1742–1751. [CrossRef]

171. Massart, J.; Sjögren, R.J.; Lundell, L.S.; Mudry, J.M.; Franck, N.; O'Gorman, D.J.; Egan, B.; Zierath, J.R.; Krook, A. Altered miR-29 Expression in Type 2 Diabetes Influences Glucose and Lipid Metabolism in Skeletal Muscle. *Diabetes* **2017**, *66*, 1807–1818. [CrossRef]

172. Guo, J.; Li, J.; Zhao, J.; Yang, S.; Wang, L.; Cheng, G.; Liu, D.; Xiao, J.; Liu, Z.; Zhao, Z. MiRNA-29c regulates the expression of inflammatory cytokines in diabetic nephropathy by targeting tristetraprolin. *Sci. Rep.* **2017**, *7*, 2314. [CrossRef]

173. Widlansky, M.E.; Jensen, D.M.; Wang, J.; Liu, Y.; Geurts, A.M.; Kriegel, A.J.; Liu, P.; Ying, R.; Zhang, G.; Casati, M.; et al. miR-29 contributes to normal endothelial function and can restore it in cardiometabolic disorders. *EMBO Mol. Med.* **2018**, *10*, e8046. [CrossRef] [PubMed]

174. Wang, H.; She, G.; Zhou, W.; Liu, K.; Miao, J.; Yu, B. Expression profile of circular RNAs in placentas of women with gestational diabetes mellitus. *Endocr. J.* **2019**, *66*, 431–441. [CrossRef] [PubMed]

175. Lakhter, A.J.; Pratt, R.E.; Moore, R.E.; Doucette, K.K.; Maier, B.F.; DiMeglio, L.A.; Sims, E.K. Beta cell extracellular vesicle miR-21-5p cargo is increased in response to inflammatory cytokines and serves as a biomarker of type 1 diabetes. *Diabetologia* **2018**, *61*, 1124–1134. [CrossRef] [PubMed]

176. Piperi, C.; Goumenos, A.; Adamopoulos, C.; Papavassiliou, A.G. AGE/RAGE signalling regulation by miRNAs: Associations with diabetic complications and therapeutic potential. *Int. J. Biochem. Cell Boil.* **2015**, *60*, 197–201. [CrossRef] [PubMed]

177. Magenta, A.; Ciarapica, R.; Capogrossi, M.C. The Emerging Role of miR-200 Family in Cardiovascular Diseases. *Circ. Res.* **2017**, *120*, 1399–1402. [CrossRef]

178. Magenta, A.; Cencioni, C.; Fasanaro, P.; Zaccagnini, G.; Greco, S.; Sarra-Ferraris, G.; Antonini, A.; Martelli, F.; Capogrossi, M.C. miR-200c is upregulated by oxidative stress and induces endothelial cell apoptosis and senescence via ZEB1 inhibition. *Cell Death Differ.* **2011**, *18*, 1628–1639. [CrossRef]

179. Belgardt, B.-F.; Ahmed, K.; Spranger, M.; Latreille, M.; Denzler, R.; Kondratiuk, N.; Von Meyenn, F.; Villena, F.N.; Herrmanns, K.; Bosco, D.; et al. The microRNA-200 family regulates pancreatic beta cell survival in type 2 diabetes. *Nat. Med.* **2015**, *21*, 619–627. [CrossRef]

180. Kitada, M.; Ogura, Y.; Monno, I.; Koya, D. Sirtuins and Type 2 Diabetes: Role in Inflammation, Oxidative Stress, and Mitochondrial Function. *Front. Endocrinol.* **2019**, *10*, 187. [CrossRef]

181. Kwon, J.J.; Factora, T.D.; Dey, S.; Kota, J. A Systematic Review of miR-29 in Cancer. *Mol. Ther. Oncol.* **2019**, *12*, 173–194. [CrossRef]

182. Slusarz, A.; Pulakat, L. The two faces of miR-29. *J. Cardiovasc. Med.* **2015**, *16*, 480–490. [CrossRef]

183. Heid, J.; Cencioni, C.; Ripa, R.; Baumgart, M.; Atlante, S.; Milano, G.; Scopece, A.; Kuenne, C.; Guenther, S.; Azzimato, V.; et al. Age-dependent increase of oxidative stress regulates microRNA-29 family preserving cardiac health. *Sci. Rep.* **2017**, *7*, 16839. [CrossRef] [PubMed]

184. David, J.A.; Rifkin, W.J.; Rabbani, P.S.; Ceradini, D.J. The Nrf2/Keap1/ARE Pathway and Oxidative Stress as a Therapeutic Target in Type II Diabetes Mellitus. *J. Diabetes Res.* **2017**, *2017*, 1–15. [CrossRef] [PubMed]

185. Matsuda, S.; Nakagawa, Y.; Kitagishi, Y.; Nakanishi, A.; Murai, T. Reactive Oxygen Species, Superoxide Dimutases, and PTEN-p53-AKT-MDM2 Signaling Loop Network in Mesenchymal Stem/Stromal Cells Regulation. *Cells* **2018**, *7*, 36. [CrossRef] [PubMed]

186. Zong, H.-Y.; Wang, E.-L.; Han, Y.-M.; Wang, Q.-J.; Wang, J.-L.; Wang, Z. Effect of miR-29b on rats with gestational diabetes mellitus by targeting PI3K/Akt signal. *Eur. Rev. Med Pharmacol. Sci.* **2019**, *23*, 2325–2331.

187. Stoll, L. Circular RNAs as novel regulators of beta-cell functions in normal and disease conditions. *Mol. Metab.* **2018**, *9*, 69–83. [CrossRef]

188. Hansen, T.B.; Jensen, T.I.; Clausen, B.H.; Bramsen, J.B.; Finsen, B.; Damgaard, C.K.; Kjems, J. Natural RNA circles function as efficient microRNA sponges. *Nature* **2013**, *495*, 384–388. [CrossRef]

4

Cross-Talk between Mitochondrial Dysfunction-Provoked Oxidative Stress and Aberrant Noncoding RNA Expression in the Pathogenesis and Pathophysiology of SLE

Chang-Youh Tsai [1,*,†], Song-Chou Hsieh [2,†], Cheng-Shiun Lu [2,3], Tsai-Hung Wu [4], Hsien-Tzung Liao [1], Cheng-Han Wu [2,3], Ko-Jen Li [2], Yu-Min Kuo [2,3], Hui-Ting Lee [5], Chieh-Yu Shen [2,3] and Chia-Li Yu [2,*]

[1] Division of Allergy, Immunology & Rheumatology, Taipei Veterans General Hospital & National Yang-Ming University, #201 Sec.2, Shih-Pai Road, Taipei 11217, Taiwan; darryliao@yahoo.com.tw

[2] Department of Internal Medicine, National Taiwan University Hospital, #7 Chung-Shan South Road, Taipei 10002, Taiwan; hsiehsc@ntu.edu.tw (S.-C.H.); b89401085@ntu.edu.tw (C.-S.L.); chenghanwu@ntu.edu.tw (C.-H.W.); dtmed170@yahoo.com.tw (K.-J.L.); 543goole@gmail.com (Y.-M.K.); tsichhl@gmail.com (C.-Y.S.)

[3] Institute of Clinical Medicine, National Taiwan University College of Medicine, #7 Chung-Shan South Road, Taipei 10002, Taiwan

[4] Division of Nephrology, Taipei Veterans General Hospital & National Yang-Ming University, #201 Sec. 2, Shih-Pai Road, Taipei 11217, Taiwan; thwu@vghtpe.gov.tw

[5] Section of Allergy, Immunology & Rheumatology, Mackay Memorial Hospital, #92 Sec. 2, Chung-Shan North Road, Taipei 10449, Taiwan; htlee1228@gmail.com

* Correspondence: cytsai@vghtpe.gov.tw (C.-Y.T.); chialiyu0717@gmail.com (C.-L.Y.)

† These authors contributed equally to this work.

Abstract: Systemic lupus erythematosus (SLE) is a prototype of systemic autoimmune disease involving almost every organ. Polygenic predisposition and complicated epigenetic regulations are the upstream factors to elicit its development. Mitochondrial dysfunction-provoked oxidative stress may also play a crucial role in it. Classical epigenetic regulations of gene expression may include DNA methylation/acetylation and histone modification. Recent investigations have revealed that intracellular and extracellular (exosomal) noncoding RNAs (ncRNAs), including microRNAs (miRs), and long noncoding RNAs (lncRNAs), are the key molecules for post-transcriptional regulation of messenger (m)RNA expression. Oxidative and nitrosative stresses originating from mitochondrial dysfunctions could become the pathological biosignatures for increased cell apoptosis/necrosis, nonhyperglycemic metabolic syndrome, multiple neoantigen formation, and immune dysregulation in patients with SLE. Recently, many authors noted that the cross-talk between oxidative stress and ncRNAs can trigger and perpetuate autoimmune reactions in patients with SLE. Intracellular interactions between miR and lncRNAs as well as extracellular exosomal ncRNA communication to and fro between remote cells/tissues via plasma or other body fluids also occur in the body. The urinary exosomal ncRNAs can now represent biosignatures for lupus nephritis. Herein, we'll briefly review and discuss the cross-talk between excessive oxidative/nitrosative stress induced by mitochondrial dysfunction in tissues/cells and ncRNAs, as well as the prospect of antioxidant therapy in patients with SLE.

Keywords: noncoding RNA; microRNA; long noncoding RNA; mitochondrial dysfunction; oxidative stress; nitrosative stress. exosome; cross-talk; systemic lupus erythematosus

1. Introduction

Systemic lupus erythematosus (SLE) is a highly heterogeneous disorder with chronic inflammatory and autoimmune reactions all over the body. It is characterized by the production of diverse autoantibodies [1,2] and chronic tissue inflammation [3–6]. There are multiple factors associated with lupus pathogenesis, including genetic predisposition [7–15], epigenetic dysregulation of gene transcription [16–21] and aberrant post-transcriptional events by noncoding (nc)RNAs [19,22–25], sex hormonal imbalance [26–29], environmental stimulation [30,31], mental/psychological stresses [28], dietary/nutritional influence [32–35], mitochondrial dysfunctions [36–39], and other yet-undefined factors [40]. Figure 1 shows the factors contributing to the pathogenesis of SLE, in which environmental factors such as infections, chemicals, heavy metals, medications, exogenous estrogens, and phthalate trigger its development in susceptible individuals. The genome-wide association study (GWAS) has identified over 100 risk loci for SLE susceptibility across populations [13]. However, functional studies have revealed that many of them fall in the category of noncoding regions of genomes, suggesting that they probably play a regulatory role. Many loci exhibit protean environmental interactions, epigenetic modifications, or association with genetic variants [10]. Nevertheless, the expression of IFN-α in tissues and circulation has been consistently found at a hereditary risk locus in patients with SLE [14]. The genetic predisposition for lupus pathogenesis is summarized in Table 1.

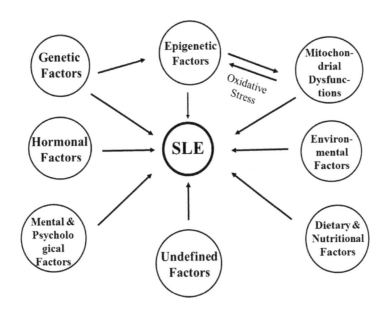

Figure 1. Factors contributing to the development of systemic lupus erythematosus. It is worthy to note that cross-talk between mitochondrial dysfunction and aberrant epigenetic regulation is mediated via excessive oxidative stress.

Recent investigations revealed that increased oxidative and/or nitrosative stress could induce structural and functional changes in different biomolecules, including proteins, lipids, nucleic acids, and glycoproteins [41,42]. The oxidative stress may also modulate proinflammatory cytokine gene expression [43–46] and cell senescence/apoptosis [47,48]. Antioxidants have been tried in the treatment of SLE with effectiveness [49–53]. Accordingly, the presence of oxidative stresses and their associated biomarkers are definitely playing a decisive role in the pathogenesis of SLE [54].

Epigenetics is an investigation of the changes in phenotypic presentation (or gene expression) that are caused by mechanisms other than the polymorphism of genome per se. It is conceivable that more than 97% of cellular RNAs are not transcribed for protein coding in nature. These ncRNAs, including microRNAs (miRs, 20–24 bp in length) and long noncoding (lnc) RNAs, which are >200 bp in length are the major molecules for post-transcriptional modifications of messenger (m)RNAs [55,56]. Interestingly, many reports have demonstrated that oxidative stress can modulate ncRNA expression in different diseases [57,58]. Conversely, ncRNAs have also been found to be regulators of oxidative stresses in different pathological conditions [59]. Furthermore, the cross-talk between miRs and lncRNAs has also been found [60,61]. Based on these facts, we hereby review and discuss briefly the molecular basis of epigenetic regulations, the underlying mechanism of mitochondrial dysfunctions, and the cross-talk between mitochondrial dysfunction-provoked oxidative stress and abnormal expression of ncRNAs during the pathologic development of SLE. At the end, a potential use of antioxidants as the therapy for SLE will also be concisely overviewed.

Table 1. Some of the genetic loci involved in the risk for SLE.

- **MHC association** [7–9]

    - MHC class II: DR_2, DR_3
    - MHC class III: C_4 null, TNF-α

- **Immune complex processing and phagocytosis** [7–15]

    - $C1_{q/r/s}$, $C_{4A/B}$, CFB
    - FCGR2A/B, CR2, CR3
    - CRP
    - ICHMs (intercellular adhesion molecules)
    - ITGAM (integrin subunit alpha M)

- **TLR and type I IFN signaling** [7–15]:

    - TLR7 (toll-like receptor 7)
    - TREX1 (three prime repair exonuclease 1)
    - DNASE1 (DNA degrading enzyme 1)
    - IRAK1/MECP2 (interleukin-receptor-associated kinase 1)
    - IRF5/7/8 (interferon regulatory factor 5, 7, 8)
    - STAT1 (signal transducer and activator of transcription 1)
    - STAT4 (signal transducer and activator of transcription 4)

- **B and T cell function and signal genes** [7–15]

    - IL10 (interleukin 10)
    - STAT4 (signal transducer and activator of transcription 4)
    - PTPN22 (protein tyrosine phosphatase non-receptor type 22)
    - PDCD1 (programmed cell death 1)
    - TNFSF4 (TNF superfamily member 4)
    - BLK (B lymphoid tyrosine kinase)
    - BANK1 (B cell scaffold protein with ankyrin repeats 1)

- **Others**

    - PXK/ABHD6 (PX domain containing serine/threonine kinase likes)
    - XKR6 (XK related 6)
    - UPF1/SMG7 (RNA helicase and ATPase)
    - NMNAT2 (nicotinamide nucleotide adenyltransferase 2)
    - UHRF1BP1 (ubiquitin like with PHD and ring finger domains 1 binding protein 1)

2. Epigenetic Regulations of Gene Expression/Silencing in Physiological Conditions

Epigenetic variation is a reversible but heritable change in gene expression without alterations in genetic code. It may include DNA methylation, histone modification, and post-transcriptional mRNA modification by ncRNAs [16]. DNA methylation is a biochemical process that involves a methyl group being added to a cytosine or adenine residue at the position of a repeated CpG dinucleotide (CpG island) in the promoter region to repress gene expression by DNA methyl- transferase (DNMT) 1, 3a, and 3b. In contrast, reactivation of DNA by demethylation to restore gene transcription can be achieved by ten-eleven translocation (TET) enzymes TET1, TET2, and TET3.

2.1. Abnormal DNA Methylation/Demethylation in SLE

DNA methylation is catalyzed by DNMT1 for gene silencing. A status of DNA hypomethylation to enhance gene expression can be found in CD4+T cells of SLE patients as a result of decreased expression of DNMT1 originating from a deficient *ras-MAPK* signature [62,63]. In addition, DNA methylation acts as a housekeeping mechanism for physiological inactivation of X-chromosomes in female [26,27,64]. Recent studies have suggested that *CD40L* demethylation is responsible for CD40L overexpression in T cells of women with SLE [64].

2.2. Abnormal Histone Modification in SLE

The degree of chromatin tightness is regulated via complex mechanisms, including structural changes in histones. Usually, double helix-chromatin coils around a protein core composed of histone octamers (H2A, H2B, H3, and H4 with two copies of each). The biochemical processes to change the 3D structure of histones include ubiquitination, phosphorylation, SUMOylation, methylation, and acetylation. The methylation and acetylation of histones are the most extensively studied [17]. These two biochemical changes are controlled by two major enzymes, histone acetyl transferase (HATs) and histone deacetylase (HDACs), that catalyze the addition/removal of an acetyl group on the lysine residues of histones. Acetylation relaxes the chromatin structures by diminishing the electric charge between histone and DNA as a result of offering an acetyl group. Conversely, deacetylation tightens the chromatin structure to silence gene expression.

The participation of histone modifications in lupus pathogenesis has been well documented. Hu et al. [65] demonstrated a global hyperacetylation of histones H3 and H4 in lupus CD4+T cells. Zhou et al. [66] reported that abnormal histone modifications within TNFSF7 promotor caused CD70 (a ligand for CD27) overexpression in SLE-T cells. Furthermore, Hedrich et al. [67] demonstrated that CREM, a transcription factor, participated in histone deacetylation in active T cells of SLE patients by way of silencing IL-2 expression, which normally recruits HDAC to cis-regulatory element (Cre) sites in IL-2 promotors. Dai et al. [68] showed in GWAS an alteration in histone H3 lysine K4 trimethylation (H3K4me3) by chromatin immunoprecipitation linked to microarray in peripheral blood mononuclear cells of some SLE patients. In addition, Zhang et al. [69] have found global H4 acetylation occurs in monocytes/macrophages in SLE subjects, which is regulated by IFN regulatory factors. The release of SLE-related cytokines such as IL-17, IL-10, and TNF-α was also abnormally increased in H3 acetylation by *stat3* [70–72]. In lupus-prone MRL/*lpr* mice, a histone deacetylation gene, *sirtuin-1* (*Sirt*-1), was found overexpressed [73], indicating a compensatory repression of gene over-reactivation. Hu et al. [73] further noted downregulation of *Sirt*-1 would transiently enhance H3 and H4 acetylation and subsequently mitigate serum levels of anti-dsDNA, as well as kidney damage in lupus mice. Javierre et al. [74] reported a global decrease in the 5-methylcytosine content in parallel with DNA hypomethylation and high expression levels of ribosomal RNA genes relevant to SLE pathogenesis. In short, abnormal histone modifications are implicated in lupus pathogenesis and immunopathological changes in these patients.

2.3. Physiological Functions of ncRNAs

Besides DNA methylation/acetylation and histone modification, the most recently discovered epigenetic mechanisms for gene expression are dependent on the class of ncRNAs that are not translated into proteins. These molecules include both housekeeping ncRNAs and regulatory ncRNA [55]. In total 50% of mRNAs are located in chromosomal regions with liability to undergo structural changes [75]. On the other hand, lncRNA can regulate gene expression by different ways, including epigenetic, transcriptional, post-transcriptional, translational, and peptide localization modifications [56]. Interestingly, the interactions between lncRNAs and miRs, as well as their pathophysiological significance, have recently been reported [60,61]. It is believed that lncRNAs mediate "sponge-like" effects on various miRs and subsequently inhibit miR-mediated functions [60,61].

The regulatory effects of intracellular and extracellular (exosomal) ncRNA on cell functions are illustrated in Figure 2.

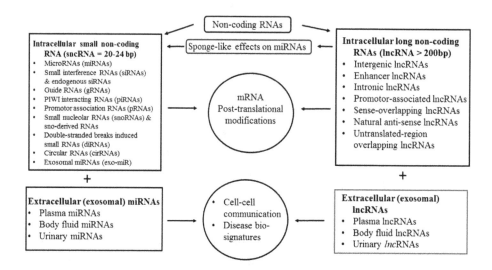

Figure 2. Different kinds of noncoding RNAs, including groups of small noncoding and long noncoding RNA, distributed in the intracellular and extracellular compartments, such as plasma, urine, and other body fluids, for regulation of messenger RNA translation and remote cell–cell communications in the body.

2.4. Aberrant Intracellular and Extracellular Exosomal ncRNA Expression in Association with Pathological Changes in Patients with SLE

It is not surprising that miRs play important roles in the regulation of innate and adaptive immunity, and the aberrantly expressed miRs are associated with autoimmune diseases [22,76–80]. Lu et al. [23,81–83] and Su et al. [84] have found various aberrantly expressed intracellular miRs implicated in the cell signaling abnormalities, deranged cytokine and chemokine release, and Th17/Treg ratio alterations in patients with SLE. Different from miRs, lncRNAs are expressed at lower levels in cells and tissues, more specifically [85–87]. These lncRNA are obviously modulating innate immunity [88] and inflammatory responses [89]. Luo et al. [90], Zhao et al. [91], and Wang et al. [92] reviewed the literature and found that lncRNA expression profiles in SLE were remarkably different from the normal.

The regulatory functions of miRNAs can be validated by transfecting miRNA mimics or antagonists using electroporator. Lu et al. [81] found increased miR-224 could target apoptosis inhibitory protein 5 (API5) and enhance T cell activation, and then activate induced cell apoptosis. Besides, the same group found decreased miR-31 in SLE T cells targeted the *Ras* homologue gene family member A (*RhoA*), which led to a decreased nuclear factor of activated T cells (NFAT) and cell apoptosis [23]. In addition, decreased miR-146a may result in upregulation of interferon regulatory factor 5 (IRF-5) and then enhanced production of IFN-α, STAT-1, IL-1 receptor associated kinase-1 (IRAK1), and TRAF6, which then increase innate immune responses, lupus disease activity, and lupus nephritis [23]. Furthermore, increased miR-524-5p that targets Jagged-1 and Hes-1mRNA may enhance IFN-γ production and then increase disease activity of SLE [82]. Su et al. [84] demonstrated that increased expression of miR-199-3p promoted ERK-mediated IL-10 production by targeting poly-(ADP-ribose) polymerase-1 (PARP-1) in SLE.

While their major functions are executed intracellularly, many miRs can be detected extracellularly in plasma/serum and urine. This extracellular form of ncRNA is protected from degradation by conjugation with carrier proteins or by being enclosed in subcellular vesicles by lipid bilayer exosomes [85]. With characteristics of the tissue- and disease-specific expression, these extracellular ncRNAs can carry out intercellular communication, signal transduction, transport of genetic information, immunomodulation, and can be taken as diagnostic biosignatures or as research tools for understanding the pathophysiology of autoimmune diseases [85–92]. Plasma circulating microRNAs exist in a rather stable form and are incorporated into distant cells to regulate protein translation and synthesis there. Carlsen et al. [87] have found plasma exosomal miR-142-3p, which targets IL-1β, and miR-181a, which targets FoxO1, are increased in active SLE patients. Kim et al. [88] demonstrated that increased plasma circulatory hsa-miR-30e-5p, hsa-miR-92a-3p, and hsa-miR-223-3p could become novel biosignatures in patients with SLE. The exosomal miRs can be found in other body fluids including breast milk, saliva, and urine, in addition to plasma [89]. Hsieh et al. [93] and Tsai et al. [94] concluded that urinary exosomal miRs could be used as biomarkers/biosignatures in lupus nephritis. Tsai et al. [94] have also noted aberrant miRNA expression in the immune-related cells could become biosignatures in correlation with pathological processes in different autoimmune and inflammatory rheumatic diseases. In addition, Perez-Hernandez et al. [95] and Xu et al. [96] have suggested the potential therapeutic application of exosomal ncRNA in different autoimmune diseases. Not only exosomal miRs, extracellularly expressed lncRNA profiles could also become potential biomarkers for human diseases [97,98]. lncRNAs are another regulatory noncoding RNA, capable of modulating many biological functions more specifically than miRs [99–102]. Aberrant expression of lncRNAs obviously induces different disease entities [99–106]. Table 2 summarizes the aberrant intracellular and circulating plasma exosomal lncRNA expression, their target mRNA, and related pathological processes in patients with SLE. Wang et al. [103] found that increased lncRNA ENST00000604411.1 expression in macrophages/dendritic cells, through targeting the X inactive specific transcript (XIST) that is normally implicated in keeping the active X chromosome in an activated state by protecting it from ectopic silencing after commencement of the silencing process of the haplotype X chromosome, could induce lupus development. Another lncRNA ENST 00000501122.2 (also known as NEAT1) overexpressed in SLE monocytes may activate CXCL-10 and IL-6 expression. Furthermore, Wu et al. [98] reported that elevated expression of plasma GAS-5, linc 0640, and linc 5150 may activate MAPK signaling pathway. The five lncRNA panels, including GAS-5, linc7074, linc 0597, linc 0640, and linc 5150 in plasma, could be regarded as biosignatures in SLE. The biochemical properties of extracellular ncRNAs and the pathophysiological roles of these aberrant exosomal ncRNAs in SLE are further discussed in the following paragraph.

Table 2. Aberrant expression of long none-coding RNAs, their target mRNAs, and related pathological processes in patients with systemic lupus erythematosus.

SLE	lnc RNA Expression	Target mRNA	Pathological Processes
	Intracellular [103–106]		
	NEAT$_1$↑*	IL-6↑, IFN↑, CXCL10↑	DNA hypomethylation
	MALAT$_1$↑	IL-21↑, SIRT$_1$↑	SLEDAI-2K↑
	Linc0597↑	TNF-α↑, IL-6↑	ESR↑, CRP↑, C3 ↓,
	Linc DC↑	STAT3↑	Th1↑
	ENST00000604411.1↑	XIST	SLEDAI score↑
	ENST0000050111222↑	NEAT$_1$	
	Linc 0949↓	TNF-α↑, IL-6↑	Inflammation↑
	Linc-HSFY2-3:3↓	-	SLEDAI score↑
	Linc-SERPIN139-1:2↓	-	
	Gas 5↓	Apoptotic gene↓	T cell apoptosis↓
	Circulating plasma exosomal [98]		
	Linc0597↑	TNF-α↑, IL-6↑	MAPK signaling↑
	Lnc0640↑	Phosphatase 4 (DUSP4)↑	Lupus pathogenesis
	Lnc5150↑	Arrestin β2 (ARRB$_2$)↑	
		Ribosomal protein S$_6$ kinase A$_5$ (RPS6KA5)↑	
	Gas 5↓	Apoptotic gene↓	T cell apoptosis↓
	Lnc 7074↓		

↑: increased expression or production; ↓: decreased expression or production; *: Oxidative stress-induced [107].

3. Increased Oxidative Stress in Patients with SLE

3.1. Causes of Excessive Oxidative Stress in SLE

Li et al. [108] have compared the reduction–oxidation (redox) capacity between normal and SLE immune cells. They found decreased plasma and intracellular glutathione (GSH) levels, and decreased intracellular GSH-peroxidase and gamma-glutamyl-transpeptidase activity in patients with SLE. Besides, the defective expression of facilitative glucose transporter (GLUT) 3 and 6 led to increased intracellular basal lactate levels, as well as decreased ATP production in SLE T cells and polymorphonuclear leukocytes. These results may indicate deranged cellular bioenergetics and defective redox capacity in immune cells that would increase oxidative stress in SLE. Lee et al. [36–39] demonstrated that mitochondrial dysfunctions in SLE patients included decreased mitochondrial DNA (mtDNA) copy number, increased mtDNA D-310 (4977 bp) heteroplasmy, and variants, as well as polymorphism of $C_{1245}G$ in *hOGG1* gene in leukocytes. Leishangthem et al. [41] found a significant decrease in enzyme activity of complex I, IV, and V in mitochondria of patients with SLE. Lee et al. [109] have extensively investigated the cause of excessive stress in patients with SLE. They reported a number of antioxidant enzyme deficiencies in SLE leukocytes, including copper/zinc superoxide dismutase (Cu/ZnSOD), catalase, glutathione peroxidase 4 (GPx-4), glutathione reductase (GR), and glutathione synthetase (GS). In addition, the mitochondrial biogenesis-related proteins, such as mtDNA-encoded ND1 peptide (ND1), ND6, nuclear respiratory factor 1(NRF-1), and pyruvate dehydrogenase E1 component alpha subunit (PDHA1), and glycolytic enzymes, including hexokinase II (HK-II), glucose 6-phosphatate isomerase (GPI), phosphofructokinase (PFK), and glyceraldehyde 3-phosphate dehydrogenase (GAPDH), are also reduced in SLE immune cells. These mitochondrial functional abnormalities may further increase oxidative stress and cell apoptosis in patients with SLE, in addition to the defective bioenergetics. Yang et al. [110] and Tsai et al. [111] concluded that enhanced oxidative stress could facilitate mitophagy, inflammatory reactions, cell senescence/apoptosis, neoantigen formation, and NETosis in SLE. The causes of mitochondrial dysfunction to induce excessive oxidative stresses and their effects on the lupus pathogenesis and pathological processes are illustrated in Figure 3.

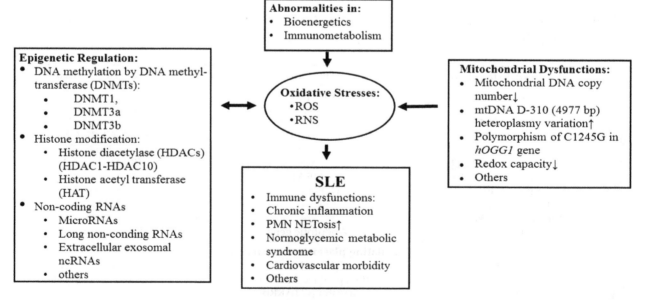

Figure 3. The origins of excessive oxidative stresses and their roles in abnormal epigenetic regulation and pathological processes in patients with SLE.

3.2. Effects of Excessive Oxidative Stress on the Pathogenesis and Pathophysiology in SLE Patients

The modifications of intra- and extracellular biomolecules by oxidative stress result in glycation and nitrosation of proteins [112], lipid peroxidation [42], as well as mitochondrial [113] and nuclear DNA strand breaks [114]. These biochemical and structural modifications of intracellular biomolecules would induce histone modification, nuclear and mitochondrial DNA damage, and aberrant ncRNA expression. As a consequence, the resulting sensitivity to environmental stress and sex hormone dysregulation [26–31] may further trigger the occurrence of lupus flare-ups. In addition, cardiovascular morbidities are enhanced due to increased glycation end products in patients with SLE [111,112,115]. The molecular basis and adverse effects of excessive oxidative stress in lupus pathogenesis and pathology are summarized in Figure 4.

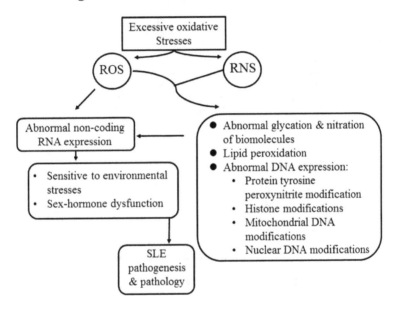

Figure 4. The molecular basis of excessive oxidative stress in the pathogenesis and pathological changes in patients with SLE.

4. Cross-Talk Between Oxidative Stress and ncRNAs in Physiological Condition

Recently, ever-increasing studies have emphasized the significance of the interactions between redox signaling and expression of ncRNAs in normal physiological conditions, as well as in disease status [44–46,57–59]. Sustained high levels of oxidative stress can cause cell senescence and even cell death, while optimal oxygen radicals are important for cell signaling. Dandekar et al. [44] and Lin et al. [116] have found mutual cross-talk among endoplasmic reticulum stress, oxidative stress, inflammatory response, and autophagy.

4.1. Excessive Oxidative Stress May Influence ncRNA Expression in Various Diseases

Many authors have demonstrated that redox-dependent signaling is essential for host's cellular decisions on differentiation, senescence, or death to maintain homeostasis of the body [117–119]. Figure 5 summarizes the aberrant miR expression resulting from excessive oxidative stress in different diseases, which include Alzheimer's disease [120], Parkinson's disease [121], hearing disorders [122], aging [123], osteoarthritis [124], cardiomyopathy in diabetes [125], and cancers [126]. However, despite the association of aberrant ncRNA expression with various pathological changes in SLE, as listed in Tables 2 and 3, there has been no literature demonstrating direct evidence for specific oxidative-induced ncRNA in patients with SLE. The combination of Table 3 and Figure 5 leads us to speculate that miR-21, miR-29b, miR-146a, and miR-126b may be induced by excessive oxidative stress in SLE as asterisked in Table 3 and its footnote.

Table 3. Aberrant expression of microRNAs, their target mRNAs, and pathological effects in patients with SLE.

SLE	miRNA	Target mRNA	Pathological Process
Intracellular [82–86]	● Increase in:		
	miR-21*	Arylamide small nucleotide inhibiors	DNA hypomethylation↑
	miR-524-5p	Jagged-1, Hes-1	IFN-γ↑, SLEDAI↑
	miR-126	KRAS	
	miR-148a	PTEN	
	● Decrease in:		
	miR-142-3p	HMGB-1	T and B activation↑
	miR-142-5p	PD-L1	
	miR-146a*	IRF-5, STAF-1	Innate immune response↑, lupus nephritis↑
	miR-224↑	API5	Type 1, IFN↑
	miR199-3p↑	PARP-1	IL-10↑
	● Decrease in:		
	miR-31	RhoA	Cell apoptosis↑
	miR-142-3p	HMGB-1	
	miR410	STAT3	
	miR-125a	STAT3, hexokinase 2, NEDDG	IL-10↑
	miR-125b*	Claudin 2, cingulin, SYVN1	
	mi-1273e		Th17/Treg ratio↑
	miR-3201		
Circulating plasma [87–94]	● Increase in:		
	miR-142-3p	IL-1β	
	miR-181a	FoxO1	
	hsa-miR-30e-5p		
	hsa-miR-92a-3p		Oral ulcer and lupus anticoagulant
	hsa-miR-223-3p		
	miR-16-5p	p38MAPK, NF-κB	
	miR-223-3p	Voltage-gated K$^+$ channel K$_{V4.2}$	
	miR-451	LKB1/AMPK	
	● Decrease in:		
	miR-106a	THBS$_2$	
	miR-17	JAB1/CSN5	
	miR-20a	IkBβ	
	miR-203	ZEB1	
	miR-92a	p63	
	miR-146a	JAK2/STAT3	
	miR-1202	cyclin dependent kinase 14	
Urinary exosomal (lupus Nephritis) [95,96]	● Increase in:		
	miR-125a	STAT3, hexokinase 2, NEDDG	Glomerulonephritis
	miR-146*	NF-κB	
	miR-150	Akt3	
	miR-155	PTEN, Wnt/β-catenin	
	● Decrease in:		
	miR-141	Tram1, GL/2, TGF-β	Glomerulonephritis
	miR-192	nin one binding protein	
	miR-200a	HMGB1/RAGE	
	miR-200c	ZEB1, Notch 1	
	miR-221	BIM-Bax/Bak, TIMP3	
	miR-222	PPP2R2A/Akt/mTOR, PCSK9	
	miR-429	TRAF6, DLC-1, HIF-1α	
	● Decrease in:		
	miR-3201		Endocapillary glomerular inflammation
	miR-1273e		

↑: increased expression or function; *: oxidative stress-induced microRNAs.

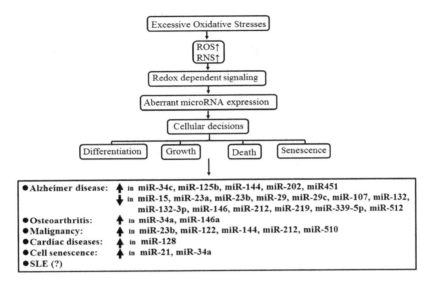

Figure 5. The effect of excessive oxidative stress on aberrant microRNA expression in various degenerative, malignant, cardiovascular, and autoimmune diseases. (?): increased miR-21, miR-29, miR-126b, and miR-146a expression induced by excessive oxidative stress is suspected in SLE patients, but no direct evidence has been published in the literature.

4.2. Aberrant ncRNA Expression Induces Oxidant/Antioxidant Imbalance in Different Pathological Processes

It has been demonstrated that excessive oxidative stress can affect ncRNA expression in Section 4.1. However, it is quite interesting that aberrant expression of ncRNAs conversely regulates redox balance in some pathological conditions. Esposti et al. [127] found miR-500a-5p could modulate oxidative stress-responsive genes in breast cancer and predict breast cancer progression as well as survival. Sangokoya et al. [128] have demonstrated that miR-144 modulates oxidative stress tolerance and, thus, is associated with changes in anemia severity in sickle cell disease. Kim et al. [129] found the roles of lncRNA and RNA-binding proteins in oxidative stress, cellular senescence, and age-related diseases. Tehrani et al. [130] further demonstrated multiple functions of lncRNAs in regulating oxidative stress, DNA damage response, and cancer progression. Mechanistically, ncRNAs can regulate enzymatic activity of different glutathione S-transferases (GSTs) to affect redox homeostasis [58]. These GSTs include microsomal GST, GST zeta l, GST mu1, GST theca 1, and sirtuin 1, superoxide dismutase 2 and thioredoxin reductase 2. In addition, the cellular oxidant/antioxidant balance can also be regulated by lncRNAs [59]. The abnormal ncRNA expression to affect the oxidant/antioxidant system is summarized in Figure 6.

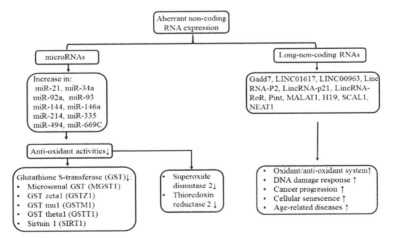

Figure 6. The effects of aberrant noncoding RNA expression on redox capacity and the induction of various age-related and malignant diseases.

5. Antioxidant Therapy and Manipulation of Epigenetic Expression to Treat Patients with SLE

In addition to increased oxygen free radicals in the plasma of SLE patients, there are other novel findings regarding the pro-oxidant/antioxidant balance in SLE. Mohan et al. [131] firstly confirmed that plasma concentrations of lipid peroxidase and nitric oxide were increased, whereas antioxidant molecules such as catalase, superoxide dismutase (SOD), GSH peroxidase, and vitamin E were decreased. Obviously, the pro-oxidant/antioxidant balance in SLE is disturbed [53]. Antioxidant therapy has been advocated for ameliorating tissue damage caused by excessive pro-oxidant radicals. Supplemented with GSH precursor, N-acetyl-cysteine (NAC) can improve disease activity in lupus-prone mice [50]. Delivering the *oxidation resistance*-1 (*OXR1*) gene to mouse kidneys by genetic manipulation can protect the kidney from damage induced by serum nephrotoxic agents, and prevent the animal from developing lupus nephritis [52]. Many authors, by administering NAC, have found remedies to ameliorate lupus activities in human SLE. Kudaravalli et al. [132] reported the improvement of endothelial dysfunction in patients with SLE by NAC and atorvastatin. Lai et al. [133] reported that NAC reduced disease activity by blocking mammalian targets of rapamycin (mTOR) in T cells of SLE patients. Tzang et al. [134] found cystamine attenuated lupus-associated apoptosis in ventricular tissue by suppressing both intrinsic and extrinsic apoptotic pathways. Nevertheless, much more clinical data are necessary to validate the efficacy of antioxidant therapy in managing patients with SLE.

Since there are so many intricate interactions among oxidative/nitrosative stress, epigenetic regulations, and gene expression in SLE, as discussed in the above sections, interference with epigenetic mechanisms such as modifying the activity of histone acetylase and/or DNA methylation, or inducing up- or downregulation of ncRNA expression may be helpful and can also be advocated to detour lupus pathogenesis and to diminish SLE disease activity in the future [135 136].

6. Conclusions

Mitochondrial dysfunction-provoked excessive oxidative stress is a crucial downstream contributory factor for lupus pathogenesis in addition to the dysregulation of upstream genetic/epigenetic functions. Recent studies have revealed that mutual interactions between oxidative stress and epigenetic regulation can perpetuate pathogenesis and pathological processes in SLE and other autoimmune diseases, as well as ageing-related diseases. In the ncRNA regulatory system, cross-talk between lncRNAs and miRs can occur for fine tuning of gene expression. Excessive oxidative stress-derived ROS and RNS may trigger autoimmune reaction and increase cell senescence/cell death in lupus-susceptible individuals. Antioxidant therapy and epigenetic modulators might become novel therapeutic strategies to treat SLE in the future.

Author Contributions: C.-L.Y. and C.-Y.T. supervised the writing project of the manuscript; C.-Y.T. and S.-C.H. prepared the manuscript and wrote the draft together; H.-T.L. prepared the figures; C.-S.L., C.-H.W., K.-J.L., Y.-M.K., H.-T.L., T.-H.W. and C.-Y.S. actively participated in the discussion and suggestions for the manuscript.

Acknowledgments: The authors thanks all of the individuals participating in the investigations.

Abbreviations

C.V	cardiovascular
DNA	deoxyribonucleic acid
DNMT	DNA methyltransfersase
FcγR	Immunoglobulin G Fragment C-gamma receptor
GLUT	glucose transporter
GSH	reduced form glutathione
GPx	glutathione peroxidase
GST	glutathione S-transferase

HAT	histone acetyltransferase
HDAC	histone deacetylase
IFN	interferon
IL	interleukin
LN	lupus nephritis
lncRNA	long noncoding ribonucleic acid
MAPK	mitogen-activated protein kinase
MHC	major histocompatibility complex
miR	microRNA
mtDNA	mitochondrial DNA
mTOR	mammalian target of rapamycin
NAC	N-acetylcysteine
ncRNA	non-coding RNA
NET	neutrophil extracellular trap
Ras	rat sarcoma protein, a superfamily of small GTPase
RNS	reactive nitrogen species
ROS	reactive oxygen species
SIRT1	sirtuin 1
SLE	systemic lupus erythematosus
SLEDAI	SLE disease activity index
SLEDAI-2K	SLEDAI in 2000 year
SOD	superoxide dismutase
TET	ten-eleven translocation DNA dioxygenase
Th	helper T cell
Treg	regulatory T cell

References

1. Wang, L.; Mohan, C.; Li, Q.Z. Arraying autoantibodies in SLE-lessons learned. *Curr. Mol. Med.* **2015**, *15*, 456–461. [CrossRef]

2. Yaniv, G.; Twig, G.; Shor, D.B.; Furer, A.; Sherer, Y.; Mozes, O.; Kimisar, O.; Slonimsky, E.; Klang, E.; Lotan, E.; et al. A volcano explosion of autoantibodies in systemic lupus erythematosus: A diversity of 180 different antibodies found in SLE patients. *Autoimmun. Rev.* **2015**, *14*, 75–79. [CrossRef]

3. Kahlenberg, J.M.; Kaplan, M.J. The inflammasome and lupus—Another innate immune mechanism contributing to disease pathogenesis? *Curr. Opin. Rheumatol.* **2014**, *26*, 475–481. [CrossRef]

4. Weidenbusch, M.; Kulkarni, O.P.; Anders, H.J. The innate immune system in human systemic lupus erythematosus. *Clin. Sci.* **2017**, *131*, 625–634. [CrossRef]

5. Tsai, C.Y.; Li, K.J.; Hsieh, S.C.; Liao, H.T.; Yu, C.L. What's wrong with neutrophils in lupus? *Clin. Exp. Rheumatol.* **2019**, *37*, 684–693. [PubMed]

6. Zharkova, O.; Celhar, T.; Crarens, P.D.; Satherthwaite, A.B.; Fairhurst, A.W.; Davis, L.S. Pathways leading to an immunological diseases: Systemic lupus erythematosus. *Rheumatology* **2017**, *56*, i55–i66. [CrossRef] [PubMed]

7. Harley, I.T.W.; Kaufman, K.M.; Langefeld, C.D.; Haraey, J.B.; Kelly, J.A. Genetic susceptibility to SLE: New insights from fine mapping and genome-wide association studies. *Nat. Rev. Genet.* **2009**, *10*, 285–290. [CrossRef] [PubMed]

8. Liu, Z.; Davidson, A. Taming lupus—A new understanding of pathogenesis is leading to clinical advances. *Nat. Med.* **2012**, *18*, 870–882. [CrossRef]

9. Ghodke-Puranick, Y.; Niewold, T.T. Immunogenetics of systemic lupus erythematosus: A comprehensive review. *J. Autoimmun.* **2015**, *64*, 125–136. [CrossRef]

10. Teruel, M.; Alacon-Riguelme, M.E. The genetic basis of systemic lupus erythematosus: What are the risk factors and what have we learned. *J. Autoimmun.* **2016**, *74*, 161–175. [CrossRef]

11. Iwamoto, T.; Niewold, T.B. Genetics of human lupus nephritis. *Clin. Immunol.* **2017**, *185*, 32–39. [CrossRef] [PubMed]

12. Hiraki, L.T.; Silverman, E.D. Genomics of systemic lupus erythematosus: Insights gained by studying monogenic young-onset systemic lupus erythematosus. *Rheum. Dis. N. Am.* **2017**, *43*, 415–434. [CrossRef] [PubMed]

13. Saeed, M. Lupus pathology based on genomics. *Immunogenetics* **2017**, *69*, 1–12. [CrossRef] [PubMed]

14. Goulielmos, G.N.; Zervou, M.I.; Vazgiourakis, V.M.; Ghodke-Puranik, Y.; Garyballos, A.; Niewold, T.B. The genetics and molecular pathogenesis of systemic lupus erythematosus (SLE) in populations of different ancestry. *Gene* **2018**, *668*, 59–72. [CrossRef] [PubMed]

15. Javinani, A.; Ashraf-Ganjouei, A.; Asloni, S.; Janshidi, A.; Mahmoudi, M. Exploring the etiopathogenesis of systemic lupus erythematosus: A genetic perspective. *Immunogenetics* **2019**, *71*, 283–297. [CrossRef] [PubMed]

16. Wu, H.; Zhao, M.; Chang, C.; Lu, Q. The real culprit in systemic lupus erythematosus: Abnormal epigenetic regulation. *Int. J. Mol. Sci.* **2015**, *16*, 11013–11033. [CrossRef] [PubMed]

17. Miceli-Richard, C. Epigenetics and lupus. *Jt. Bone Spine* **2015**, *82*, 90–93. [CrossRef]

18. Hedrick, C.M.; Mabert, K.; Rouen, T.; Tsokos, G.C. DNA methylation in systemic lupus erythematosus. *Epigenomics* **2017**, *9*, 505–525. [CrossRef]

19. Zhan, Y.; Guo, Y.; Lu, Q. Aberrant epigenetic regulation in the pathogenesis of systemic lupus erythematosus and its implications in precision medicine. *Cytogent. Genome Res.* **2016**, *149*, 141–155. [CrossRef]

20. Wang, Z.; Chang, C.; Peng, M.; Lu, Q. Translating epigenetics into clinic: Focus on lupus. *Clin. Epigenet.* **2017**, *9*, 78. [CrossRef]

21. Ren, J.; Panther, E.; Liao, X.; Grammer, A.C.; Lipsky, P.E.; Reilly, C.M. The impact of protein acetylation/deacetylation on systemic lupus erythematosus. *Int. J. Mol. Sci.* **2018**, *19*, 4007. [CrossRef] [PubMed]

22. Long, H.; Yin, H.; Wang, L.; Gershwin, M.E.; Lu, Q. The critical role of epigenetics in systemic lupus erythematosus and autoimmunity. *J. Autoimmun.* **2016**, *74*, 118–138. [CrossRef] [PubMed]

23. Lai, N.-S.; Koo, M.; Yu, C.-L.; Lu, M.C. Immunopathogenesis of systemic lupus erythematosus and rheumatoid arthritis: The role of aberrant expression of non-coding RNAs in T cells. *Clin. Exp. Immunol.* **2017**, *187*, 327–336. [CrossRef] [PubMed]

24. Zununi Vahed, S.; Nakhjarvani, M.; Etemadi, J.; Jamashidi, N.; Pourlak, T.; Abedizar, S. Altered levels of immune-regulatory microRNAs in plasma samples of patients with lupus nephritis. *Bioimpacts* **2018**, *8*, 177–183. [CrossRef] [PubMed]

25. Honarpisheh, M.; Kohler, P.; von Rauchhaupt, E.; Lech, M. The involvement of microRNAs in modulation of innate and adaptive immunity in systemic lupus erythematosus and lupus nephritis. *J. Immunol. Res.* **2018**, 4126106. [CrossRef] [PubMed]

26. McMurray, R.W. Sex-hormones in the pathogenesis in systemic lupus erythematosus. *Front Biosci.* **2001**, *6*, E193–E206. [CrossRef] [PubMed]

27. Khan, D.; Dai, R.; Ahmed, S.A. Sex differences and estrogen regulation of miRNA in lupus, a prototypical autoimmune disease. *Cell Immunol.* **2015**, *294*, 70–79. [CrossRef]

28. Assad, S.; Khan, H.H.; Ghazanfar, H.; Khan, Z.-H.; Mansor, S.; Rahman, M.A.; Khan, G.H.; Zafar, B.; Tariz, U.; Malik, S.A. Role of sex-hormone levels and psychological stress in the pathogenesis of autoimmune diseases. *Cureus* **2017**, *9*, E1315. [CrossRef]

29. Christou, E.A.A.; Banos, A.; Kosmara, D.; Bertsias, G.K.; Boumpas, D.T. Sexual dimorphism in SLE; above and beyond sex hormones. *Lupus* **2019**, *28*, 3–10. [CrossRef]

30. Sari-Puttini, P.; Atzeni, F.; Laccarino, L.; Doria, A. Environment and systemic lupus erythematosus: An overview. *Autoimmunity* **2005**, *38*, 465–472. [CrossRef]

31. Parks, C.G.; de Souza Espinodola Santos, A.; Barbhaiya, M.; Costenbader, K.H. Understanding the role of environmental factors in the development of systemic lupus erythematosus. *Best Pract. Res. Clin. Rheumatol.* **2017**, *31*, 306–320. [CrossRef] [PubMed]

32. Brown, A.C. Lupus erythematosus and nutrition: A review of the literature. *J. Ren. Nutr.* **2000**, *10*, 170–183. [CrossRef] [PubMed]

33. Minami, Y.; Sasaki, T.; Arai, Y.; Kurisu, Y.; Hisamichi, S. Diet and systemic lupus erythematosus: A 4 year prospective study of Japanese patients. *J. Rheumatol.* **2003**, *30*, 747–754. [PubMed]

34. Hsieh, C.-C.; Lin, B.-F. Dietary factors regulate cytokines in murine models of systemic lupus erythematosus. *Autoimmun. Rev.* **2011**, *11*, 22–27. [CrossRef]

35. Klack, K.; Bonfa, E.; Borba Neto, E.F. Diet and nutritional aspects in systemic lupus erythematosus. *Rev. Bras. Rheumatol.* **2012**, *52*, 384–408.

36. Lee, H.-T.; Lin, C.-S.; Chen, W.-S.; Liao, H.-T.; Tsai, C.-Y.; Wei, Y.-H. Leukocyte mitochondrial DNA alteration in systemic lupus erythematosus and its relevance to the susceptibility to lupus nephritis. *Int. J. Mol. Sci.* **2012**, *13*, 8853–8868. [CrossRef]

37. Lee, H.-T.; Wu, T.-H.; Lin, C.-S.; Lee, C.-S.; Wei, Y.-H.; Tsai, C.-Y.; Chang, D.-M. The pathogenesis of systemic lupus erythematosus—From the viewpoint of oxidative stress and mitochondrial dysfunction. *Mitochondrion* **2016**, *30*, 1–7. [CrossRef]

38. Lee, H.-T.; Wu, T.-H.; Lin, C.-S.; Lee, C.-S.; Pan, S.-C.; Chang, D.-M.; Wei, Y.-H.; Tsai, C.-Y. Oxidative DNA and mitochondrial DNA change in patients with SLE. *Front Biosci. Landmark.* **2017**, *22*, 493–503.

39. Lee, H.-T.; Lin, C.-S.; Pan, S.-C.; Wu, T.-H.; Lee, C.-S.; Chang, D.-M.; Tsai, C.-Y.; Wei, Y.-H. Alterations of oxygen consumption and extracellular acidification rates by glutamine in PBMCs of SLE patients. *Mitochondrion* **2019**, *44*, 65–74. [CrossRef]

40. Marion, T.N.; Postlethwaite, A.E. Chance, genetics, and the heterogeneity of disease and pathogenesis in systemic lupus erythematosus. *Sem. Immunopathol.* **2014**, *36*, 495–517. [CrossRef]

41. Leishangthem, B.D.; Sharma, A.; Bhatnagar, A. Role of altered mitochondria functions in the pathogenesis of systemic lupus erythematosus. *Lupus* **2016**, *25*, 272–281. [CrossRef] [PubMed]

42. Kuren, B.T.; Scofield, R.H. Lipid peroxidation in systemic lupus erythematosus. *Indian J. Exp. Biol.* **2006**, *44*, 349–356.

43. Das, U.N. Oxidative, anti-oxidants, essential fatty acids, eicosanoids, cytokines, gene/oncogene expression and apoptosis in systemic lupus erythematosus. *J. Assoc. Physicians India* **1998**, *46*, 630–634.

44. Dandekar, A.; Mendez, R.; Zhang, K. Cross talk between ER stress, oxidative stress, and inflammation in health and disease. *Methods Mol. Biol.* **2015**, *1292*, 205–214.

45. Hussain, T.; Tan, B.; Yin, Y.; Blachier, F.; Tossou, M.C.; Rahu, N. Oxidative stress and inflammation: What polyphenols can do for us? *Oxidative Med. Cell Longev.* **2016**, *2016*, 7432797. [CrossRef]

46. Guzik, T.J.; Touyz, R.M. Oxidative stress, inflammation, and vascular aging in hypertension. *Hypertension* **2017**, *70*, 660–667. [CrossRef]

47. Plotnikov, E.; Losenkov, I.; Epimakhova, E.; Bohan, N. Protective effects of pyruvic acid salt against lithium toxicity and oxidative damage in human blood mononuclear cells. *Adv. Pharm. Bull.* **2019**, *9*, 302–306. [CrossRef]

48. Lee, D.; Lee, S.H.; Noh, I.; Oh, E.; Ryu, H.; Ha, J.; Jeong, S.; Yoo, J.; Jeon, T.J.; Yun, C.O.; et al. A helical polypeptide-based potassium ionophore induces endoplasmic reticulum stress-mediated apoptosis by perturbing ion homeostasis. *Adv. Sci. (Weinheim)* **2019**, *6*, 1801995. [CrossRef]

49. Das, U.N. Current and emerging strategies for the treatment and management of systemic lupus erythematosus based on molecular signatures of acute and chronic inflammation. *J. Inflamm. Res.* **2010**, *3*, 143–170. [CrossRef]

50. Perl, A. Oxidative stress in the pathology and treatment of systemic lupus erythematosus. *Nat. Rev. Rheumatol.* **2013**, *9*, 674–686. [CrossRef]

51. Su, Y.J.; Cheng, T.T.; Chen, C.J.; Chiu, W.C.; Chang, W.N.; Tsai, N.W.; Kung, C.T.; Lin, W.C.; Huang, C.C.; Chang, Y.T.; et al. The association among antioxidant enzymes, autoantibodies, and disease severity score in systemic lupus erythematosus: Comparison of neuropsychiatric and nonneuropsychiatric groups. *BioMed Res. Int.* **2014**, *2014*, 137231. [CrossRef] [PubMed]

52. Li, Y.; Li, W.; Liu, C.; Yan, M.; Raman, I.; Du, Y.; Fang, X.; Zhou, X.J.; Mohan, C.; Li, Q.Z. Delivering oxidation resistance-1 (OXR1) to mouse kidney by genetic modified mesenchymal stem cells exhibited enhanced protection against nephrotoxic serum induced renal injury and lupus nephritis. *J. Stem Cell Res. Ther.* **2014**, *4*, 231. [PubMed]

53. Jafari, S.M.; Salimi, S.; Nakhaee, A.; Kalani, H.; Tavallaie, S.; Farajian-Mashhkadi, F.; Zakerj, Z.; Sandoughi, M. Prooxidant-antioxidant balance in patients with systemic lupus erythematosus and its relationship with clinical and laboratory findings. *Autoimmune Dis.* **2016**, *2016*, 4343514. [CrossRef] [PubMed]

54. Shah, D.; Mahajan, N.; Sah, S.; Nath, S.K.; Paudyal, B. Oxidative stress and its biomarkers in systemic lupus erythematosus. *J. Biomed. Sci.* **2014**, *21*, 23. [CrossRef] [PubMed]

55. Wei, J.-W.; Huang, K.; Yang, C.; Kang, C.-S. Non-coding RNAs as regulators in epigenetics. *Oncol. Rep.* **2017**, *37*, 3–9. [CrossRef] [PubMed]

56. Li, J.; Liu, C. Coding or noncoding, the converging concepts of RNAs. *Front. Genet.* **2019**, *10*, 496. [CrossRef]

57. Banerjee, J.; Khanna, S.; Bhattacharya, A. MicroRNA regulation of oxidative stress. *Oxid. Med. Cell Longev.* **2017**, 2872156. [CrossRef]

58. Bu, H.; Wedel, S.; Cavinato, M.; Jansen-Dürr, P. MicroRNA regulation of oxidative stress-induced cellular senescence. *Oxid. Med. Cell Longev.* **2017**, *2017*, 2398696. [CrossRef]

59. Wang, X.; Shen, C.; Zhu, J.; Shen, G.; Li, Z.; Dong, J. Long non-coding RNAs in the regulation of oxidative stress. *Oxid. Med. Cell Longev.* **2019**, 1318795. [CrossRef]

60. Bayoumi, A.S.; Sayed, A.; Broskova, Z.; Teoh, J.-P.; Wilson, J.; Su, H.; Tang, Y.-L.; Kim, I. Crosstalk between long noncoding RNAs and microRNAs in health and disease. *Int. J. Mol. Sci.* **2016**, *17*, 356. [CrossRef]

61. Yamamura, S.; Imai-Sumida, M.; Tanaka, Y.; Dahiya, R. Interaction and cross-talk between non-coding RNAs. *Cell Mol. Life Sci.* **2018**, *75*, 467–484. [CrossRef] [PubMed]

62. Deng, C.; Yang, J.; Scott, J.; Hanash, S.; Richardson, B.C. Role of the ras-MAPK signaling pathway in the DNA methyltransferase response to DNA hypomethylation. *Biol. Chem.* **1998**, *379*, 1113–1120. [CrossRef] [PubMed]

63. Sawalha, A.H.; Jeffries, M.; Webb, R.; Lu, Q.; Gorelik, G.; Ray, D.; Osban, J.; Knowlion, N.; Johnson, K.; Richardsonm, B. Defective T cell ERK signaling induces interferon-regulated gene expression and overexpression of methylation sensitive genes similar to lupus patients. *Genes Immun.* **2008**, *9*, 368–378. [CrossRef] [PubMed]

64. Lu, Q.; Wu, A.; Tesmer, L.; Ray, D.; Yousif, N.; Richardson, B. Demethylation of CD40LG on the inactive X in T cells from women with lupus. *J. Immunol.* **2007**, *179*, 6352–6358. [CrossRef] [PubMed]

65. Hu, N.; Qiu, X.; Luo, Y.; Yuan, J.; Li, Y.; Lei, W.; Zhang, G.; Zhou, Y.; Su, Y.; Lu, Q. Abnormal histone modification patterns in lupus CD4+T cells. *J. Rheumatol.* **2008**, *35*, 804–810. [PubMed]

66. Zhou, Y.; Qiu, X.; Luo, Y.; Yuan, J.; Li, Y.; Zhong, Q.; Zhao, M.; Lu, Q. Histone modifications and methyl-CpG binding domain protein levels at the TNFSF7 (CD70) promoter in SLE CD4+ T cells. *Lupus* **2011**, *20*, 1365–1371. [CrossRef] [PubMed]

67. Hedrich, C.M.; Tsokos, G.C. Epigenetic mechanisms in systemic lupus erythematosus and other autoimmune diseases. *Trends Mol. Med.* **2011**, *17*, 714–724. [CrossRef]

68. Dai, Y.; Zhang, L.; Hu, C.; Zhang, Y. Genome-wide analysis of histone H3 lysine 4 trimethylation by CHIP-chip in peripheral blood mononuclear cells of systemic lupus erythematosus patients. *Clin. Exp. Rheumatol.* **2010**, *28*, 158–168.

69. Zhang, Z.; Song, L.; Maurer, K.; Petri, M.A.; Sullivan, K.E. Global H4 acetylation analysis by CHIP-chip in SLE monocytes. *Genes Immun.* **2010**, *11*, 124–133. [CrossRef]

70. Apostolidis, S.A.; Rauren, T.; Hedrich, C.M.; Tsokos, G.C.; Crispin, J.C. Protein phosphatase 2A enables expression of interleukin 17 (IL-17) through chromatin remodeling. *J. Biol. Chem.* **2013**, *288*, 26775–26784. [CrossRef]

71. Hedrich, C.M.; Rauen, J.; Apostolidis, S.A.; Grammatikos, A.P.; Rodriguez Rodrigues, N.; Ioannidis, C.; Kyttaris, V.C.; Crispin, J.C.; Tsokos, G.C. Stat3 promotes IL-10 expression in lupus T cells through trans-activation and chromatin remodeling. *Proc. Natl. Acad. Sci. USA* **2014**, *111*, 13457–13462. [CrossRef] [PubMed]

72. Sullivan, K.E.; Suriano, A.; Dietzmann, K.; Lin, J.; Goldman, D.; Petri, M.A. The TNF-alpha locus is altered in monocytes from patients with systemic lupus erythematosus. *Clin. Immunol.* **2007**, *123*, 74–81. [CrossRef] [PubMed]

73. Hu, N.; Long, H.; Zhao, M.; Yin, H.; Lu, Q. Aberrant expression pattern of histone acetylation modifier and mitigation of lupus by SIRT1-siRNA in MRL/*lpr* mice. *Scand. J. Rheumatol.* **2009**, *38*, 464–471. [CrossRef] [PubMed]

74. Javierre, B.M.; Fernandez, A.F.; Richter, J.; Al-Shahrour, F.; Martin-Subero, J.I.; Rodriguez-Ubreva, J.; Berdasco, M.; Fraga, M.F.; O'Hanlon, T.P.; Rider, L.G.; et al. Changes in the pattern of DNA methylation associate with twin discordance in systemic lupus erythematosus. *Genome Res.* **2010**, *20*, 170–179. [CrossRef]

75. Ruvkun, G. Molecular biology. Glimpses of a tiny RNA world. *Science* **2001**, *294*, 797–799. [CrossRef]

76. Dai, R.; Ahmed, S.A. MicroRNA, a new paradigm for understanding immunoregulation, inflammation, and autoimmune diseases. *Transl. Res.* **2011**, *157*, 163–179. [CrossRef]

77. Qu, B.; Shen, N. miRNAs in the pathogenesis of systemic lupus erythematosus. *Int. J. Mol. Sci.* **2015**, *16*, 9557–9572. [CrossRef]

78. Chen, J.-Q.; Papp, G.; Szodoray, P.; Zeher, M. The role of microRNAs in the pathogenesis of autoimmune diseases. *Autoimmun. Rev.* **2016**, *15*, 1171–1180. [CrossRef]

79. Le, X.; Yu, X.; Shen, N. Novel insights of microRNAs in the development of systemic lupus erythematosus. *Curr. Opin. Rheumatol.* **2017**, *29*, 450–457. [CrossRef]

80. Long, H.; Wang, X.; Chen, Y.; Wang, L.; Zhao, M.; Lu, Q. Dysregulation of microRNAs in autoimmune diseases: Pathogenesis, biomarkers and potential therapeutic targets. *Cancer Lett.* **2018**, *428*, 90–103. [CrossRef]

81. Lu, M.C.; Lai, N.S.; Chen, H.C.; Yu, H.C.; Huang, K.Y.; Tung, C.H.; Huang, H.B.; Yu, C.L. Decreased microRNA (miR)-145 and increased miR-224 expression in T cells from patients with systemic lupus erythematosus involved in lupus immunopathogenesis. *Clin. Exp. Immunol.* **2013**, *171*, 91–99. [CrossRef] [PubMed]

82. Lu, M.C.; Yu, C.L.; Chen, H.C.; Yu, H.C.; Huang, H.B.; Lai, N.S. Aberrant T cell expression of Ca2+ influx-regulated miRNA in patients with systemic lupus erythematosus promotes lupus pathogenesis. *Rheumatology (Oxford)* **2015**, *54*, 343–348. [CrossRef] [PubMed]

83. Tsai, C.Y.; Hsieh, S.-C.; Lu, M.-C.; Yu, C.-L. Aberrant non-coding RNA expression profiles as biomarker/biosignature in autoimmune and inflammatory rheumatic diseases. *J. Lab. Preci. Med.* **2018**, *3*, 51. [CrossRef]

84. Su, X.; Ye, L.; Chen, X.; Zhang, H.; Zhou, Y.; Ding, X.; Chen, D.; Lin, Q.; Chen, C. MiR-199-3p promotes ERK-mediated IL-10 production by targeting poly(ADP-ribose)polymerase-1 in patients with systemic lupus eruthematosus. *Chemo-Biol. Interact.* **2019**, *306*, 110–116. [CrossRef]

85. Heegaard, N.H.H.; Carlsen, A.L.; Skovgaard, K.; Heegaard, P.M.H. Circulating extracellular microRNA in systemic autoimmunity. *Exp. Suppl.* **2015**, *106*, 171–195.

86. Turpin, D.; Truchetet, M.E.; Faustin, B.; Augusto, J.F.; Contin-Bordes, C.; Brisson, A.; Blanco, P.; Duffau, P. Role of extracellular vesicles in autoimmune diseases. *Autoimmun. Rev.* **2016**, *15*, 174–183. [CrossRef]

87. Carlsen, A.L.; Schetter, A.J.; Nielsen, C.T.; Lood, C.; Knudsen, S.R.; Voss, A.; Harris, C.C.; Hellmark, T.; Segelmark, M.; Jacobsen, S.; et al. Circulating microRNA expression profiles associated with systemic lupus erythematosus. *Arthritis Rheum.* **2013**, *65*, 1324–1334. [CrossRef]

88. Kim, B.-S.; Jung, J.-Y.; Jeon, J.-Y.; Kim, H.-A.; Suh, C.-H. Circulating hsa-miR-30e-5p, hsa-miR-92a-3p, and hsa-miR-223-3p may be novel biomarkers in systemic lupus erythematosus. *HLA* **2016**, *88*, 187–193. [CrossRef]

89. Ishibe, Y.; Kusaoi, M.; Murayama, G.; Nemoto, T.; Kon, T.; Ogasawara, M.; Kempe, K.; Yamaji, K.; Tamura, N. Changes in the expression of circulating microRNAs in systemic lupus erythematosus patient blood plasma after passing through a plasma absorption membrane. *Ther. Apher. Dial.* **2018**, *22*, 278–289. [CrossRef]

90. Natasha, G.; Gundogan, B.; Tan, A.; Farhatnia, Y.; Wu, W.; Rajadas, J.; Seifalian A.M. Exosomes as immunotheranostic nanoparticles. *Clin. Ther.* **2014**, *36*, 820–829. [CrossRef]

91. Tan, L.; Wu, H.; Liu, Y.; Zhao, M.; Li, D.; Lu, Q. Recent advances of exosomes in immune modulation and autoimmune diseases. *Autoimmunity* **2016**, *49*, 357–365. [CrossRef] [PubMed]

92. Rekker, K.; Saare, M.; Roost, A.M.; Kubo, A.-L.; Zarovni, N.; Chiesi, A.; Salumets, A.; Peters, M. Comparison of serum exosome isolation methods for microRNA profiling. *Clin. Biochem.* **2014**, *47*, 135–138. [CrossRef] [PubMed]

93. Hsieh, S.C.; Tsai, C.Y.; Yu, C.L. Potential serum and urine biomarkers in patients with lupus nephritis and the unsolved problems. *Open Access Rheumatol.* **2016**, *8*, 81–91. [PubMed]

94. Tsai, C.Y.; Lu, M.C.; Yu, C.L. Can urinary exosomal micro-RNA detection become a diagnostic and prognostic gold standard for patients with lupus nephritis and diabetic nephropathy? *J. Lab. Precis. Med.* **2017**, *2*, 91. [CrossRef]

95. Perez-Hernandez, J.; Redon, J.; Cortes, R. Extracellular vesicles as therapeutic agents in systemic lupus erythemaotusus. *Int. J. Mol. Sci.* **2017**, *18*, E717. [CrossRef]

96. Xu, H.; Jia, S.; Xu, H. Potential therapeutic applications of exosomes in different autoimmune diseases. *Clin. Immunol.* **2019**, *205*, 116–124. [CrossRef]

97. Kelemen, E.; Danis, J.; Göblös, A.; Bata-Csörgö, Z.; Szell, M. Exosomal long non-coding RNAs as biomarkers in human diseases. *J. Int. Fed. Clin. Chem. Lab. Med.* **2019**, *30*, 224–236.

98. Wu, G.C.; Hu, Y.; Guan, S.Y.; Ye, D.Q.; Pan, H.F. Differential plasma expression profiles of long non-coding RNAs, reveal potential biomarkers for systemic lupus erythematosus. *Biomolecules* **2019**, *9*, E206. [CrossRef]

99. Gloss, B.S.; Dinger, M.E. The specificity of long noncoding RNA expression. *Biochim. Biophys. Acta* **2016**, *1859*, 16–22. [CrossRef]

100. Derrien, T.; Johnson, R.; Bussotti, G.; Tanzer, A.; Djebali, S.; Tilgner, H.; Guermec, G.; Martin, D.; Merkel, A.; Knowles, D.G.; et al. The GENCODE v7 catalog of human lung noncoding RNAs: Analysis of their gene structure, evolution, and expression. *Genome Res.* **2012**, *22*, 1775–1789. [CrossRef]

101. Cabili, M.N.; Trapnell, C.; Goff, L.; Koziol, M.; Tazon-Vega, B.; Regev, A.; Rinn, J.L. Integrative annotation of human large intergenic noncoding RNAs reveals global properties and specific subclasses. *Genes Dev.* **2011**, *25*, 1915–1927. [CrossRef] [PubMed]

102. Hadjicharalambous, M.R.; Lindsay, M.A. Long non-coding RNAs and the innate immune response. *Noncoding RNA* **2019**, *5*, E34. [CrossRef] [PubMed]

103. Wang, Y.; Chen, S.; Chen, S.; Du, J.; Lin, J.; Qin, H.; Wang, J.; Liang, J.; Xu, J. Long noncoding RNA expression profile and association with SLEDAI score in monocyte-derived dendritic cells from patients with systemic lupus erythematosus. *Arthritis Res. Ther.* **2018**, 20–138. [CrossRef]

104. Mathy, N.W.; Chen, X.-M. Long non-coding RNAs (lncRNAs) and their transcriptional control of inflammatory responses. *J. Biol. Chem.* **2017**, *292*, 12375–12382. [CrossRef] [PubMed]

105. Luo, Q.; Li, X.; Xu, C.; Zeng, L.; Ye, J.; Guo, Y.; Huang, Z.; Li, J. Integrative analysis of long non-coding RNAs and messenger RNA expression profiles in systemic lupus erythematosus. *Mol. Med. Rep.* **2018**, *17*, 3489–3496. [PubMed]

106. Zhao, C.N.; Mao, Y.M.; Liu, L.N.; Li, X.M.; Wang, D.G.; Pan, H.F. Emerging role of lncRNAs in systemic lupus erythematosus. *Biomed. Pharm.* **2018**, *106*, 584–592. [CrossRef]

107. Simchovitz, A.; Hanan, M.; Niederhoffer, N.; Madrer, N.; Yayon, N.; Bennett, E.R.; Greenberg, D.S.; Kadener, S.; Soreq, H. NEAT1 is overexpressed in Parkinson's disease substantia nigra and confers drug-inducible neuroprotection from oxidative stress. *FASEB J.* **2019**, *33*, 11223–11234. [CrossRef]

108. Li, K.J.; Wu, C.H.; Hsieh, S.C.; Lu, M.C.; Tsai, C.Y.; Yu, C.L. Deranged bioenergetics and defective redox capacity in T-lymphocytes and neutrophils are related to cellular dysfunction and increased oxidative stress in patients with active systemic lupus erythematosus. *Clin. Dev. Immunol.* **2012**, *2012*, 548516. [CrossRef]

109. Lee, H.-T.; Lin, C.-S.; Lee, C.-S.; Tsai, C.-Y.; Wei, Y.-H. Increase 8-hydroxy-2'-deoxyguanosine in plasma and decreased mRNA expression of human 8-oxoguanine DNA glycosylase 1, anti-oxidant enzymes, mitochondrial biogenesis-related proteins and glycolytic enzymes in leukocytes in patients with systemic lupus erythematosus. *Clin. Exp. Immunol.* **2014**, *176*, 66–77.

110. Yang, S.K.; Zhang, H.R.; Shi, S.P.; Zhu, Y.Q.; Song, N.; Dai, Q.; Zhang, W.; Gui, M.; Zhang, H. The role of mitochondria in systemic lupus erythematosus: A glimpse of various pathogenetic mechanisms. *Curr. Med. Chem.* **2018**. [CrossRef]

111. Tsai, C.Y.; Shen, C.Y.; Liao, H.T.; Li, K.J.; Lee, H.T.; Lu, C.S.; Wu, C.H.; Kuo, Y.M.; Hsieh, S.C.; Yu, C.L. Molecular and cellular bases of immunosenescence, inflammation, and cardiovascular complications mimicking "inflammaging" in patients with systemic lupus erythematosus. *Int. J. Mol. Sci.* **2019**, *20*, E3878. [CrossRef] [PubMed]

112. Vlassopoulos, A.; Lean, M.E.J.; Combet, E. Oxidative stress, protein glycation and nutrition-interactions relevant to health and disease throughout the lifecycle. *Proc. Nutr. Soc.* **2014**, *73*, 430–438. [CrossRef] [PubMed]

113. McGuire, P.J. Mitochondrial dysfunction and the aging immune system. *Biology (Basel)* **2019**, *8*, E26. [CrossRef] [PubMed]

114. Ye, B.; Hou, N.; Xiao, L.; Xu, Y.; Xu, H.; Li, F. Dynamic monitoring of oxidative DNA double strand break and repair in cardiomyocytes. *Cardiovasc. Pathol.* **2016**, *25*, 93–100. [CrossRef] [PubMed]

115. de Leeuw, K.; Graaff, R.; de Vries, R.; Dullaart, R.P.; Smit, A.J.; Kallenberg, C.G.; Bjil, M. Accumulation of advanced glycation endproducts in patients with systemic lupus erythematosus. *Rheumatology* **2007**, *46*, 1551–1556. [CrossRef] [PubMed]

116. Lin, Y.; Jiang, M.; Chen, W.; Zhao, T.; Wei, Y. Cancer and ER stress: Mutual crosstalk between autophagy, oxidative stress and inflammatory response. *Biomed. Pharmacother.* **2019**, *118*, 109249. [CrossRef] [PubMed]

117. Zhang, Y.; Du, Y.; Le, W.; Wang, K.; Kieffer, N.; Zhang, J. Redox control of the survival of healthy and diseased cells. *Antioxid. Redox Signal.* **2011**, *15*, 2867–2908. [CrossRef]

118. Cao, S.S.; Kaufman, R.J. Endoplasmic reticulum stress and oxidative stress in cell fate decision and human disease. *Antioxid. Redox Signal.* **2014**, *21*, 396–413. [CrossRef]

119. Pervaiz, S. Redox dichotomy in cell fate decision: Evasive mechanism or Achilles heel? *Antioxid. Redox Signal.* **2018**, *29*, 1191–1195. [CrossRef]

120. Prasad, K.N. Oxidative stress and pro-inflammatory cytokines may act as one of the signals for regulating microRNAs expression in Alzheimer's disease. *Mech. Ageing Dev.* **2017**, *162*, 63–71. [CrossRef]

121. Prasad, K.N. Oxidative stress, pro-inflammatory cytokines, and antioxidants regulate expression levels of microRNAs in Parkinson's disease. *Curr. Aging Sci.* **2017**, *10*, 177–184. [CrossRef] [PubMed]

122. Prasad, K.N.; Bondy, S.C. MicroRNAs in hearing disorders: Their regulation by oxidative stress, inflammation and antioxidants. *Front Cell Neurosci.* **2017**, *11*, 276. [CrossRef] [PubMed]

123. Guillaumet-Adkins, A.; Yañez, Y.; Peris-Diaz, M.D.; Calabria, I.; Palanca-Ballester, C.; Sandoval, J. Epigenetics and oxidative stress in aging. *Oxid. Med. Cell Longev.* **2017**, *2017*, 9175806. [CrossRef]

124. Cheleschi, S.; De Palma, A.; Pascarelli, N.A.; Giordano, N.; Galeazzi, M.; Tenti, S.; Fioravanti, A. Could oxidative stress regulate the expression of microRNA-146a and microRNA-34a in human osteoarthritic chondrocyte cultures? *Int. J. Mol. Sci.* **2017**, *18*, E2660. [CrossRef] [PubMed]

125. Zhang, W.; Xu, W.; Feng, Y.; Zhou, X. Non-coding RNA involvement in the pathogenesis of diabetic cardiomyopathy. *J. Cell. Mol. Med.* **2019**, *23*, 5859–5867. [CrossRef]

126. Lan, J.; Huang, Z.; Han, J.; Shao, J.; Huang, C. Redox regulation of microRNAs in cancer. *Cancer Lett.* **2018**, *418*, 250–259. [CrossRef]

127. Esposti, D.D.; Aushev, V.N.; Lee, E.; Cros, M.-P.; Zhu, J.; Herceg, Z.; Chen, J.; Hernandez-Vargas, H. miR-500a-5p regulates oxidative stress response genes in breast cancer and predicts cancer survival. *Sci. Rep.* **2017**, *7*, 15966. [CrossRef]

128. Sangokoya, C.; Telen, M.J.; Chi, J.-T. microRNA miR-144 modulates oxidative stress tolerance and associates with anemia severity in sickle cell disease. *Blood* **2010**, *116*, 4338–4348. [CrossRef]

129. Kim, C.; Kang, D.; Lee, E.K.; Lee, J.S. Long noncoding RNAs and RNA binding proteins in oxidative stress, cellular senescence, and age-related diseases. *Oxid. Med. Cell Longev.* **2017**, *2017*, 2062384. [CrossRef]

130. Tehrani, S.S.; Karimian, A.S.; Parsian, H.; Majidinia, M.; Yousefi, G. Multiple functions of long non-coding RNAs in oxidative stress, DNA damage response and cancer progression. *J. Cell. Biochem.* **2018**, *119*, 223–236. [CrossRef]

131. Mohan, I.K.; Das, U.N. Oxidant stress, anti-oxidants and essential fatty acids in systemic lupus erythematosus. *Prostaglandins Leukot. Essent Fatty Acids* **1997**, *56*, 193–198. [CrossRef]

132. Kudaravalli, J. Improvement in endothelial dysfunction in patients with systemic lupus erythematosus with N-acetylcysteine and atorvastatin. *Ind. J. Pharmacol.* **2011**, *43*, 311–315. [CrossRef] [PubMed]

133. Lai, Z.W.; Hanczko, R.; Bonilla, E.; Caza, T.N.; Clair, B.; Bartos, A.; Miklossy, G.; Jimah, J.; Deherty, E.; Tily, H.; et al. N-aceylcysteine reduces disease activity by blocking mammalian target of rapamycin in T cells from systemic lupus erythematosus patients: A randomized double blind, placebo-controlled trial. *Arthritis Rheum.* **2012**, *64*, 2937–2946. [CrossRef] [PubMed]

134. Tzang, B.S.; Hsu, T.C.; Kuo, C.Y.; Chen, T.Y.; Chiang, S.Y.; Li, S.L.; Kao, S.H. Cytamine attenuates lupus-associated apoptosis of ventricular tissue by suppressing both intrinsic and extrinsic pathways. *J. Cell. Mol. Med.* **2012**, *16*, 2104–2111. [CrossRef] [PubMed]

135. Portal-Nùñez, S.; Esbrit, P.; Alcaraz, M.J.; Largo, R. Oxidative stress, autophagy, epigenetic changes and regulation by miRNAs as potential therapeutic targets in osteoarthritis. *Biochem. Pharmacol.* **2016**, *108*, 1–10. [CrossRef]

136. Dong, D.; Zhang, Y.; Reece, E.A.; Wang, L.; Harman, C.R.; Yang, P. microRNA expression profiling and functional annotation analysis of their targets modulated by oxidative stress during embryonic heart development in diabetic mice. *Reprod. Toxicol.* **2016**, *65*, 365–374. [CrossRef]

5

MicroRNAs Tune Oxidative Stress in Cancer Therapeutic Tolerance and Resistance

Wen Cai Zhang

Burnett School of Biomedical Sciences, College of Medicine, University of Central Florida, 6900 Lake Nona Blvd, Orlando, FL 32827, USA; wencai.zhang@ucf.edu

Abstract: Relapsed disease following first-line therapy remains one of the central problems in cancer management, including chemotherapy, radiotherapy, growth factor receptor-based targeted therapy, and immune checkpoint-based immunotherapy. Cancer cells develop therapeutic resistance through both intrinsic and extrinsic mechanisms including cellular heterogeneity, drug tolerance, bypassing alternative signaling pathways, as well as the acquisition of new genetic mutations. Reactive oxygen species (ROSs) are byproducts originated from cellular oxidative metabolism. Recent discoveries have shown that a disabled antioxidant program leads to therapeutic resistance in several types of cancers. ROSs are finely tuned by dysregulated microRNAs, and vice versa. However, mechanisms of a crosstalk between ROSs and microRNAs in regulating therapeutic resistance are not clear. Here, we summarize how the microRNA–ROS network modulates cancer therapeutic tolerance and resistance and direct new vulnerable targets against drug tolerance and resistance for future applications.

Keywords: microRNA; cancer; oxidative stress; reactive oxygen species; redox signaling; hypoxia; therapeutic tolerance; therapeutic resistance

---

1. Reactive Oxygen Species (ROSs)

There are many types of free radicals including oxygen- and nitrogen-based species. ROSs or reactive oxygen metabolites are free radicals containing oxygen metabolites such as single oxygen, the superoxide anion, hydrogen peroxide, and the hydroxyl radical [1]. ROSs are generated from cellular oxidative metabolism, including mitochondrial oxidative phosphorylation and electron transfer reactions, and optimal levels of ROSs play a pivotal role in many cellular functions [2]. At physiological levels, ROSs are considered signaling molecules or secondary messengers that participate in cell signal transduction, a process known as redox signaling [3]. In addition, the production of ROSs by phagocytic cells is recognized as an important part of innate immunity that kills invading pathogens [4].

The coordination between ROS generation and scavenging ensures that ROS levels are tightly controlled and fine-tuned so as to act as secondary messengers for cell signaling [5]. However, the aberrant production of ROSs, or the failure of the capacity to scavenge excessive ROSs, results in an imbalance in the redox environment of the cell [6]. High levels of ROSs have deleterious effects including nucleic acid (DNA and RNA), lipid, and protein oxidation, as well as membrane destruction by lipid peroxide formation, leading to the development of various diseases such as cancer [7]. Using antioxidant-based strategies [8] to decrease ROS levels or inhibit oxidative damage may prevent ROS-induced cell damage. For example, peroxisome proliferator-activated receptor-gamma coactivator 1 alpha upregulates expression levels of superoxide dismutase enzymes (SOD2/SOD3) and catalase to protect cells from oxidative damage via detoxification and DNA repair [9].

Aberrantly regulated metabolic pathways lead to tumorigenesis [10] and preferential survival of tumor cells [11]. Accumulating evidence suggests that tumorigenesis is dependent on mitochondrial metabolism [12], especially the tricarboxylic acid (TCA) cycle [13]. The TCA cycle is a central pathway in the metabolism of sugars, lipids, and amino acids [14]. Dysregulation of the TCA cycle can induce oncogenesis by activating pseudohypoxia responses, which result in the expression of hypoxia-associated proteins irrespective of oxygen status [15]. For example, succinate accumulation caused by functional loss of the TCA cycle enzyme succinate dehydrogenase complex stabilizes hypoxia-inducible factor (HIF)-1α via inhibition of prolyl hydroxylase (PHD) [16]. In addition, loss of function of the von Hippel–Lindau (VHL) protein [17] also induces pseudohypoxia responses through decreased ubiquitination and proteasomal degradation of HIF-1α [18]. Among the 1158 mitochondrial genes discovered in MitoCarta2.0 (Broad Institute) [19,20], the succinate dehydrogenase complex [21] inclusive of succinate dehydrogenase A [22], succinate dehydrogenase B [23], succinate dehydrogenase C [24], and succinate dehydrogenase D [25], as well as glycine decarboxylase [26–29] and glutaminase [30], is especially critical for tumorigenesis. Hypoxia, acting through HIF-1α, results in a low production of ROSs and high antioxidant defense in cancers such as leukemia [31]. It suggests that targeting key enzymes of hypoxia metabolism pathways might provide a new way to eradicate tumor formation [32].

2. microRNAs (miRNAs)

miRNAs are important regulators of mRNA expression [33] and play critical roles in regulating tumor initiation and progression [34]. Importantly, single miRNAs have been shown to regulate entire cell signaling networks in a cell-context dependent manner [35] and may also be utilized as biomarkers [36–38] for both invasive [39,40] and non-invasive [41–43] detection. Dysregulated expressions of miRNAs may function as oncogenes (oncomiRs) [44] such as miR-21 [45], miR-31 [46], miR-155 [47,48], and miR-10b [49] or as tumor suppressors such as *let-7* [50] and miR-34 [51,52] in many cancers.

ROSs are finely tuned by dysregulated miRNAs, and vice versa. Many studies are focused on regulatory interactions between miRNAs and ROSs attributing to oxidative stress-related tissue [53]. It is important for a well-regulated cellular ROS level, and miRNAs fill in the role of maintaining this homeostasis. A dysregulation of normal physiological miRNA levels can thus lead to oxidative damage and the development of diseases such as cancer. For example, oncogenic miR-21 enhances both KRAS [54] and epidermal growth factor receptor (EGFR) signaling [55] and promotes tumorigenesis through stimulation of mitogen-activated protein kinase (MAPK)-mediated ROS production by downregulation of SOD2/SOD3 [56]. On the other hand, oxidative stress can alter the expression level of many miRNAs [57–59]. For instance, oxidative stress such as hydrogen peroxide elevates miR-34a with concomitant reduction of sirtuin-1 and sirtuin-6 in bronchial epithelial cells [60], which is associated with chronic obstructive pulmonary disease and tumorigenesis [61]. However, oxidative stress decreases expression levels of the *let-7* family [62] in a p53-dependent manner in a variety of tumor cells [63]. These findings suggest that ROSs may exert a pivotal role in the regulation of microRNA expression in a cell-context-dependent manner.

miRNA-based monotherapy has not been developed well in clinical settings [64–66]. For example, a first-in-man, phase 1 clinical trial of miR-16-loaded nanoparticles as a treatment for recurrent malignant pleural mesothelioma patients has been completed [67]. Delivery of tumor suppressive miR-16 in 22 patients led to 5% objective response, 68% stable disease, and 27% progressive disease. Possible mechanisms of low objective response include miRNA sequestration through leaky cancer blood vessels as well as endocytosis by cancer cells [68]. Nevertheless, miR-16 expression levels in patients should be detected prior to receiving miR-16 treatment in future clinical trials [69]. Furthermore, miRNA-based treatment may combine with other current or potential therapeutics in combating cancer [70,71]. In addition, increasing evidence has revealed that miRNAs can be directly linked to therapeutic resistance in some cancers. For instance, overexpressing miR-205 sensitizes radioresistant breast cancer cells to radiation in a xenograft model [72]. Similarly, administration of miR-24 sensitizes radioresistant nasopharyngeal carcinoma cells to radiation in vitro [73]. miRNA-mediated regulation of signaling pathways involved in tumorigenesis as well as therapeutic tolerance and resistance is summarized in Table 1. It is revealed that miRNAs may serve both as drug targets and as therapeutic agents to eradicate cancer cells and sensitize therapeutic resistant cells [74].

Table 1. miRNA-mediated regulation of signaling pathways involved in tumorigenesis as well as therapeutic tolerance and resistance.

miRNA	Signaling Involved in Tumorigenesis	Signaling Involved in Therapeutic Tolerance and Resistance
miR-1246 and miR-1290 ↑	(+) tumorigenesis via repressing metallothioneins in human non-small cell lung cancer [75]	(+) resistance to EGFR tyrosine kinase inhibitor gefitinib via repressing metallothioneins in human non-small cell lung cancer [75]
miR-147b ↑	N.A.	(+) tolerance to EGFR tyrosine kinase inhibitor osimertinib through activating pseudohypoxia signaling pathways via repressing VHL and succinate dehydrogenase in human non-small cell lung cancer [76]
miR-155 ↑	(+) tumorigenesis in mouse miR155 transgenic B cell lymphomas [77]	(+) chemoresistance to gemcitabine through decreasing apoptosis in human pancreatic cancer [78]
miR-21 ↑	(+) Ras/MEK/ERK signaling via repressing negative regulators of the Ras/MEK/ERK pathway and inhibition of apoptosis in mouse KRAS transgenic non-small cell lung cancer [54]	(+) chemoresistance to gemcitabine through decreasing apoptosis and activating Akt phosphorylation in human pancreatic cancer [79,80] (+) radioresistance through upregulation of hypoxia-inducible factor 1α in human non-small cell lung cancer [81] (+) resistance to EGFR tyrosine kinase inhibitors through activating PI3K-AKT signaling pathway in human non-small cell lung cancer [82]
miR-31 ↑	(+) tumorigenesis through activating RAS/MAPK signaling via repressing negative regulators of RAS/MAPK signaling in mouse KRAS transgenic non-small cell lung cancer [46]	N.A.
let-7 family ↓	(+) tumorigenesis in human breast cancer through repressing H-RAS and high mobility group AT-hook 2 [83]	(+) resistance to EGFR tyrosine kinase inhibitor gefitinib through upregulation of MYC in human non-small cell lung cancer [84]
miR-30 ↓	• (+) tumor initiation and (−) apoptosis by repressing ubiquitin-conjugating enzyme 9 and integrin beta3, respectively, in human breast cancer [85] • (+) mTOR/AKT-signaling pathway through repressing transmembrane 4 super family member 1 in human non-small cell lung cancer [86]	• (−) resistance to EGFR tyrosine kinase inhibitor gefitinib through repressing BCL2-like 11 and apoptotic peptidase activating factor 1 in human non-small cell lung cancer [87] • (+) chemoresistance to cisplatin through activating autophagy in human gastric cancer [88]
miR-34a/b/c ↓	• (+) tumor initiation in mouse Kras; Trp53 transgenic lung cancer [51] • (+) tumor initiation by repressing inhibin subunit beta B and AXL in mouse Apc transgenic colorectal cancer [89]	(+) chemoresistance to fludarabine through p53 inactivation and apoptosis resistance in human chronic lymphocytic leukemia [90]

EGFR: epidermal growth factor receptor; Akt: Akt Serine/Threonine Kinase; MAPK: mitogen-activated protein kinase; MEK: Mitogen-activated protein kinase kinase; ERK: extracellular-signal-regulated kinase; PI3K: phosphatidylinositol 3-kinase; AXL: AXL receptor tyrosine kinase; Apc: adenomatous polyposis coli; VHL: Von Hippel–Lindau; mTOR: mammalian target of rapamycin; ↑: upregulation; ↓: downregulation; (+): promotion; (−): repression; N.A.: not available.

3. Therapeutic Tolerance and Resistance

The discovery of genetic mutation on tyrosine kinase, such as *EGFR* mutations including exon 19 deletion (*Del19*) and exon 21 Leu858Arg substitution (*L858R*), that confer sensitivity to EGFR-targeted tyrosine kinase inhibitors in lung adenocarcinomas heralded the beginning of the era of precision medicine for lung cancer [91,92]. However, the success of EGFR-based therapy was compromised by therapeutic resistance following initial treatment response in most cancer patients [93]. Exon 20 Thr790Met substitution (*T790M*), affecting the ATP binding pocket of the EGFR kinase domain, accounts for approximately half of all lung cancer cases with acquired resistance to the current first generation EGFR tyrosine kinase inhibitors, erlotinib and gefitinib [94]. In erlotinib- and gefitinib-resistant lung tumors with $EGFR^{T790M}$, rociletinib and osimertinib are highly active [95]. However, resistance to the third generation EGFR tyrosine kinase inhibitor osimertinib is now emerging clinically [96]. In addition to genetic mutations, intratumor heterogeneity also drives neoplastic progression and therapeutic resistance [97]. Recently, it has been found that $EGFR^{T790M}$-positive drug-resistant cells are derived from $EGFR^{T790M}$-negative drug-tolerant persister cells that survive initial EGFR tyrosine kinase inhibitors treatment [98,99]. It is therefore crucial to identify molecular changes that drive drug tolerance.

Consistently, Zhang et al. have revealed that lung tumor cells protect themselves with a drug-tolerance mechanism when the cells are treated with osimertinib [76]. These findings align with previous data showing that tumor cells enter into a tolerant state when they are treated with tyrosine kinase inhibitors in lung and other cancers [100–102]. These tolerant persister cells precede and evolve into resistant cells over time by acquiring *EGFR*-resistant mutations [98,99]. These tolerant cells are slow cycling and are enriched in the expression of stem-associated genes in the WNT/planar cell polarity signaling pathway, such as *WNT5A*, *FZD2*, and *FZD7*. These findings are conceptually similar to a recent report that post-drug transition to stable resistance consists of dedifferentiation [103].

Excessive ROSs produced by damaged mitochondria can trigger mitophagy, a process that can scavenge impaired mitochondria and reduce ROS levels to maintain a stable mitochondrial function in cells [104]. Therefore, mitophagy helps maintain cellular homeostasis under oxidative stress. For example, protein kinase inhibitor sorafenib shows activities against many protein kinases, including vascular endothelial growth factor receptor (VEGFR), platelet-derived growth factor receptor (PDGFR), and rapidly accelerated fibrosarcoma (RAF) kinases [105]. Resistance to sorafenib in cancers such as hepatocellular carcinoma is frequent [106] partially due to antiangiogenic effects-mediated hypoxia [107]. Administration of tryptophan-derived metabolites such as melatonin [108] increased ROS production and mitophagy, resulting in increased sensitivity to sorafenib in hepatocellular carcinoma cells [109]. Additionally, melatonin downregulated the HIF-1α protein synthesis through inhibition of the mammalian target of rapamycin complex 1 (mTORC1)-mediated pathway [110]. Most recently, it was shown that drug-tolerant persister cancer cells were vulnerable to inhibition of the glutathione peroxidase 4, owing to a disabled antioxidant program [102]. It suggests that increasing ROS levels may re-sensitize therapeutic resistant cancer cells to current treatments.

4. miRNA–ROS Interaction Regulates Therapeutic Tolerance/Resistance at the Phenotypic Level

The miRNA–ROS network in a scenario of therapeutic tolerance/resistance is grouped at three levels including phenotype, signaling/metabolism, and genetics/epigenetics (Figure 1). Phenotypic changes include the enrichment of tumor-initiating cells, the histological transformation from *EGFR*-mutant non-small cell lung cancer to small cell lung cancer, and epithelial–mesenchymal transition resulting in therapeutic tolerance/resistance.

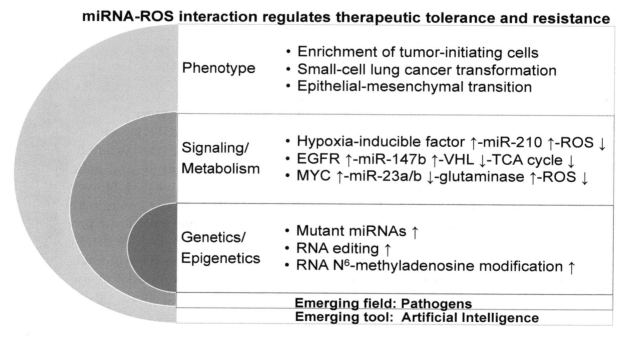

Figure 1. miRNA–ROS interaction regulates cancer therapeutic tolerance and resistance through heterogeneous mechanisms. The mechanisms at hierarchy levels include phenotypic, signaling/metabolic, and genetic/epigenetic changes. ROS: reactive oxygen species; HIF: hypoxia-inducible factor; EGFR: epidermal growth factor receptor; VHL: von Hippel–Lindau; TCA: tricarboxylic acid; ↑: upregulation; ↓: downregulation.

4.1. Enrichment of Tumor-Initiating Cells

Therapeutic resistance is frequent after primary and adjuvant cancer therapy, often evolving into a lethal relapse disease [111]. These observations may be attributed to the highly heterogeneous nature of tumors that contain distinct tumoral and microenvironment cells, all of which contribute in varying degrees toward self-renewal, drug resistance, and relapse [112]. The tumor-initiating cell or cancer stem cell model provides one explanation for the phenotypic and functional diversity among cancer cells in some tumors [113]. Tumor-initiating cells have been demonstrated to be more resistant to conventional therapeutic interventions [114] and are key drivers of relapse in many types of cancers including leukemia [115], lung cancer [116], breast cancer [117], brain cancer [118], colon cancer [119], and nasopharyngeal carcinoma [120]. There is, therefore, increasing interest in developing strategies that can specifically target tumor-initiating cells with novel and emerging therapeutic modalities, thereby halting cancer progression and improving disease outcome [121]. Tumor-initiating cells protect their genomes from ROS-mediated damage [122] via increased production of free radical scavengers [123] leading to low ROS levels [124]. Thus, heterogeneity of ROS levels in cancers such as glioma may influence the extent to which tumor-initiating cell-enriched populations are resistant to therapies such as ionizing radiation [125]. Tumor-initiating cells display heterogeneous phenotypes due to different genotypes in tumors [126]. Thus, the genetic backgrounds, such as mutant *EGFR* and *RAS*, need to be taken into consideration to better understand the association between tumor-initiating cells and therapeutic resistance in the future.

In non-small cell lung cancer, a panel of tumor-initiating cell-relevant miRNAs is enriched when assessed by a miRNA microarray [75]. Those top upregulated miRNAs include miR-1290 and miR-1246 (Table 1). The top downregulated miRNAs comprise miR-23a and *let-7b/c/d/i*. Further analysis showed that miR-1246 and miR-1290 regulate tumor-initiating cells via repressing cysteine-rich metal-binding proteins (metallothioneins) [75]. The reduced expression of metallothioneins has been implicated as biomarkers of low ROSs, which is consistent with the previous finding that pharmacological anti-oxidants such as N-acetyl cysteine or the knock-down of *nuclear respiratory factor 2* (*NRF2*) prevented the induction of metallothionein-1 induced by tyrosine kinase inhibitor sorafenib [127].

Another direct target of miR-1290, glioma pathogenesis-related protein 1, promotes apoptosis through upregulating ROS production by activating the c-Jun-NH(2) kinase signaling cascade in cancer cells [128]. Other evidence has shown that extracellular miR-1246 could enhance radioresistance of lung cancer cells [129]. In addition, miR-21 is enriched in tumor-initiating cells in many types of cancers such as gastric and breast cancers [130]. Functional loss of miR-21 reduces a frequency of tumor-initiating cells, consistently with decreased capacity of therapeutic resistance against EGFR tyrosine kinase inhibitors [82] (Table 1). Whether these miRNAs regulate ROSs resulting in therapeutic tolerance and resistance still needs further study. Thus, targeting enriched tumor-initiating cells might overcome miRNA–ROS-mediated therapeutic tolerance/resistance.

4.2. Small Cell Lung Cancer Transformation

Small cell lung cancer is a highly aggressive disease that exhibits rapid growth and genetic instability including inactivated *tumor suppressor retinoblastoma 1 (RB1)* and amplified *MYC proto-oncogene (MYC)* [131]. Histologic transformation of *EGFR* mutant non-small cell lung cancer to small cell lung cancer is an important mechanism of resistance to EGFR tyrosine kinase inhibitors that occurs in approximately 3–10% of *EGFR* mutant non-small cell lung cancers [132]. Transformation to small cell lung cancer occurs in a subpopulation of *EGFR* mutant non-small cell lung cancer patients and is frequently associated with mutant *RB1, TP53,* and *PIK3CA* [133,134]. Future studies might help define which subsets of non-small cell lung cancer are most prone to small cell lung cancer transformation.

Frequent overexpression of the miR-17~92 cluster in small cell lung cancer [135] is a fine-tuner to reduce excessive ROS-induced DNA damage in RB1-inactivated small cell lung cancer cells [136]. Therefore, miR-17~92 may be excellent therapeutic target candidates to overcome small cell lung cancer transformation.

4.3. Epithelial–Mesenchymal Transition

An epithelial–mesenchymal transition is a biologic process that allows a polarized epithelial cell to undergo multiple biochemical changes that enable it to assume a mesenchymal cell phenotype, which includes increased resistance to apoptosis [137]. Epithelial–mesenchymal transition is tightly regulated by microRNAs. For example, downregulation of miR-200 family members is linked to enhanced epithelial–mesenchymal transition and tumor-initiating cell acquisition [138,139] in many cancers [140]. Reduced miR-200s directly increase p38α [141], leading to decreased levels of ROSs and subsequent inactivation of the NRF2 oxidative stress response pathway [142]. The decreased ROSs, in turn, inhibit expression of the miR-200s [143], thus establishing a miR-200s-activated stress signature, which strongly correlates with shorter patient survival caused by chemotherapeutic resistance. In addition, miR-30b/c and miR-222 mediate gefitinib-induced apoptosis and the epithelial–mesenchymal transition leading to therapeutic resistance in non-small cell lung cancer [87]. These discoveries collectively indicate potential roles of the miRNA family in the regulation of ROS homeostasis in tumor-initiating cells and therapeutic resistance.

5. miRNA–ROS Interaction Regulates Therapeutic Tolerance/Resistance at a Signaling/Metabolic Level

5.1. HIF-miR-210-ROS

Under hypoxic conditions, upregulated HIF-1α directly binds to a hypoxia-responsive element on the proximal miR-210 promoter and induces miR-210 expression in cancer cells [144]. miR-210 activates generation of ROSs [145] via suppressing iron–sulfur cluster assembly enzyme [146,147] and cytochrome c oxidase assembly protein [148] in the mitochondria electron transport chain and the TCA cycle. miR-210 knockdown decreased resistance to radiotherapy in hypoxic glioma stem cells and hepatoma cells [149,150]. These discoveries suggest that the HIF-miR-210-ROS [151] pathway might be a target to overcome therapeutic resistance (Figure 1).

5.2. EGFR-miR-147b-VHL-TCA Cycle

Increasing evidence suggests that the metabolic enzymes and the catalyzed metabolites, such as isocitrate dehydrogenase, succinate dehydrogenase, and succinate [16,152] in the TCA cycle, are involved in not only tumorigenesis but also therapeutic resistance. A hypoxia response is linked to tumor cell survival and drug-resistance in many cancers [153,154]. Dysregulated cancer metabolism has recently gained attention for its potential role in promoting therapeutic resistance by a therapeutic tolerance strategy in a novel manner [102]. Furthermore, Zhang et al. discovered that lung cancer cells adopt a tolerance strategy to protect from EGFR tyrosine kinase inhibitors by modulating miR-147b-dependent pseudohypoxia signaling pathways [76]. The study revealed that VHL [155] and succinate dehydrogenase play roles in tolerance-mediated cancer progression. Decreasing miR-147b and reactivation of the TCA cycle pathway provides a promising strategy to prevent therapeutic tolerance-mediated tumor relapse (Figure 1).

In addition, VHL regulates Akt activity [156], suggesting that miR-147b-VHL axis might confer therapeutic tolerance through activating Akt activity. In addition, other upstream transcription factors such as the inhibitor of DNA binding 2 might regulate VHL levels [157]. The interaction between miR-147b and other transcription factors controlling VHL needs to be investigated in the future.

Furthermore, the reciprocal changes of metabolites in the TCA cycle such as increased levels of succinate and 2-oxoglutarate (also known as α-ketoglutarate) [158] as well as decreased levels of malate and fumarate in osimertinib-tolerant cells indicate that silenced activity for succinate dehydrogenase is linked to therapeutic tolerance. In addition, small molecule inhibitor R59949 silencing succinate dehydrogenase activity enhances therapeutic tolerance, which is comparable to the function of miR-147b overexpression in tolerant persister cells. It is not surprising that accumulated succinate due to a loss of function of succinate dehydrogenase could activate the pseudohypoxia signaling pathway by repressing PHD2 as reported previously [16]. This is consistent with the findings that the miR-147b/succinate dehydrogenase axis could increase the gene expression for pseudohypoxia signaling pathways. In addition to inactivated VHL and succinate dehydrogenase, other factors such as reduced nicotinamide adenine dinucleotide (NAD$^+$) and decreased glutathione [159] might also activate pseudohypoxia responses leading to therapeutic tolerance. In addition, these pseudohypoxia responses may further perturb the TCA cycle and cooperatively regulate therapeutic tolerance.

These discoveries suggests that miR-147b may promote drug-tolerance to EGFR tyrosine kinase inhibitors either through reactivation of the EGFR downstream signaling pathway or through bypass by another receptor tyrosine kinase that sustains downstream signaling despite inhibition of EGFR [160,161].

5.3. Myc-miR-23a/b-Glutaminase-ROS

Cancer cells depend on both glycolysis and glucose oxidation to support their growth [162,163] as well as glutaminolysis that catabolizes glutamine to generate ATP and lactate [164]. Oncogenic c-Myc represses miR-23a and miR-23b, resulting in increased levels of mitochondrial glutaminase in cancer cells [30]. Glutaminase converts glutamine to glutamate, which is further catabolized through the TCA cycle for the production of adenosine triphosphate (ATP) or serves as substrate for glutathione synthesis [165]. Glutamine withdrawal or glutaminase knockdown resulted in increased levels of ROSs. Thus, the Myc-miR-23-glutaminase axis provides a new mechanism for regulating ROS homeostasis in cancer cells. Considering that downregulated miR-23a is enriched in tumor-initiating cells [75], it is of great interest to explore a link between miR-23 and ROSs in therapeutic tolerance/resistance (Figure 1).

6. miRNA–ROS Interaction Regulates Therapeutic Tolerance/Resistance at a Genetic/Epigenetic Level

6.1. Mutant miRNAs

The whole genome sequencing analysis of lung adenocarcinomas showed noncoding somatic mutational hotspots near *vacuolar membrane protein 1/MIR21* [166]. Samples harboring indels or single nucleotide variants in this locus demonstrated significantly higher levels of *MIR21* expression. miR-21 high levels are linked to therapeutic resistance to several treatments, including EGFR tyrosine kinase inhibitors [167] and chemotherapeutic agents [168]. Thus, it is valuable to predict therapeutic response by detecting the sequence of miR-21 in biopsies from cancer patients before they receive treatments such as EGFR tyrosine kinase inhibitors (Figure 1).

6.2. RNA Editing

Adenosine deaminases acting on RNA (ADARs) convert adenosine to inosine in double-stranded RNA including both protein-coding [169] non-coding RNAs [170]. ADAR editase activation has been associated with progression of a broad array of malignancies including therapeutic resistance [171]. ADAR1 promotes tumor-initiating cell activity [172] and resistance to BCR-ABL1 inhibitor or janus kinase 2 inhibitor in chronic myeloid leukemia through inactivating biogenesis of the *let-7* [173] or pri-miR-26a maturation [174]. In addition, most cancer patients either do not respond to the immune checkpoint blockade or develop resistance to it, often because of acquired mutations [175] that impair antigen presentation [176]. Loss of function of ADAR1 in tumor cells profoundly sensitizes tumors to immunotherapy and overcomes resistance to the programmed cell death protein 1 (PD-1) checkpoint blockade [177]. It is of interest to further study how the ADAR-miRNA axis regulates therapeutic tolerance/resistance through controlling potential genes encoding ROS scavengers [178] such as *Drosophila homolog of the mammalian protein thioredoxin-1* and *cytochrome P450 4g1* (Figure 1).

6.3. RNA m⁶A Modification

N^6-methyladenosine (m⁶A) modification of mRNA (RNA m⁶A modification) is the most abundant RNA modification in eukaryotes and highly conserved among multiple species [179]. RNA m⁶A modification is emerging as an important regulator of gene expression that affects different developmental and biological processes [180], and altered m⁶A homeostasis is linked to cancer [181–183]. RNA m⁶A modification is catalyzed by the dynamic regulation of methyltransferases and demethylases. Methyltransferase include methyltransferase-like 3 (METTL3), METTL14, and Wilms' tumor 1-associating protein, and the demethylases include fat mass- and obesity-associated protein and ALKB homolog 5 [184]. Upregulation of METTL3 is associated with poor prognosis in tumorigenesis and increased chemo- and radio-resistance in cancers such as glioblastomas [185] and pancreatic cancer [186]. Developing resistant phenotypes during tyrosine kinase inhibitor therapy is controlled by m⁶A modification [187]. Leukemia cells with mRNA m⁶A demethylation are more tolerant to tyrosine kinase inhibitor treatment. Recovery of m⁶A methylation re-sensitizes therapeutic resistant cells towards tyrosine kinase inhibitors. The findings identify a novel function for the m⁶A methylation in regulating reversible tyrosine kinase inhibitor-tolerance state, providing a mechanistic paradigm for drug resistance in cancer. In addition, METTL3 plays roles in the maturation process of miRNAs against ROSs in an m⁶A-dependent manner [188]. For example, METTL3-mediated miR-873 upregulation controls the kelch-like ECH associated protein 1 (KEAP1)-NRF2 [142] pathway against ROSs. These studies revealed that RNA m⁶A might regulate therapeutic tolerance/resistance through miRNA–ROS pathways (Figure 1).

7. Emerging Fields and Tools in Preventing and Overcoming Therapeutic Tolerance/Resistance

7.1. Artificial Intelligence (AI)

AI is an area of computer science that emphasizes the creation of intelligent machines that work and react like humans and that uses labeled big data along with markedly enhanced computing power and cloud storage [189]. The most common applications of AI in drug treatment have to do with matching patients to their optimal drug or combination of drugs, predicting drug–target or drug–drug interactions and optimizing treatment protocols [190]. AI-based models have been developed for predicting synergistic treatment combinations in many diseases such as infectious diseases [191] and cancers [192,193]. One challenge is determining how AI-based technology may design tools which improve identification of therapeutic tolerance and resistance and develop new treatment combinations against tolerant and resistant cancers. The success of this AI-based approach may provide earlier and targeted anticancer treatment, which would prevent therapeutic tolerance/resistance emerging and cure cancer patients more effectively (Figure 1).

7.2. Pathogens

Pathogens such as microbiomes and viruses are becoming increasingly recognized for their effects on tumorigenesis and therapeutic resistance to cancer treatment [194]. Bacterial dysbiosis accompanies carcinogenesis in several malignancies such as gastric [195], colon [196], liver [197], and pancreatic [198] cancers by affecting metabolism and impairing immune functions [199]. Additionally, fungi [200] and viruses [201] also induce carcinogenesis in several cancers. Furthermore, intratumoral bacteria induced therapeutic resistance through breakdown of chemotherapy gemcitabine into inactive metabolites via bacterial enzymes such as cytidine deaminase [202] and via impairing response to immune checkpoint blockade [198]. Gut microbiota plays a critical role in mediating colorectal cancer chemoresistance in response to chemotherapeutics via a selective target loss of miR-18a* (miR-18a-3p) and miR-4802, and via activation of the autophagy pathway [203]. In addition, miR-18a* is a tumor suppressor that inhibits KRAS expression [204]. Activating *KRAS* mutations confer both primary [205] and acquired [206] resistance to anti-EGFR cetuximab therapy in colorectal cancer. Thus, targeting intratumoral pathogens provide a new angle in cancer treatment to overcome therapeutic tolerance/resistance. Some intracellular pathogens interact directly with receptor tyrosine kinases, and this interaction is critical for pathogen entry [207]. This establishes that pathogen-encoded receptor tyrosine kinase-interacting epitopes represent promising candidates for the development of novel therapeutic and prophylactic vaccines and of small-molecule interaction disruptors [208]. It would be of great interest to investigate whether those pathogens will confer therapeutic tolerance/resistance in host tumor cells by regulating miRNA–ROS interaction (Figure 1).

8. Concluding Remarks and Future Directions

Therapeutic tolerance/resistance raise major problems for the successful treatment of cancer, including conventional therapy and recent molecular therapy. There is an increasing importance of studying the role of ROS-relevant miRNAs to identify more effective biomarkers and develop better therapeutic targets against therapeutic tolerance/resistance. The interaction between miRNAs and ROSs fits in with the opportunities and challenges of studying mechanisms by which cancer cells resist therapy and ways by which therapeutic tolerance/resistance can be overcome. New concepts and emerging research tools bring potential to overcome therapeutic tolerance/resistance. However, some major challenges should be addressed properly. First, cancer relapse is driven by a small subpopulation of drug-tolerant persister cells, known as minimal residual disease in clinic. Single cell-relevant technologies, such as single-cell sequencing [209] might be applied to track single tolerant persister cells to gain insights into drug tolerance dynamics and heterogeneity [210]. In addition, preventative strategies using potential agents targeting those therapeutic tolerant cells at early stages in combination with molecular therapeutics will help prevent therapeutic tolerance and the resulting therapeutic resistance [211]. Second, new ex vivo models such as the organoid have been widely applied in cancer treatment response and therapeutic tolerance/resistance [212,213]. One of the advantages of the three-dimensional organoid model compared to a conventional two-dimensional monolayer is that tumor microenvironments established in organoids are similar to those found in vivo. For example, cancer organoids show heterogeneous hypoxic regions and show their enriched tumor-initiating cells and relevant metabolism pathway [214]. The organoid model may be used for large-scale screening, especially when incorporated with AI-based technology, to optimize the best drug combinations and thus reduce therapeutic tolerance/resistance. However, lacking immune cells and other types of cells has challenged this model [215]. Thus, incorporating immune cells will help better understand tolerance and resistance to immunotherapy [216]. Third, applications of non-invasive biomarkers to predict drug response represents a future direction in clinical settings. For example, cell-free circulating miRNAs have been successfully combined with low dose computed tomography scanning for diagnoses of early-stage lung cancer patients [217]. It is reasonable to incorporate cell-free circulating miRNAs signature together with cell-free DNAs signature [218] to predict and track the emergence of therapeutic tolerance/resistance. However, microRNAs predicting therapeutic tolerance/resistance might be dependent on specific mutant driver genes. For instance, increased miR-147b is relevant to mutant *EGFR* [76], and downregulated miR-23a is relevant to mutant *MYC* [30]. Thus, genetic mutation background and specific treatment agents should be considered comprehensively. Ultimately, early intervention on genetic/epigenetic, signaling/metabolic, and phenotypic changes in the miRNA–ROS network should be considered comprehensively to prevent and overcome therapeutic tolerance/resistance.

Acknowledgments: This work is supported by the Burnett School of Biomedical Sciences, College of Medicine, University of Central Florida grant 25400714, the NIH-Yale SPORE in Lung Cancer Career Development Program Award, and NRSA grant 5T32HL007893 awarded to W.C.Z. We thank Joshua Roney for critical reading and comments. We apologize to all researches whose work could not be cited due to reference limitations.

References

1. Harman, S.M.; Liang, L.; Tsitouras, P.D.; Gucciardo, F.; Heward, C.B.; Reaven, P.D.; Ping, W.; Ahmed, A.; Cutler, R.G. Urinary excretion of three nucleic acid oxidation adducts and isoprostane $F_2\alpha$ measured by liquid chromatography-mass spectrometry in smokers, ex-smokers, and nonsmokers. *Free Radic. Biol Med.* **2003**, *1301*, 35–1309. [CrossRef]

2. Apel, K.; Hirt, H. Reactive oxygen species: Metabolism, oxidative stress, and signal transduction. *Annu. Rev. Plant. Biol.* **2004**, *55*, 373–399. [CrossRef]

3. Finkel, T. Signal transduction by reactive oxygen species. *J. Cell Biol.* **2011**, *194*, 7–15. [CrossRef]

4. Dupré-Crochet, S.; Erard, M.; Nüße, O. ROS production in phagocytes: Why, when, and where? *J. Leukoc. Biol.* **2013**, *94*, 657–670. [CrossRef]

5. Li, R.; Jia, Z.; Trush, M.A. Defining ROS in biology and medicine. *React. Oxyg. Species* **2016**, *1*, 9–21.

6. Sabharwal, S.S.; Schumacker, P.T. Mitochondrial ROS in cancer: Initiators, amplifiers or an Achilles' heel? *Nat. Rev. Cancer* **2014**, *14*, 709–721. [CrossRef]

7. Zorov, D.B.; Juhaszova, M.; Sollott, S.J. Mitochondrial Reactive Oxygen Species (ROS) and ROS-Induced ROS Release. *Physiol. Rev.* **2014**, *94*, 909–950. [CrossRef]

8. Yeung, A.W.K.; Tzvetkov, N.T.; El-Tawil, O.S.; Bungau, S.G.; Abdel-Daim, M.M.; Atanasov, A.G. Antioxidants: Scientific literature landscape analysis. *Oxid. Med. Cell Longev.* **2019**, *8278*, 2019454. [CrossRef]

9. Valle, I.; Alvarez-Barrientos, A.; Arza, E.; Lamas, S.; Monsalve, M. Pgc-1α regulates the mitochondrial antioxidant defense system in vascular endothelial cells. *Cardiovasc. Res.* **2005**, *66*, 562–573. [CrossRef]

10. Ward, P.S.; Thompson, C.B. Metabolic reprogramming: A cancer hallmark even Warburg did not anticipate. *Cancer Cell* **2012**, *21*, 297–308. [CrossRef]

11. Vander Heiden, M.G.; Cantley, L.C.; Thompson, C.B. Understanding the Warburg effect: The metabolic requirements of cell proliferation. *Science* **2009**, *324*, 1029–1033. [CrossRef]

12. Vyas, S.; Zaganjor, E.; Haigis, M.C. Mitochondria and cancer. *Cell* **2016**, *166*, 555–566. [CrossRef]

13. Raimundo, N.; Baysal, B.E.; Shadel, G.S. Revisiting the TCA cycle: Signaling to tumor formation. *Trends Mol. Med.* **2011**, *17*, 641–649. [CrossRef]

14. Fernie, A.R.; Carrari, F.; Sweetlove, L.J. Respiratory metabolism: Glycolysis, the TCA cycle and mitochondrial electron transport. *Curr. Opin. Plant Biol.* **2004**, *7*, 254–261. [CrossRef]

15. Dahia, P.L. Pheochromocytoma and paraganglioma pathogenesis: Learning from genetic heterogeneity. *Nat. Rev. Cancer* **2014**, *14*, 108–119. [CrossRef]

16. Selak, M.A.; Armour, S.M.; MacKenzie, E.D.; Boulahbel, H.; Watson, D.G.; Mansfield, K.D.; Pan, Y.; Simon, M.C.; Thompson, C.B.; Gottlieb, E. Succinate links TCA cycle dysfunction to oncogenesis by inhibiting HIF-α prolyl hydroxylase. *Cancer Cell* **2005**, *7*, 77–85. [CrossRef]

17. Kaelin, W.G., Jr. Molecular basis of the VHL hereditary cancer syndrome. *Nat. Rev. Cancer* **2002**, *2*, 673–682. [CrossRef]

18. Krieg, M.; Haas, R.; Brauch, H.; Acker, T.; Flamme, I.; Plate, K.H. Up-regulation of hypoxia-inducible factors HIF-1α and HIF-2α under normoxic conditions in renal carcinoma cells by von Hippel-Lindau tumor suppressor gene loss of function. *Oncogene* **2000**, *19*, 5435–5443. [CrossRef]

19. Pagliarini, D.J.; Calvo, S.E.; Chang, B.; Sheth, S.A.; Vafai, S.B.; Ong, S.E.; Walford, G.A.; Sugiana, C.; Boneh, A.; Chen, W.K.; et al. A mitochondrial protein compendium elucidates complex I disease biology. *Cell* **2008**, *134*, 112–123. [CrossRef]

20. Calvo, S.E.; Clauser, K.R.; Mootha, V.K. MitoCarta2.0: An updated inventory of mammalian mitochondrial proteins. *Nucleic Acids Res.* **2016**, *44*, D1251–D1257. [CrossRef]

21. Gottlieb, E.; Tomlinson, I.P.M. Mitochondrial tumour suppressors: A genetic and biochemical update. *Nat. Rev. Cancer* **2005**, *5*, 857–866. [CrossRef]

22. Wagner, A.J.; Remillard, S.P.; Zhang, Y.X.; Doyle, L.A.; George, S.; Hornick, J.L. Loss of expression of SDHA predicts *SDHA* mutations in gastrointestinal stromal tumors. *Mod. Pathol.* **2013**, *26*, 289–294. [CrossRef] [PubMed]

23. Astuti, D.; Latif, F.; Dallol, A.; Dahia, P.L.M.; Douglas, F.; George, E.; Skoldberg, F.; Husebye, E.S.; Eng, C.; Maher, E.R. Gene mutations in the succinate dehydrogenase subunit SDHB cause susceptibility to familial pheochromocytoma and to familial paraganglioma. *Am. J. Hum. Genet.* **2001**, *69*, 49–54. [CrossRef] [PubMed]

24. Niemann, S.; Muller, U. Mutations in *SDHC* cause autosomal dominant paraganglioma, type 3. *Nat. Genet.* **2000**, *26*, 268–270. [CrossRef]

25. Baysal, B.E.; Ferrell, R.E.; Willett-Brozick, J.E.; Lawrence, E.C.; Myssiorek, D.; Bosch, A.; van der Mey, A.; Taschner, P.E.M.; Rubinstein, W.S.; Myers, E.N.; et al. Mutations in *SDHD*, a mitochondrial complex II gene, in hereditary paraganglioma. *Science* **2000**, *287*, 848–851. [CrossRef]

26. Zhang, W.C.; Shyh-Chang, N.; Yang, H.; Rai, A.; Umashankar, S.; Ma, S.; Soh, B.S.; Sun, L.L.; Tai, B.C.; Nga, M.E.; et al. Glycine decarboxylase activity drives non-small cell lung cancer tumor-initiating cells and tumorigenesis. *Cell* **2012**, *148*, 259–272. [CrossRef]

27. Zhang, W.C.; Lim, B. Targeting metabolic enzyme with locked nucleic acids in non-small cell lung cancer. *Cancer Res.* **2014**, *74*, 1438.

28. Lim, B.; Zhang, W. Targeting Metabolic Enzymes in Human Cancer. U.S. Patent 9,297,813,B2, 11 November 2011.

29. Go, M.K.; Zhang, C W.; Lim, B.; Yew, W.S. Glycine decarboxylase is an unusual amino acid decarboxylase involved in tumorigenesis. *Biochemistry* **2014**, *53*, 947–956. [CrossRef]

30. Gao, P.; Tchernyshyov, I.; Chang, T.C.; Lee, Y.S.; Kita, K.; Ochi, T.; Zeller, K.I.; De Marzo, A.M.; Van Eyk, J.E.; Mendell, J.T.; et al. C-Myc suppression of miR-23a/B enhances mitochondrial glutaminase expression and glutamine metabolism. *Nature* **2009**, *458*, 762–765. [CrossRef]

31. Testa, U.; Labbaye, C.; Castelli, G.; Pelosi, E. Oxidative stress and hypoxia in normal and leukemic stem cells. *Exp. Hematol.* **2016**, *44*, 540–560. [CrossRef]

32. Hanahan, D.; Weinberg, R.A. Hallmarks of cancer: The next generation. *Cell* **2011**, *144*, 646–674. [CrossRef]

33. Kasinski, A.L.; Slack, F.J. MicroRNAs en route to the clinic: Progress in validating and targeting microRNAs for cancer therapy. *Nat. Rev. Cancer* **2011**, *11*, 849–864. [CrossRef]

34. Anastasiadou, E.; Jacob, L.S.; Slack, F.J. Non-coding RNA networks in cancer. *Nat. Rev. Cancer* **2018**, *18*, 5–18. [CrossRef]

35. Fan, M.; Krutilina, R.; Sun, J.; Sethuraman, A.; Yang, H.C.; Wu, H.Z.; Yue, J.; Pfeffer, L.M. Comprehensive analysis of microRNA (miRNA) targets in breast cancer cells. *J. Biol. Chem.* **2013**, *288*, 27480–27493. [CrossRef]

36. Calin, G.A.; Croce, C.M. MicroRNA signatures in human cancers. *Nat. Rev. Cancer* **2006**, *6*, 857–866. [CrossRef]

37. Lu, J.; Getz, G.; Miska, A.E.; Alvarez-Saavedra, E.; Lamb, J.; Peck, D.; Sweet-Cordero, A.; Ebert, B.L.; Mak, R.H.; Ferrando, A.A.; et al. MicroRNA expression profiles classify human cancers. *Nature* **2005**, *435*, 834–838. [CrossRef]

38. Nouraee, N.; Calin, G.A. MicroRNAs as cancer biomarkers. *MicroRNA* **2013**, *2*, 102–117. [CrossRef]

39. Tentori, A.M.; Nagarajan, M.B.; Kim, J.J.; Zhang, W.C.; Slack, F.J.; Doyle, P.S. Quantitative and multiplex microRNA assays from unprocessed cells in isolated nanoliter well arrays. *Lab Chip* **2018**, *18*, 2410–2424. [CrossRef]

40. Nagarajan, M.B.; Tentori, A.M.; Zhang, W.C.; Slack, F.J.; Doyle, P.S. Nonfouling, encoded hydrogel microparticles for multiplex microRNA profiling directly from formalin-fixed, paraffin-embedded tissue. *Anal. Chem.* **2018**, *90*, 10279–10285. [CrossRef]

41. Anfossi, S.; Babayan, A.; Pantel, K.; Calin, G.A. Clinical utility of circulating non-coding RNAs—An update. *Nat. Rev. Clin. Oncol.* **2018**, *15*, 541–563. [CrossRef]

42. Schwarzenbach, H.; Nishida, N.; Calin, G.A.; Pantel, K. clinical relevance of circulating cell-free microRNAs in cancer. *Nat. Rev. Clin. Oncol.* **2014**, *11*, 145–156. [CrossRef]

43. Cortez, M.A.; Bueso-Ramos, C.; Ferdin, J.; Lopez-Berestein, G.; Sood, A.K.; Calin, G.A. MicroRNAs in body fluids—The mix of hormones and biomarkers. *Nat. Rev. Clin. Oncol.* **2011**, *8*, 467–477. [CrossRef]

44. Esquela-Kerscher, A.; Slack, F.J. Oncomirs—MicroRNAs with a role in cancer. *Nat. Rev. Cancer* **2006**, *6*, 259–269. [CrossRef]

45. Medina, P.P.; Nolde, M.; Slack, F.J. Oncomir addiction in an in vivo model of microRNA-21-induced pre-B-cell lymphoma. *Nature* **2010**, *467*, 86–90. [CrossRef]

46. Edmonds, M.D.; Boyd, K.L.; Moyo, T.; Mitra, R.; Duszynski, R.; Arrate, M.P.; Chen, X.; Zhao, Z.; Blackwell, T.S.; Andl, T.; et al. MicroRNA-31 initiates lung tumorigenesis and promotes mutant KRAS-driven lung cancer. *J. Clin. Investig.* **2016**, *126*, 349–364. [CrossRef] [PubMed]

47. Cheng, C.J.; Bahal, R.; Babar, I.A.; Pincus, Z.; Barrera, F.; Liu, C.; Svoronos, A.; Braddock, D.T.; Glazer, P.M.; Engelman, D.M.; et al. MicroRNA silencing for cancer therapy targeted to the tumour microenvironment. *Nature* **2015**, *518*, 107–110. [CrossRef] [PubMed]

48. Babar, I.A.; Cheng, C.J.; Booth, C.J.; Liang, X.; Weidhaas, J.B.; Saltzman, W.M.; Slack, F.J. Nanoparticle-based therapy in an in vivo microRNA-155 (miR-155)-dependent mouse model of lymphoma. *Proc. Natl. Acad. Sci. USA* **2012**, *109*, E1695–E1704. [CrossRef]

49. Ma, L.; Reinhardt, F.; Pan, E.; Soutschek, J.; Bhat, B.; Marcusson, E.G.; Teruya-Feldstein, J.; Bell, G.W.; Weinberg, R.A. Therapeutic silencing of miR-10b inhibits metastasis in a mouse mammary tumor model. *Nat. Biotechnol.* **2010**, *28*, 341–347. [CrossRef]

50. Johnson, S.M.; Grosshans, H.; Shingara, J.; Byrom, M.; Jarvis, R.; Cheng, A.; Labourier, E.; Reinert, K.L.; Brown, D.; Slack, F.J. RAS is regulated by the let-7 microRNA family. *Cell* **2005**, *120*, 635–647. [CrossRef]

51. Kasinski, A.L.; Slack, F.J. MiRNA-34 prevents cancer initiation and progression in a therapeutically resistant k-RAS and P53-induced mouse model of lung adenocarcinoma. *Cancer Res.* **2012**, *72*, 5576–5587. [CrossRef]

52. Krzeszinski, J.Y.; Wei, W.; Huynh, H.; Jin, Z.; Wang, X.; Chang, T.C.; Xie, X.J.; He, L.; Mangala, L.S.; Lopez-Berestein, G.; et al. MiR-34a blocks osteoporosis and bone metastasis by inhibiting osteoclastogenesis and Tgif2. *Nature* **2014**, *512*, 431–435. [CrossRef]

53. Banerjee, J.; Khanna, S.; Bhattacharya, A. MicroRNA regulation of oxidative stress. *Oxid. Med. Cell Longev.* **2017**, *2872*, 2017156. [CrossRef]

54. Hatley, M.E.; Patrick, D.M.; Garcia, M.R.; Richardson, J.A.; Bassel-Duby, R.; van Rooij, E.; Olson, E.N. Modulation of K-Ras-dependent lung tumorigenesis by microRNA-21. *Cancer Cell* **2010**, *18*, 282–293. [CrossRef]

55. Seike, M.; Goto, A.; Okano, T.; Bowman, D.E.; Schetter, J.A.; Horikawa, I.; Mathe, A.E.; Jen, J.; Yang, P.; Sugimura, H.; et al. MiR-21 is an EGFR-regulated anti-apoptotic factor in lung cancer in never-smokers. *Proc. Natl. Acad. Sci. USA* **2009**, *106*, 12085–12090. [CrossRef]

56. Zhang, X.; Ng, W.L.; Wang, P.; Tian, L.; Werner, E.; Wang, H.; Doetsch, P.; Wang, Y. MicroRNA-21 modulates the levels of reactive oxygen species by targeting Sod3 Tnfα. *Cancer Res.* **2012**, *72*, 4707–4713. [CrossRef]

57. Engedal, N.; Zerovnik, E.; Rudov, A.; Galli, F.; Olivieri, F.; Procopio, A.D.; Rippo, M.R.; Monsurro, V.; Betti, M.; Albertini, M.C. From oxidative stress damage to pathways, networks, and autophagy via microRNAs. *Oxid. Med. Cell Longev.* **2018**, *2018*, 4968321. [CrossRef]

58. Poyil, P.; Son, Y.-O.; Divya, S.P.; Wang, L.; Zhang, Z.; Shi, H. Oncogenic transformation of human lung bronchial epithelial cells induced by arsenic involves ROS-dependent activation of STAT3-miR-21-PDCD4 mechanism. *Sci. Rep.* **2016**, *6*, 37227.

59. Thulasingam, S.; Massilamany, C.; Gangaplara, A.; Dai, H.; Yarbaeva, S.; Subramaniam, S.; Riethoven, J.J.; Eudy, J.; Lou, M.; Reddy, J. MiR-27b*, an oxidative stress-responsive microRNA modulates nuclear factor-kB pathway in Raw 264.7 cells. *Mol. Cell Biochem.* **2011**, *352*, 181–188. [CrossRef]

60. Baker, J.R.; Vuppusetty, C.; Colley, T.; Papaioannou, A.I.; Fenwick, P.; Donnelly, L.; Ito, K.; Barnes, P.J. Oxidative stress dependent microRNA-34a activation via PI3Kα reduces the expression of sirtuin-1 and sirtuin-6 in epithelial cells. *Sci. Rep.* **2016**, *6*, 35871. [CrossRef]

61. Lin, Z.; Fang, D. The roles of SIRT1 in cancer. *Genes Cancer* **2013**, *4*, 97–104. [CrossRef] [PubMed]

62. Bussing, I.; Slack, F.J.; Grosshans, H. Let-7 microRNAs in development, stem cells and cancer. *Trends Mol. Med.* **2008**, *14*, 400–409. [CrossRef] [PubMed]

63. Saleh, A.D.; Savage, J.E.; Cao, L.; Soule, B.P.; Ly, D.; DeGraff, W.; Harris, C.C.; Mitchell, J.B.; Simone, N.L. Cellular stress induced alterations in microRNA let-7a and let-7b expression are dependent on P53. *PLoS ONE* **2011**, *6*, e24429. [CrossRef] [PubMed]

64. Ling, H.; Fabbri, M.; Calin, G.A. MicroRNAs and other non-coding RNAs as targets for anticancer drug development. *Nat. Rev. Drug Discov.* **2013**, *12*, 847–865. [CrossRef]

65. Adams, B.D.; Parsons, C.; Walker, L.; Zhang, W.C.; Slack, F.J. Targeting noncoding RNAs in disease. *J. Clin. Investig.* **2017**, *127*, 761–771. [CrossRef]

66. Rupaimoole, R.; Slack, F.J. MicroRNA therapeutics: Towards a new era for the management of cancer and other diseases. *Nat. Rev. Drug Discov.* **2017**, *16*, 203–222. [CrossRef]

67. Van Zandwijk, N.; Pavlakis, N.; Kao, S.C.; Linton, A.; Boyer, M.J.; Clarke, S.; Huynh, Y.; Chrzanowska, A.; Fulham, M.J.; Bailey, D.L.; et al. Safety and activity of microRNA-loaded minicells in patients with recurrent malignant pleural mesothelioma: A first-in-man, phase 1, open-label, dose-escalation study. *Lancet Oncol.* **2017**, *18*, 1386–1396. [CrossRef]

68. Van Zandwijk, N.; McDiarmid, J.; Brahmbhatt, H.; Reid, G. Response to an innovative mesothelioma treatment based on miR-16 mimic loaded EGFR targeted minicells (TargomiRs). *Transl. Lung Cancer Res.* **2018**, *7*, S60–S61. [CrossRef]

69. Fennell, D. MiR-16: Expanding the range of molecular targets in mesothelioma. *Lancet Oncol.* **2017**, *18*, 1296–1297. [CrossRef]

70. Slack, F.J.; Chinnaiyan, A.M. The role of non-coding RNAs in oncology. *Cell* **2019**, *179*, 1033–1055. [CrossRef]

71. Garzon, R.; Marcucci, G.; Croce, C.M. Targeting microRNAs in cancer: Rationale, strategies and challenges. *Nat. Rev. Drug Discov.* **2010**, *9*, 775–789. [CrossRef]

72. Zhang, P.; Wang, L.; Rodriguez-Aguayo, C.; Yuan, Y.; Debeb, B.G.; Chen, D.; Sun, Y.; You, M.J.; Liu, Y.; Dean, C.D.; et al. MiR-205 acts as a tumour radiosensitizer by targeting ZEB1 and Ubc13. *Nat. Commun.* **2014**, *5*, 5671. [CrossRef] [PubMed]

73. Wang, S.; Zhang, R.; Claret, F.X.; Yang, H. Involvement of microRNA-24 and DNA methylation in resistance of nasopharyngeal carcinoma to ionizing radiation. *Mol. Cancer Ther.* **2014**, *13*, 3163–3174. [CrossRef] [PubMed]

74. Babar, I.A.; Czochor, J.; Steinmetz, A.; Weidhaas, J.B.; Glazer, P.M.; Slack, F.J. Inhibition of hypoxia-induced miR-155 radiosensitizes hypoxic lung cancer cells. *Cancer Biol. Ther.* **2011**, *12*, 908–914. [CrossRef]

75. Zhang, W.C.; Chin, T.M.; Yang, H.; Nga, M.E.; Lunny, D.P.; Lim, E.K.; Sun, L.L.; Pang, Y.H.; Leow, Y.N.; Malusay, S.R.; et al. Tumour-initiating cell-specific miR-1246 and miR-1290 expression converge to promote non-small cell lung cancer progression. *Nat. Commun.* **2016**, *7*, 11702. [CrossRef]

76. Zhang, W.C.; Wells, J.M.; Chow, K.H.; Huang, H.; Yuan, M.; Saxena, T.; Melnick, M.A.; Politi, K.; Asara, J.M.; Costa, D.B.; et al. MiR-147b-mediated TCA cycle dysfunction and pseudohypoxia initiate drug tolerance to EGFR inhibitors in lung adenocarcinoma. *Nat. Metab.* **2019**, *1*, 460–474. [CrossRef]

77. Costinean, S.; Zanesi, N.; Pekarsky, Y.; Tili, E.; Volinia, S.; Heerema, N.; Croce, M.C. Pre-B Cell Proliferation and Lymphoblastic Leukemia/High-Grade Lymphoma in E(Mu)-Mir155 Transgenic Mice. *Proc. Natl. Acad. Sci. USA* **2006**, *103*, 7024–7029. [CrossRef]

78. Mikamori, M.; Yamada, D.; Eguchi, H.; Hasegawa, S.; Kishimoto, T.; Tomimaru, Y.; Asaoka, T.; Noda, T.; Wada, H.; Kawamoto, K.; et al. MicroRNA-155 controls exosome synthesis and promotes gemcitabine resistance in pancreatic ductal adenocarcinoma. *Sci. Rep.* **2017**, *7*, 42339. [CrossRef]

79. Moriyama, T.; Ohuchida, K.; Mizumoto, K.; Yu, J.; Sato, N.; Nabae, T.; Takahata, S.; Toma, H.; Nagai, E.; Tanaka, M. MicroRNA-21 modulates biological functions of pancreatic cancer cells including their proliferation, invasion, and chemoresistance. *Mol. Cancer Ther.* **2009**, *8*, 1067–1074. [CrossRef]

80. Giovannetti, E.; Funel, N.; Peters, J.G.; Del Chiaro, M.; Erozenci, L.A.; Vasile, E.; Leon, L.G.; Pollina, L.E.; Groen, A.; Falcone, A.; et al. MicroRNA-21 in pancreatic cancer: correlation with clinical outcome and pharmacologic aspects underlying its role in the modulation of gemcitabine activity. *Cancer Res.* **2010**, *70*, 4528–4538. [CrossRef]

81. Jiang, S.; Wang, R.; Yan, H.; Jin, L.; Dou, X.; Chen, D. MicroRNA-21 modulates radiation resistance through upregulation of hypoxia-inducible factor-1α-promoted glycolysis in non-small cell lung cancer cells. *Mol. Med. Rep.* **2016**, *13*, 4101–4107. [CrossRef]

82. Zhang, W.C.; Slack, F.J. MicroRNA-21 mediates resistance to EGFR tyrosine kinase inhibitors in lung cancer. *J. Thorac. Oncol.* **2017**, *12*, S1536. [CrossRef]

83. Yu, F.; Yao, H.; Zhu, P.; Zhang, X.; Pan, Q.; Gong, C.; Huang, Y.; Hu, X.; Su, F.; Lieberman, J.; et al. Let-7 regulates self renewal and tumorigenicity of breast cancer cells. *Cell* **2007**, *131*, 1109–1123. [CrossRef] [PubMed]

84. Yin, J.; Hu, W.; Pan, L.; Fu, W.; Dai, L.; Jiang, Z.; Zhang, F.; Zhao, J. Let-7 and miR-17 promote self-renewal and drive gefitinib resistance in non-small cell lung cancer. *Oncol. Rep.* **2019**, *42*, 495–508. [CrossRef] [PubMed]

85. Yu, F.; Deng, H.; Yao, H.; Liu, Q.; Su, F.; Song, E. MiR-30 reduction maintains self-renewal and inhibits apoptosis in breast tumor-initiating cells. *Oncogene* **2010**, *29*, 4194–4204. [CrossRef] [PubMed]

86. Ma, Y.S.; Yu, F.; Zhong, X.M.; Lu, G.X.; Cong, X.L.; Xue, S.B.; Xie, W.T.; Hou, L.K.; Pang, L.J.; Wu, W.; et al. MiR-30 family reduction maintains self-renewal and promotes tumorigenesis in NSCLC-initiating cells by targeting oncogene TM4SF1. *Mol. Ther.* **2018**, *26*, 2751–2765. [CrossRef] [PubMed]

87. Garofalo, M.; Romano, G.; Di Leva, G.; Nuovo, G.; Jeon, Y.J.; Ngankeu, A.; Sun, J.; Lovat, F.; Alder, H.; Condorelli, G.; et al. EGFR and MET receptor tyrosine kinase-altered microRNA expression induces tumorigenesis and gefitinib resistance in lung cancers. *Nat. Med.* **2012**, *18*, 74–82. [CrossRef]

88. Du, X.; Liu, B.; Luan, X.; Cui, Q.; Li, L. MiR-30 decreases multidrug resistance in human gastric cancer cells by modulating cell autophagy. *Exp. Ther. Med.* **2018**, *15*, 599–605. [CrossRef]

89. Jiang, L.; Hermeking, H. MiR-34a and miR-34b/c suppress intestinal tumorigenesis. *Cancer Res.* **2017**, *77*, 2746–2758. [CrossRef]

90. Zenz, T.; Mohr, J.; Eldering, E.; Kater, A.P.; Bühler, A.; Kienle, D.; Winkler, D.; Dürig, J.; van Oers, M.H.; Mertens, D.; et al. MiR-34a as part of the resistance network in chronic lymphocytic leukemia. *Blood* **2009**, *113*, 3801–3808. [CrossRef]

91. Pao, W.; Miller, V.; Zakowski, M.; Doherty, J.; Politi, K.; Sarkaria, I.; Singh, B.; Heelan, R.; Rusch, V.; Fulton, L.; et al. EGF receptor gene mutations are common in lung cancers from "never smokers" and are associated with sensitivity of tumors to gefitinib and erlotinib. *Proc. Natl. Acad. Sci. USA* **2004**, *101*, 13306–13311. [CrossRef]

92. Sordella, R.; Bell, D.W.; Haber, D.A.; Settleman, J. Gefitinib-sensitizing EGFR mutations in lung cancer activate anti-apoptotic pathways. *Science* **2004**, *305*, 1163–1167. [CrossRef]

93. Politi, K.; Herbst, R.S. Lung cancer in the era of precision medicine. *Clin. Cancer Res.* **2015**, *21*, 2213–2220. [CrossRef]

94. Kobayashi, S.; Boggon, T.J.; Dayaram, T.; Janne, P.A.; Kocher, O.; Meyerson, M.; Johnson, B.E.; Eck, M.J.; Tenen, D.G.; Halmos, B. EGFR mutation and resistance of non-small-cell lung cancer to gefitinib. *N. Engl. J. Med.* **2005**, *352*, 786–792. [CrossRef] [PubMed]

95. Politi, K.; Ayeni, D.; Lynch, T. The next wave of EGFR tyrosine kinase inhibitors enter the clinic. *Cancer Cell* **2015**, *27*, 751–753. [CrossRef] [PubMed]

96. Thress, K.S.; Paweletz, C.P.; Felip, E.; Cho, B.C.; Stetson, D.; Dougherty, B.; Lai, Z.; Markovets, A.; Vivancos, A.; Kuang, Y.; et al. Acquired EGFR C797S mutation mediates resistance to AZD9291 in non-small cell lung cancer harboring EGFR T790M. *Nat. Med.* **2015**, *21*, 560–562. [CrossRef]

97. Greaves, M.; Maley, C.C. Clonal evolution in cancer. *Nature* **2012**, *481*, 306–313. [CrossRef]

98. Hata, A.N.; Niederst, M.J.; Archibald, H.L.; Gomez-Caraballo, M.; Siddiqui, F.M.; Mulvey, H.E.; Maruvka, Y.E.; Ji, F.; Bhang, H.E.; Krishnamurthy Radhakrishna, V.; et al. Tumor cells can follow distinct evolutionary paths to become resistant to epidermal growth factor receptor inhibition. *Nat. Med.* **2016**, *22*, 262–269. [CrossRef]

99. Ramirez, M.; Rajaram, S.; Steininger, R.J.; Osipchuk, D.; Roth, M.A.; Morinishi, L.S.; Evans, L.; Ji, W.; Hsu, C.H.; Thurley, K.; et al. Diverse drug-resistance mechanisms can emerge from drug-tolerant cancer persister cells. *Nat. Commun.* **2016**, *7*, 10690. [CrossRef]

100. Sharma, S.V.; Lee, D.Y.; Li, B.; Quinlan, M.P.; Takahashi, F.; Maheswaran, S.; McDermott, U.; Azizian, N.; Zou, L.; Fischbach, M.A.; et al. A chromatin-mediated reversible drug-tolerant state in cancer cell subpopulations. *Cell* **2010**, *141*, 69–80. [CrossRef]

101. Smith, M.P.; Brunton, H.; Rowling, E.J.; Ferguson, J.; Arozarena, I.; Miskolczi, Z.; Lee, J.L.; Girotti, M.R.; Marais, R.; Levesque, M.P.; et al. Inhibiting drivers of non-mutational drug tolerance is a salvage strategy for targeted melanoma therapy. *Cancer Cell* **2016**, *29*, 270–284. [CrossRef]

102. Hangauer, M.J.; Viswanathan, V.S.; Ryan, J.M.; Bole, D.; Eaton, J.K.; Matov, A.; Galeas, J.; Dhruv, H.D.; Berens, M.E.; Schreiber, L.S.; et al. Drug-tolerant persister cancer cells are vulnerable to GPX4 inhibition. *Nature* **2017**, *551*, 247–250. [CrossRef] [PubMed]

103. Shaffer, S.M.; Dunagin, M.C.; Torborg, S.R.; Torre, E.A.; Emert, B.; Krepler, C.; Beqiri, M.; Sproesser, K.; Brafford, P.A.; Xiao, M.; et al. Rare cell variability and drug-induced reprogramming as a mode of cancer drug resistance. *Nature* **2017**, *546*, 431–435. [CrossRef] [PubMed]

104. Pavlides, S.; Vera, I.; Gandara, R.; Sneddon, S.; Pestell, R.G.; Mercier, I.; Martinez-Outschoorn, U.E.; Whitaker-Menezes, D.; Howell, A.; Sotgia, F.; et al. Warburg meets autophagy: Cancer-associated fibroblasts accelerate tumor growth and metastasis via oxidative stress, mitophagy, and aerobic glycolysis. *Antioxid. Redox Signal.* **2012**, *16*, 1264–1284. [CrossRef]

105. Sebolt-Leopold, J.S.; English, J.M. Mechanisms of drug inhibition of signalling molecules. *Nature* **2006**, *441*, 457–462. [CrossRef] [PubMed]

106. Zhu, Y.-j.; Zheng, B.; Wang, H.-y.; Chen, L. New knowledge of the mechanisms of sorafenib resistance in liver cancer. *Acta Pharmacol. Sin.* **2017**, *38*, 614–622. [CrossRef]

107. Mendez-Blanco, C.; Fondevila, F.; Garcia-Palomo, A.; Gonzalez-Gallego, J.; Mauriz, L.J. Sorafenib resistance in hepatocarcinoma: Role of hypoxia-inducible factors. *Exp. Mol. Med.* **2018**, *50*, 134. [CrossRef]

108. Vanecek, J. Cellular mechanisms of melatonin action. *Physiol. Rev.* **1998**, *78*, 687–721. [CrossRef]

109. Prieto-Domínguez, N.; Ordóñez, R.; Fernández, A.; Méndez-Blanco, C.; Baulies, A.; Garcia-Ruiz, C.; Fernández-Checa, J.C.; Mauriz, J.L.; González-Gallego, J. Melatonin-induced increase in sensitivity of human hepatocellular carcinoma cells to sorafenib is associated with reactive oxygen species production and mitophagy. *J. Pineal Res.* **2016**, *61*, 396–407. [CrossRef]

110. Prieto-Dominguez, N.; Mendez-Blanco, C.; Carbajo-Pescador, S.; Fondevila, F.; Garcia-Palomo, A.; Gonzalez-Gallego, J.; Mauriz, J.L. Melatonin enhances sorafenib actions in human hepatocarcinoma cells by inhibiting mTORC1/p70S6K/HIF-1α and hypoxia-mediated mitophagy. *Oncotarget* **2017**, *8*, 91402–91414. [CrossRef]

111. Herbst, R.S.; Heymach, J.V.; Lippman, S.M. Lung cancer. *N. Engl. J. Med.* **2008**, *359*, 1367–1380. [CrossRef]

112. Chen, Z.; Fillmore, C.M.; Hammerman, P.S.; Kim, C.F.; Wong, K.K. Non-small-cell lung cancers: A heterogeneous set of diseases. *Nat. Rev. Cancer* **2014**, *14*, 535–546. [CrossRef]

113. Nguyen, L.V.; Vanner, R.; Dirks, P.; Eaves, C.J. Cancer stem cells: An evolving concept. *Nat. Rev. Cancer* **2012**, *12*, 133–143. [CrossRef]

114. Tang, D.G. Understanding cancer stem cell heterogeneity and plasticity. *Cell Res.* **2012**, *22*, 457–472. [CrossRef] [PubMed]

115. Bonnet, D.; Dick, J.E. Human acute myeloid leukemia is organized as a hierarchy that originates from a primitive hematopoietic cell. *Nat. Med.* **1997**, *3*, 730–737. [CrossRef]

116. Zhang, W.C.; Yang, H.; Soh, B.S.; Sun, L.L.; Chin, T.M.; Lim, E.H.; Lim, B. Abstract 487: Evidence for tumor initiating stem cells in lung cancer. *Cancer Res.* **2011**, *71*, 487.

117. Tam, W.L.; Lu, H.; Buikhuisen, J.; Soh, B.S.; Lim, E.; Reinhardt, F.; Wu, Z.J.; Krall, J.A.; Bierie, B.; Guo, W.; et al. Protein kinase C α is a central signaling node and therapeutic target for breast cancer stem cells. *Cancer Cell* **2013**, *24*, 347–364. [CrossRef] [PubMed]

118. Singh, S.K.; Hawkins, C.; Clarke, I.D.; Squire, J.A.; Bayani, J.; Hide, T.; Henkelman, R.M.; Cusimano, M.D.; Dirks, P.B. Identification of human brain tumour initiating cells. *Nature* **2004**, *432*, 396–401. [CrossRef]

119. Ricci-Vitiani, L.; Lombardi, D.G.; Pilozzi, E.; Biffoni, M.; Todaro, M.; Peschle, C.; De Maria, R. Identification and expansion of human colon-cancer-initiating cells. *Nature* **2007**, *445*, 111–115. [CrossRef]

120. Hoe, S.L.L.; Tan, L.P.; Jamal, J.; Peh, S.C.; Ng, C.C.; Zhang, W.C.; Ahmad, M.; Khoo, A.S.B. Evaluation of stem-like side population cells in a recurrent nasopharyngeal carcinoma cell line. *Cancer Cell Int.* **2014**, *14*, 101. [CrossRef] [PubMed]

121. Gupta, P.B.; Onder, T.T.; Jiang, G.; Tao, K.; Kuperwasser, C.; Weinberg, R.A.; Lander, E.S. Identification of selective inhibitors of cancer stem cells by high-throughput screening. *Cell* **2009**, *138*, 645–659. [CrossRef] [PubMed]

122. Ito, K.; Hirao, A.; Arai, F.; Matsuoka, S.; Takubo, K.; Hamaguchi, I.; Nomiyama, K.; Hosokawa, K.; Sakurada, K.; Nakagata, N.; et al. Regulation of oxidative stress by ATM is required for self-renewal of haematopoietic stem cells. *Nature* **2004**, *431*, 997–1002. [CrossRef]

123. Diehn, M.; Cho, R.W.; Lobo, N.A.; Kalisky, T.; Dorie, M.J.; Kulp, A.N.; Qian, D.; Lam, J.S.; Ailles, L.E.; Wong, M.; et al. Association of reactive oxygen species levels and radioresistance in cancer stem cells. *Nature* **2009**, *458*, 780–783. [CrossRef]

124. Dando, I.; Cordani, M.; Dalla Pozza, E.; Biondani, G.; Donadelli, M.; Palmieri, M. Antioxidant mechanisms and ROS-related microRNAs in cancer stem cells. *Oxid. Med. Cell Longev.* **2015**, *4257*, 201508. [CrossRef]

125. Bao, S.; Wu, Q.; McLendon, R.E.; Hao, Y.; Shi, Q.; Hjelmeland, A.B.; Dewhirst, M.W.; Bigner, D.D.; Rich, J.N. Glioma stem cells promote radioresistance by preferential activation of the DNA damage response. *Nature* **2006**, *444*, 756–760. [CrossRef]

126. Curtis, S.J.; Sinkevicius, K.W.; Li, D.; Lau, A.N.; Roach, R.R.; Zamponi, R.; Woolfenden, A.E.; Kirsch, D.G.; Wong, K.K.; Kim, C.F. Primary tumor genotype is an important determinant in identification of lung cancer propagating cells. *Cell Stem Cell* **2010**, *7*, 127–133. [CrossRef]

127. Houessinon, A.; Francois, C.; Sauzay, C.; Louandre, C.; Mongelard, G.; Godin, C.; Bodeau, S.; Takahashi, S.; Saidak, Z.; Gutierrez, L.; et al. Metallothionein-1 as a biomarker of altered redox metabolism in hepatocellular carcinoma cells exposed to sorafenib. *Mol. Cancer* **2016**, *15*, 38. [CrossRef]

128. Li, L.; Abdel Fattah, E.; Cao, G.; Ren, C.; Yang, G.; Goltsov, A.A.; Chinault, A.C.; Cai, W.W.; Timme, T.L.; Thompson, T.C. Glioma pathogenesis-related protein 1 exerts tumor suppressor activities through proapoptotic reactive oxygen species-c-Jun-NH$_2$ kinase signaling. *Cancer Res.* **2008**, *68*, 434–443. [CrossRef]

129. Yuan, D.; Xu, J.; Wang, J.; Pan, Y.; Fu, J.; Bai, Y.; Zhang, J.; Shao, C. Extracellular miR-1246 promotes lung cancer cell proliferation and enhances radioresistance by directly targeting DR5. *Oncotarget* **2016**, *7*, 32707–32722. [CrossRef]

130. Golestaneh, A.F.; Atashi, A.; Langroudi, L.; Shafiee, A.; Ghaemi, N.; Soleimani, M. MiRNAs expressed differently in cancer stem cells and cancer cells of human gastric cancer cell line MKN-45. *Cell Biochem. Funct.* **2012**, *30*, 411–418. [CrossRef]

131. Jackman, D.M.; Johnson, B.E. Small-cell lung cancer. *Lancet* **2005**, *366*, 1385–1396. [CrossRef]

132. Oser, M.G.; Niederst, M.J.; Sequist, L.V.; Engelman, J.A. Transformation from non-small-cell lung cancer to small-cell lung cancer: molecular drivers and cells of origin. *Lancet Oncol.* **2015**, *16*, e165–e172. [CrossRef]

133. Sequist, L.V.; Waltman, B.A.; Dias-Santagata, D.; Digumarthy, S.; Turke, A.B.; Fidias, P.; Bergethon, K.; Shaw, A.T.; Gettinger, S.; Cosper, A.K.; et al. Genotypic and histological evolution of lung cancers acquiring resistance to EGFR inhibitors. *Sci. Transl. Med.* **2011**, *3*, 75ra26. [CrossRef]

134. Marcoux, N.; Gettinger, S.N.; O'Kane, G.; Arbour, K.C.; Neal, J.W.; Husain, H.; Evans, T.L.; Brahmer, J.R.; Muzikansky, A.; Bonomi, P.D.; et al. EGFR-mutant adenocarcinomas that transform to small-cell lung cancer and other neuroendocrine carcinomas: clinical outcomes. *J. Clin. Oncol.* **2018**, *37*, 278–285. [CrossRef]

135. Hayashita, Y.; Osada, H.; Tatematsu, Y.; Yamada, H.; Yanagisawa, K.; Tomida, S.; Yatabe, Y.; Kawahara, K.; Sekido, Y.; Takahashi, T. A polycistronic microRNA cluster, miR-17-92, is overexpressed in human lung cancers and enhances cell proliferation. *Cancer Res.* **2005**, *65*, 9628–9632. [CrossRef]

136. Ebi, H.; Sato, T.; Sugito, N.; Hosono, Y.; Yatabe, Y.; Matsuyama, Y.; Yamaguchi, T.; Osada, H.; Suzuki, M.; Takahashi, T. Counterbalance between RB inactivation and miR-17-92 overexpression in reactive oxygen species and DNA damage induction in lung cancers. *Oncogene* **2009**, *28*, 3371–3379. [CrossRef]

137. Kalluri, R.; Neilson, E.G. Epithelial-mesenchymal transition and its implications for fibrosis. *J. Clin. Investig.* **2003**, *112*, 1776–1784. [CrossRef]

138. Iliopoulos, D.; Lindahl-Allen, M.; Polytarchou, C.; Hirsch, H.A.; Tsichlis, P.N.; Struhl, K. Loss of miR-200 inhibition of Suz12 leads to polycomb-mediated repression required for the formation and maintenance of cancer stem cells. *Mol. Cell* **2010**, *39*, 761–772. [CrossRef]

139. Shimono, Y.; Zabala, M.; Cho, R.W.; Lobo, N.; Dalerba, P.; Qian, D.; Diehn, M.; Liu, H.; Panula, S.P.; Chiao, E.; et al. Downregulation of miRNA-200c links breast cancer stem cells with normal stem cells. *Cell* **2009**, *138*, 592–603. [CrossRef]

140. Gregory, P.A.; Bert, A.G.; Paterson, E.L.; Barry, S.C.; Tsykin, A.; Farshid, G.; Vadas, M.A.; Khew-Goodall, Y.; Goodall, G.J. The miR-200 family and miR-205 regulate epithelial to mesenchymal transition by targeting ZEB1 and SIP1. *Nat. Cell Biol.* **2008**, *10*, 593–601. [CrossRef]

141. Mateescu, B.; Batista, L.; Cardon, M.; Gruosso, T.; de Feraudy, Y.; Mariani, O.; Nicolas, A.; Meyniel, J.-P.; Cottu, P.; Sastre-Garau, X.; et al. MiR-141 and miR-200a act on ovarian tumorigenesis by controlling oxidative stress response. *Nat. Med.* **2011**, *17*, 1627–1635. [CrossRef]

142. DeNicola, G.M.; Karreth, F.A.; Humpton, T.J.; Gopinathan, A.; Wei, C.; Frese, K.; Mangal, D.; Yu, K.H.; Yeo, C.J.; Calhoun, E.S.; et al. Oncogene-induced Nrf2 transcription promotes ROS detoxification and tumorigenesis. *Nature* **2011**, *475*, 106–109. [CrossRef] [PubMed]

143. Magenta, A.; Cencioni, C.; Fasanaro, P.; Zaccagnini, G.; Greco, S.; Sarra-Ferraris, G.; Antonini, A.; Martelli, F.; Capogrossi, M.C. MiR-200c is upregulated by oxidative stress and induces endothelial cell apoptosis and senescence via Zeb1 inhibition. *Cell Death Differ.* **2011**, *18*, 1628–1639. [CrossRef] [PubMed]

144. Huang, X.; Ding, L.; Bennewith, K.L.; Tong, R.T.; Welford, S.M.; Ang, K.K.; Story, M.; Le, Q.T.; Giaccia, A.J. Hypoxia-inducible miR-210 regulates normoxic gene expression involved in tumor initiation. *Mol. Cell* **2009**, *35*, 856–867. [CrossRef] [PubMed]

145. Guzy, R.D.; Schumacker, P.T. Oxygen sensing by mitochondria at complex III: The paradox of increased reactive oxygen species during hypoxia. *Exp. Physiol.* **2006**, *91*, 807–819. [CrossRef]

146. Chan, S.Y.; Zhang, Y.Y.; Hemann, C.; Mahoney, C.E.; Zweier, J.L.; Loscalzo, J. MicroRNA-210 controls mitochondrial metabolism during hypoxia by repressing the iron-sulfur cluster assembly proteins ISCU1/2. *Cell Metab.* **2009**, *10*, 273–284. [CrossRef] [PubMed]

147. Favaro, E.; Ramachandran, A.; McCormick, R.; Gee, H.; Blancher, C.; Crosby, M.; Devlin, C.; Blick, C.; Buffa, F.; Li, J.L.; et al. MicroRNA-210 regulates mitochondrial free radical response to hypoxia and krebs cycle in cancer cells by targeting iron sulfur cluster protein ISCU. *PLoS ONE* **2010**, *5*, e10345. [CrossRef]

148. Chen, Z.; Li, Y.; Zhang, H.; Huang, P.; Luthra, R. Hypoxia-regulated microRNA-210 modulates mitochondrial function and decreases ISCU and COX10 expression. *Oncogene* **2010**, *29*, 4362–4368. [CrossRef]

149. Yang, W.; Sun, T.; Cao, J.; Liu, F.; Tian, Y.; Zhu, W. Downregulation of miR-210 expression inhibits proliferation, induces apoptosis and enhances radiosensitivity in hypoxic human hepatoma cells in vitro. *Exp. Cell Res.* **2012**, *318*, 944–954. [CrossRef]

150. Yang, W.; Wei, J.; Guo, T.; Shen, Y.; Liu, F. Knockdown of miR-210 decreases hypoxic glioma stem cells stemness and radioresistance. *Exp. Cell Res.* **2014**, *326*, 22–35. [CrossRef]

151. Ivan, M.; Huang, X. MiR-210: Fine-tuning the hypoxic response. *Adv. Exp. Med. Biol.* **2014**, *772*, 205–227.

152. Calvert, A.E.; Chalastanis, A.; Wu, Y.; Hurley, L.A.; Kouri, F.M.; Bi, Y.; Kachman, M.; May, J.L.; Bartom, E.; Hua, Y.; et al. Cancer-associated IDH1 promotes growth and resistance to targeted therapies in the absence of mutation. *Cell Rep.* **2017**, *19*, 1858–1873. [CrossRef]

153. Keith, B.; Johnson, R.S.; Simon, M.C. HIF1α and HIF2α: Sibling rivalry in hypoxic tumour growth and progression. *Nat. Rev. Cancer* **2011**, *12*, 9–22. [CrossRef] [PubMed]

154. Samanta, D.; Gilkes, D.M.; Chaturvedi, P.; Xiang, L.; Semenza, G.L. Hypoxia-inducible factors are required for chemotherapy resistance of breast cancer stem cells. *Proc. Natl. Acad. Sci. USA* **2014**, *111*, E5429–E5438. [CrossRef]

155. Kaelin, W.G., Jr. The von Hippel-Lindau tumour suppressor protein: O_2 sensing and cancer. *Nat. Rev. Cancer* **2008**, *8*, 865–873. [CrossRef]

156. Guo, J.; Chakraborty, A.A.; Liu, P.; Gan, W.; Zheng, X.; Inuzuka, H.; Wang, B.; Zhang, J.; Zhang, L.; Yuan, M.; et al. pVHL suppresses kinase activity of Akt in a proline-hydroxylation-dependent manner. *Science* **2016**, *353*, 929–932. [CrossRef]

157. Lee, S.B.; Frattini, V.; Bansal, M.; Castano, A.M.; Sherman, D.; Hutchinson, K.; Bruce, J.N.; Califano, A.; Liu, G.; Cardozo, T.; et al. An ID2-dependent mechanism for VHL inactivation in cancer. *Nature* **2016**, *529*, 172–177. [CrossRef]

158. MacKenzie, E.D.; Selak, M.A.; Tennant, D.A.; Payne, L.J.; Crosby, S.; Frederiksen, C.M.; Watson, D.G.; Gottlieb, E. Cell-permeating α-ketoglutarate derivatives alleviate pseudohypoxia in succinate dehydrogenase-deficient cells. *Mol. Cell Biol.* **2007**, *27*, 3282–3289. [CrossRef]

159. Gomes, A.P.; Price, N.L.; Ling, A.J.; Moslehi, J.J.; Montgomery, M.K.; Rajman, L.; White, J.P.; Teodoro, J.S.; Wrann, C.D.; Hubbard, B.P.; et al. Declining NAD^+ induces a pseudohypoxic state disrupting nuclear-mitochondrial communication during aging. *Cell* **2013**, *155*, 1624–1638. [CrossRef]

160. Wilson, T.R.; Fridlyand, J.; Yan, Y.; Penuel, E.; Burton, L.; Chan, E.; Peng, J.; Lin, E.; Wang, Y.; Sosman, J.; et al. Widespread potential for growth-factor-driven resistance to anticancer kinase inhibitors. *Nature* **2012**, *487*, 505–509. [CrossRef]

161. Niederst, M.J.; Engelman, J.A. Bypass mechanisms of resistance to receptor tyrosine kinase inhibition in lung cancer. *Sci. Signal.* **2013**, *6*, re6. [CrossRef]

162. DeBerardinis, R.J.; Chandel, N.S. Fundamentals of cancer metabolism. *Sci. Adv.* **2016**, *2*, e1600200. [CrossRef] [PubMed]

163. Hensley, C.T.; Faubert, B.; Yuan, Q.; Lev-Cohain, N.; Jin, E.; Kim, J.; Jiang, L.; Ko, B.; Skelton, R.; Loudat, L.; et al. Metabolic heterogeneity in human lung tumors. *Cell* **2016**, *164*, 681–694. [CrossRef] [PubMed]

164. DeBerardinis, R.J.; Mancuso, A.; Daikhin, E.; Nissim, I.; Yudkoff, M.; Wehrli, S.; Thompson, C.B. Beyond aerobic glycolysis: transformed cells can engage in glutamine metabolism that exceeds the requirement for protein and nucleotide synthesis. *Proc. Natl. Acad. Sci. USA* **2007**, *104*, 19345–19350. [CrossRef]

165. Yuneva, M.; Zamboni, N.; Oefner, P.; Sachidanandam, R.; Lazebnik, Y. Deficiency in glutamine but not glucose induces MYC-dependent apoptosis in human cells. *J. Cell Biol.* **2007**, *178*, 93–105. [CrossRef]

166. Imielinski, M.; Guo, G.; Meyerson, M. Insertions and deletions target lineage-defining genes in human cancers. *Cell* **2017**, *168*, 460.e14–472.e14. [CrossRef] [PubMed]

167. Li, B.; Ren, S.; Li, X.; Wang, Y.; Garfield, D.; Zhou, S.; Chen, X.; Su, C.; Chen, M.; Kuang, P.; et al. MiR-21 overexpression is associated with acquired resistance of EGFR-TKI in non-small cell lung cancer. *Lung Cancer* **2014**, *83*, 146–153. [CrossRef]

168. Zheng, G.; Li, N.; Jia, X.; Peng, C.; Luo, L.; Deng, Y.; Yin, J.; Song, Y.; Liu, H.; Lu, M.; et al. MYCN-mediated miR-21 overexpression enhances chemo-resistance via targeting CADM1 in tongue cancer. *J. Mol. Med. (Berl.)* **2016**, *94*, 1129–1141. [CrossRef]

169. Nishikura, K. A-to-I editing of coding and non-coding RNAs by ADARs. *Nat. Rev. Mol. Cell Biol.* **2016**, *17*, 83–96. [CrossRef]

170. Yang, W.; Chendrimada, T.P.; Wang, Q.; Higuchi, M.; Seeburg, P.H.; Shiekhattar, R.; Nishikura, K. Modulation of microRNA processing and expression through RNA editing by ADAR deaminases. *Nat. Struct. Mol. Biol.* **2006**, *13*, 13–21. [CrossRef]

171. Lazzari, E.; Crews, L.A.; Wu, C.; Leu, H.; Ali, S.; Chiaramonte, R.; Minden, M.; Costello, C.; Jamieson, C.H.M. Abstract 2414: ADAR1-dependent RNA editing is a mechanism of therapeutic resistance in human plasma cell malignancies. *Cancer Res.* **2016**, *76*, 2414.

172. Zhang, W.C.; Slack, F.J. ADARs edit microRNAs to promote leukemic stem cell activity. *Cell Stem Cell* **2016**, *19*, 141–142. [CrossRef] [PubMed]

173. Zipeto, M.A.; Court, A.C.; Sadarangani, A.; Delos Santos, N.P.; Balaian, L.; Chun, H.J.; Pineda, G.; Morris, S.R.; Mason, C.N.; Geron, I.; et al. ADAR1 activation drives leukemia stem cell self-renewal by impairing Let-7 biogenesis. *Cell Stem Cell* **2016**, *19*, 177–191. [CrossRef]

174. Jiang, Q.; Isquith, J.; Zipeto, M.A.; Diep, R.H.; Pham, J.; Delos Santos, N.; Reynoso, E.; Chau, J.; Leu, H.; Lazzari, E.; et al. Hyper-editing of cell-cycle regulatory and tumor suppressor RNA promotes malignant progenitor propagation. *Cancer Cell* **2019**, *35*, 81.e7–94.e7. [CrossRef] [PubMed]

175. Zaretsky, J.M.; Garcia-Diaz, A.; Shin, D.S.; Escuin-Ordinas, H.; Hugo, W.; Hu-Lieskovan, S.; Torrejon, D.Y.; Abril-Rodriguez, G.; Sandoval, S.; Barthly, L.; et al. Mutations associated with acquired resistance to PD-1 blockade in melanoma. *N. Engl. J. Med.* **2016**, *375*, 819–829. [CrossRef]

176. Sade-Feldman, M.; Jiao, Y.J.; Chen, J.H.; Rooney, M.S.; Barzily-Rokni, M.; Eliane, J.-P.; Bjorgaard, S.L.; Hammond, M.R.; Vitzthum, H.; Blackmon, S.M.; et al. Resistance to checkpoint blockade therapy through inactivation of antigen presentation. *Nat. Commun.* **2017**, *8*, 1136. [CrossRef]

177. Ishizuka, J.J.; Manguso, R.T.; Cheruiyot, C.K.; Bi, K.; Panda, A.; Iracheta-Vellve, A.; Miller, B.C.; Du, P.P.; Yates, K.B.; Dubrot, J.; et al. Loss of ADAR1 in tumours overcomes resistance to immune checkpoint blockade. *Nature* **2019**, *565*, 43–48. [CrossRef]

178. Chen, L.; Rio, D.C.; Haddad, G.G.; Ma, E. Regulatory role of dADAR in ROS metabolism in drosophila CNS. *Brain Res. Mol. Brain Res.* **2004**, *131*, 93–100. [CrossRef]

179. Dominissini, D.; Moshitch-Moshkovitz, S.; Schwartz, S.; Salmon-Divon, M.; Ungar, L.; Osenberg, S.; Cesarkas, K.; Jacob-Hirsch, J.; Amariglio, N.; Kupiec, M.; et al. Topology of the human and mouse m⁶A RNA methylomes revealed by m⁶A-seq. *Nature* **2012**, *485*, 201–206. [CrossRef] [PubMed]

180. Geula, S.; Moshitch-Moshkovitz, S.; Dominissini, D.; Mansour, A.A.; Kol, N.; Salmon-Divon, M.; Hershkovitz, V.; Peer, E.; Mor, N.; Manor, Y.S.; et al. M⁶A mRNA methylation facilitates resolution of naive pluripotency toward differentiation. *Science* **2015**, *347*, 1002–1006. [CrossRef]

181. Lin, S.; Choe, J.; Du, P.; Triboulet, R.; Gregory, R.I. The m⁶A methyltransferase METTL3 promotes translation in human cancer cells. *Mol. Cell* **2016**, *62*, 335–345. [CrossRef]

182. Choe, J.; Lin, S.; Zhang, W.; Liu, Q.; Wang, L.; Ramirez-Moya, J.; Du, P.; Kim, W.; Tang, S.; Sliz, P.; et al. MRNA circularization by METTL3-eIF3h enhances translation and promotes oncogenesis. *Nature* **2018**, *561*, 556–560. [CrossRef] [PubMed]

183. Lan, Q.; Liu, P.Y.; Haase, J.; Bell, J.L.; Huttelmaier, S.; Liu, T. The critical role of RNA m⁶A methylation in cancer. *Cancer Res.* **2019**, *79*, 1285–1292. [CrossRef] [PubMed]

184. Shi, H.; Wei, J.; He, C. Where, when, and how: Context-dependent functions of RNA methylation writers, readers, and erasers. *Mol. Cell* **2019**, *74*, 640–650. [CrossRef] [PubMed]
185. Visvanathan, A.; Patil, V.; Arora, A.; Hegde, S.A.; Arivazhagan, A.; Santosh, V.; Somasundaram, K. Essential Role of Mettl3-Mediated M6a Modification in Glioma Stem-Like Cells Maintenance and Radioresistance. *Oncogene* **2018**, *37*, 522–533. [CrossRef]
186. Taketo, K.; Konno, M.; Asai, A.; Koseki, J.; Toratani, M.; Satoh, T.; Doki, Y.; Mori, M.; Ishii, H.; Ogawa, K. The Epitranscriptome M6a Writer Mettl3 Promotes Chemo- and Radioresistance in Pancreatic Cancer Cells. *Int. J. Oncol.* **2018**, *52*, 621–629. [CrossRef]
187. Yan, F.; Al-Kali, A.; Zhang, Z.; Liu, J.; Pang, J.; Zhao, N.; He, C.; Litzow, M.R.; Liu, S. A dynamic N 6-methyladenosine methylome regulates intrinsic and acquired resistance to tyrosine kinase inhibitors. *Cell Res.* **2018**, *28*, 1062–1076. [CrossRef]
188. Wang, J.; Ishfaq, M.; Xu, L.; Xia, C.; Chen, C.; Li, J. METTL3/m^{6}A/miRNA-873–5p attenuated oxidative stress and apoptosis in colistin-induced kidney injury by modulating Keap1/Nrf2 pathway. *Front. Pharmacol.* **2019**. [CrossRef]
189. Topol, E.J. High-performance medicine: The convergence of human and artificial intelligence. *Nat. Med.* **2019**, *25*, 44–56. [CrossRef]
190. Romm, E.L.; Tsigelny, I.F. Artificial intelligence in drug treatment. *Annu. Rev. Pharmacol. Toxicol.* **2020**, *60*. [CrossRef]
191. Mason, D.J.; Eastman, R.T.; Lewis, R.P.I.; Stott, I.P.; Guha, R.; Bender, A. Using machine learning to predict synergistic antimalarial compound combinations with novel structures. *Front. Pharmacol.* **2018**. [CrossRef]
192. Li, X.; Xu, Y.; Cui, H.; Huang, T.; Wang, D.; Lian, B.; Li, W.; Qin, G.; Chen, L.; Xie, L. Prediction of synergistic anti-cancer drug combinations based on drug target network and drug induced gene expression profiles. *Artif. Intell. Med.* **2017**, *83*, 35–43. [CrossRef]
193. Xia, F.; Shukla, M.; Brettin, T.; Garcia-Cardona, C.; Cohn, J.; Allen, J.E.; Maslov, S.; Holbeck, S.L.; Doroshow, J.H.; Evrard, Y.A.; et al. Predicting tumor cell line response to drug pairs with deep learning. *BMC Bioinform.* **2018**, *19*, 486. [CrossRef] [PubMed]
194. McQuade, J.L.; Daniel, C.R.; Helmink, B.A.; Wargo, J.A. Modulating the microbiome to improve therapeutic response in cancer. *Lancet Oncol.* **2019**, *20*, e77–e91. [CrossRef]
195. Peek, R.M., Jr.; Blaser, M.J. Helicobacter pylori and gastrointestinal tract adenocarcinomas. *Nat. Rev. Cancer* **2002**, *2*, 28–37. [CrossRef] [PubMed]
196. Bullman, S.; Pedamallu, C.S.; Sicinska, E.; Clancy, T.E.; Zhang, X.; Cai, D.; Neuberg, D.; Huang, K.; Guevara, F.; Nelson, T.; et al. Analysis of fusobacterium persistence and antibiotic response in colorectal cancer. *Science* **2017**, *358*, 1443–1448. [CrossRef] [PubMed]
197. Ma, C.; Han, M.; Heinrich, B.; Fu, Q.; Zhang, Q.; Sandhu, M.; Agdashian, D.; Terabe, M.; Berzofsky, J.A.; Fako, V.; et al. Gut microbiome-mediated bile acid metabolism regulates liver cancer via NKT cells. *Science* **2018**, *360*, eaan5931. [CrossRef]
198. Pushalkar, S.; Hundeyin, M.; Daley, D.; Zambirinis, C.P.; Kurz, E.; Mishra, A.; Mohan, N.; Aykut, B.; Usyk, M.; Torres, L.E.; et al. The pancreatic cancer microbiome promotes oncogenesis by induction of innate and adaptive immune suppression. *Cancer Discov.* **2018**, *8*, 403–416. [CrossRef]
199. Sam, Q.H.; Chang, M.W.; Chai, L.Y. The fungal mycobiome and its interaction with gut bacteria in the host. *Int. J. Mol. Sci.* **2017**, *18*, 330. [CrossRef]
200. Aykut, B.; Pushalkar, S.; Chen, R.; Li, Q.; Abengozar, R.; Kim, J.I.; Shadaloey, S.A.; Wu, D.; Preiss, P.; Verma, N.; et al. The fungal mycobiome promotes pancreatic oncogenesis via activation of MBL. *Nature* **2019**, *574*, 264–267. [CrossRef]
201. Moore, P.S.; Chang, Y. Why do viruses cause cancer? Highlights of the first century of human tumour virology. *Nat. Rev. Cancer* **2010**, *10*, 878–889. [CrossRef]
202. Geller, L.T.; Barzily-Rokni, M.; Danino, T.; Jonas, O.H.; Shental, N.; Nejman, D.; Gavert, N.; Zwang, Y.; Cooper, Z.A.; Shee, K.; et al. Potential role of intratumor bacteria in mediating tumor resistance to the chemotherapeutic drug gemcitabine. *Science* **2017**, *357*, 1156–1160. [CrossRef]
203. Yu, T.; Guo, F.; Yu, Y.; Sun, T.; Ma, D.; Han, J.; Qian, Y.; Kryczek, I.; Sun, D.; Nagarsheth, N.; et al. Fusobacterium nucleatum promotes chemoresistance to colorectal cancer by modulating autophagy. *Cell* **2017**, *170*, 548.e16–563.e16. [CrossRef] [PubMed]

204. Tsang, W.P.; Kwok, T.T. The miR-18a* microRNA functions as a potential tumor suppressor by targeting on K-Ras. *Carcinogenesis* **2009**, *30*, 953–959. [CrossRef]

205. Lievre, A.; Bachet, J.B.; Le Corre, D.; Boige, V.; Landi, B.; Emile, J.F.; Cote, J.F.; Tomasic, G.; Penna, C.; Ducreux, M.; et al. KRAS mutation status is predictive of response to cetuximab therapy in colorectal cancer. *Cancer Res.* **2006**, *66*, 3992–3995. [CrossRef]

206. Misale, S.; Yaeger, R.; Hobor, S.; Scala, E.; Janakiraman, M.; Liska, D.; Valtorta, E.; Schiavo, R.; Buscarino, M.; Siravegna, G.; et al. Emergence of KRAS mutations and acquired resistance to anti-EGFR therapy in colorectal cancer. *Nature* **2012**, *486*, 532–536. [CrossRef] [PubMed]

207. Haqshenas, G.; Doerig, C. Targeting of host cell receptor tyrosine kinases by intracellular pathogens. *Sci. Signal.* **2019**, *12*, eaau9894. [CrossRef]

208. Schor, S.; Einav, S. Combating intracellular pathogens with repurposed host-targeted drugs. *ACS Infect. Dis.* **2018**, *4*, 88–92. [CrossRef] [PubMed]

209. Stewart, A.C.; Gay, C.M.; Xi, Y.; Fujimoto, J.; Kalhor, N.; Hartsfield, P.M.; Tran, H.; Fernandez, L.; Lu, D.; Wang, Y.; et al. Abstract 2899: Single-cell analyses reveal increasing intratumoral heterogeneity as an essential component of treatment resistance in small cell lung cancer. *Cancer Res.* **2019**, *79*, 2899.

210. Rambow, F.; Rogiers, A.; Marin-Bejar, O.; Aibar, S.; Femel, J.; Dewaele, M.; Karras, P.; Brown, D.; Chang, Y.H.; Debiec-Rychter, M.; et al. Toward minimal residual disease-directed therapy in melanoma. *Cell* **2018**, *174*, 843.e19–855.e19. [CrossRef]

211. Bivona, T.G.; Doebele, R.C. A framework for understanding and targeting residual disease in oncogene-driven solid cancers. *Nat. Med.* **2016**, *22*, 472–478. [CrossRef]

212. Ponz-Sarvise, M.; Corbo, V.; Tiriac, H.; Engle, D.D.; Frese, K.K.; Oni, T.E.; Hwang, C.I.; Ohlund, D.; Chio, I.I.C.; Baker, L.A.; et al. Identification of resistance pathways specific to malignancy using organoid models of pancreatic cancer. *Clin. Cancer Res.* **2019**, *25*, 6742–6755. [CrossRef]

213. Usui, T.; Sakurai, M.; Umata, K.; Elbadawy, M.; Ohama, T.; Yamawaki, H.; Hazama, S.; Takenouchi, H.; Nakajima, M.; Tsunedomi, R.; et al. Hedgehog signals mediate anti-cancer drug resistance in three-dimensional primary colorectal cancer organoid culture. *Int. J. Mol. Sci.* **2018**, *19*, 1098. [CrossRef]

214. Hubert, C.G.; Rivera, M.; Spangler, L.C.; Wu, Q.; Mack, S.C.; Prager, B.C.; Couce, M.; McLendon, R.E.; Sloan, A E.; Rich, N.J. A three-dimensional organoid culture system derived from human glioblastomas recapitulates the hypoxic gradients and cancer stem cell heterogeneity of tumors found in vivo. *Cancer Res.* **2016**, *76*, 2465–2477. [CrossRef] [PubMed]

215. Chakrabarti, J.; Holokai, L.; Syu, L.; Steele, N.; Chang, J.; Dlugosz, A.; Zavros, Y. Mouse-derived gastric organoid and immune cell co-culture for the study of the tumor microenvironment. *Methods Mol. Biol.* **2018**, *1817*, 157–168. [PubMed]

216. Neal, J.T.; Li, X.; Zhu, J.; Giangarra, V.; Grzeskowiak, C.L.; Ju, J.; Liu, I.H.; Chiou, S.H.; Salahudeen, A.A.; Smith, A.R.; et al. Organoid modeling of the tumor immune microenvironment. *Cell* **2018**, *175*, 1972.e16–1988.e16. [CrossRef] [PubMed]

217. Sozzi, G.; Boeri, M.; Rossi, M.; Verri, C.; Suatoni, P.; Bravi, F.; Roz, L.; Conte, D.; Grassi, M.; Sverzellati, N.; et al. Clinical utility of a plasma-based miRNA signature classifier within computed tomography lung cancer screening: A correlative mild trial study. *J. Clin. Oncol.* **2014**, *32*, 768–773. [CrossRef]

218. Siravegna, G.; Bardelli, A. Genotyping cell-free tumor DNA in the blood to detect residual disease and drug resistance. *Genome Biol.* **2014**, *15*, 449. [CrossRef]

6

Development and Clinical Trials of Nucleic Acid Medicines for Pancreatic Cancer Treatment

Keiko Yamakawa, Yuko Nakano-Narusawa, Nozomi Hashimoto, Masanao Yokohira and Yoko Matsuda *

Oncology Pathology, Department of Pathology and Host-Defense, Faculty of Medicine, Kagawa University, 1750-1 Ikenobe, Miki-cho, Kita-gun, Kagawa 761-0793, Japan
* Correspondence: youkoh@med.kagawa-u.ac.jp

Abstract: Approximately 30% of pancreatic cancer patients harbor targetable mutations. However, there has been no therapy targeting these molecules clinically. Nucleic acid medicines show high specificity and can target RNAs. Nucleic acid medicine is expected to be the next-generation treatment next to small molecules and antibodies. There are several kinds of nucleic acid drugs, including antisense oligonucleotides, small interfering RNAs, microRNAs, aptamers, decoys, and CpG oligodeoxynucleotides. In this review, we provide an update on current research of nucleic acid-based therapies. Despite the challenging obstacles, we hope that nucleic acid drugs will have a significant impact on the treatment of pancreatic cancer. The combination of genetic diagnosis using next generation sequencing and targeted therapy may provide effective precision medicine for pancreatic cancer patients.

Keywords: nucleic acid medicine; pancreatic cancer; clinical trial; siRNA; antisense oligonucleotide

---

1. Introduction

Despite advances in diagnostics and therapeutics, the prognosis of pancreatic cancer remains poor with an overall five-year survival rate of 6%, due in part to difficulties in treating carcinoma at an advanced stage. Mutations of *KRAS*, *CDKN2a*, *TP53*, and *SMAD4* are driver mutations in pancreatic cancer; however, a targeted approach for those molecules has not been successful yet. Precision medicine for individual patient has been greatly expected to improve pancreatic cancer patients' outcomes. Recent advances of comprehensive gene analysis using next-generation sequencers can provide a wealth of information of genetic abnormalities of cancers [1,2]. There have been several candidates for treatment targets in pancreatic cancer. Approximately 30% of pancreatic cancer patients harbor druggable mutations; for example, *KRAS*, *BRCA1* and 2, *PALB2*, *ATM*, *HER2*, *MET*, *MLH1*, *MSH2*, *MSH6*, *PMS2*, *PI3CA*, *PTEN*, *CDKN2A*, *BRAF*, and *FGFR1* [2]. However, there has been no clinical therapy targeting these molecules, because it is difficult to inhibit target RNA in humans.

RNA interference (RNAi) is a biological process in which RNA molecules inhibit gene expression or translation by neutralizing targeted mRNA molecules. Nucleic acid medicine consists of natural or chemically modified nucleotides that can act directly without changes in gene expression [3]. These drugs show high specificity and can target mRNA and noncoding RNAs. Nucleic acid medicine is considered the next-generation treatment next to small molecules and antibodies. There are several aspects of nucleic acid therapy that are potentially advantageous over traditional drugs. These include the ability to generate specific inhibitors of targets that were previously inaccessible, with the only limit being the genetic information available. Inhibition of mRNA expression has the potential to produce faster and longer-lasting responses than protein inhibition by conventional targeted therapy. Moreover, the side-effects of nucleic acid medicine might be less than those of conventional therapy [4]. Lastly, oligonucleotides can be chemically synthesized and thus their development duration is relatively short compared to antibodies.

There are several kinds of nucleic acid drugs, including antisense oligonucleotides (ASOs), small interfering RNAs (siRNAs), microRNAs (miRNAs), aptamers, decoys, and CpG oligodeoxynucleotides (CpG oligos) (Table 1). They can be classified as either extracellular or intracellular according to their site of function; ASOs, siRNAs, miRNAs, and decoys act in the nucleus or cytoplasm, while aptamers bind to extracellular proteins and CpG oligos act on Toll-like receptor 9 (TLR9) in the endosome. The drugs also have different targets; ASOs, miRNAs, and siRNAs target RNA, whilst aptamers, decoys, and CpG oligos target proteins. Nucleic acid drugs are suited for coextinction or therapeutic synergy, which may represent an important step to overcome compensatory effects typically observed in cancer cells following knockdown of a single target. In this review, we provide an update on the current research of nucleic acid-based therapies, focusing on ASO and siRNA for pancreatic cancer, and summarize the outcomes from published data.

Table 1. Nucleic acid medicines.

	Antisense Oligonucleotides	siRNAs	Antisense miRNAs	miRNA Mimics	Decoys	Aptamers	CpG Oligodeoxynucleotides
Structure	Single strand DNA/RNA	Double strand RNA	Single strand DNA/RNA	Double strand RNA	Double strand DNA	Single strand DNA/RNA	Single strand DNA
Length (base pairs)	12–21 20–30	20–25	12–16	20–25	20	26–45	20
Site	Intracellular (nucleus, cytoplasm)	Intracellular (cytoplasm)	Intracellular (cytoplasm)	Intracellular (cytoplasm)	Intracellular (nucleus)	Extracellular	Extracellular (endosome)
Target	mRNA pre-mRNA miRNA	mRNA	miRNA	mRNA	Protein (transcription factor)	Protein	Protein (TLR9)
Function	mRNA degradation Translational inhibition miRNA inhibition Splicing inhibition	mRNA degradation	miRNA degradation	mRNA degradation Translational inhibition	Transcriptional inhibition	Inhibition of protein function	Activation of natural immunity via TLR9
Drug delivery system	Modified or unnecessary	Necessary	Necessary	Necessary	Necessary	PEGylation	Antigen

TLR9, toll like receptor 9.

2. Functions

2.1. Antisense Oligonucleotides

ASOs are single strands of DNA or RNA that are complementary to a chosen sequence. In the case of antisense RNA, they prevent protein translation of certain messenger RNA strands by binding to them [5].

Antisense DNA can be used to target a specific, complementary (coding or noncoding RNA). If binding takes place, this DNA/RNA hybrid can be degraded by the enzyme RNase H. After crossing the cell membrane, ASOs target mRNA directly in the nucleus or cytosol, thus blocking and neutralizing the targeted miRNA, with the help of the enzyme RNase H1. Furthermore, ASOs have various functions, including the inhibition of translation, miRNA, and splicing. ASOs have been investigated for more than 20 years and their use is now a standard technique in developmental biology and they are used to study altered gene expression and gene function. Recently, several ASOs have been modified for an unnecessary drug delivery system (DDS).

2.2. siRNAs

siRNAs are double-stranded RNAs with a length of 20–25 base pairs. siRNAs can suppress the gene expression via sequence specific inhibition of RNA expression (RNA interference, RNAi). The cellular process of RNAi occurs in almost all eukaryotic organisms [6]. After being processed by the ribonuclease III-like DICER enzyme, siRNA interacts with RNA-induced silencing complex to block and neutralize the target mRNA [7]. siRNA libraries have been created to dissect the function of independent genes since they show high sequence specificity. The application of siRNAs allows researchers to discover novel targets and pathway mediators.

2.3. Aptamers

Nucleic acid aptamers are short single-stranded DNA or RNA oligonucleotides that fold into unique three-dimensional structures and bind to a wide range of targets, including proteins, small molecules, metal ions, viruses, bacteria, and whole cells [8]. Aptamers have high specificity and binding affinities (in the low nanomolar to picomolar range) similar to those of antibodies and are frequently referred to as 'chemical antibodies'. Proteins constitute by far the largest class of aptamer targets. The high stability of aptamer–protein complexes, frequently characterized by a Kd in the low nanomolar range, combined with an excellent specificity of interaction make aptamers valuable tools for various applications, such as affinity purification, bio-sensing, imaging, and enzyme inhibition [9].

2.4. Decoys

Decoys are double-stranded molecules that mimic the consensus DNA binding site of a specific transcription factor in the promoter region of its target genes [10]. The regulation of transcription of disease-related genes in vivo has important therapeutic potential. Gene expression controlled by the transcription factor is effectively prevented, thereby effectively silencing gene expression and preventing protein production. Therefore, being less specific in comparison with the siRNA or ASO method, the decoy technique can be considered a gene silencing approach.

2.5. CpG Oligos

CpG oligodeoxynucleotides (CpG oligos) are short single-stranded synthetic DNA molecules that contain cytosine triphosphate deoxynucleotide followed by a guanine triphosphate deoxynucleotide [11]. Synthetic phosphorothioate oligodeoxynucleotides bearing unmethylated CpG motifs can mimic the immune-stimulatory effects of bacterial DNA and are recognized by Toll-like receptor 9 (TLR9), which is constitutively expressed only in B cells and plasmacytoid dendritic cells. Nucleotide modifications at positions at or near the CpG dinucleotides can severely affect immune modulation. CpG oligos induce type I interferon, cytokines, B cell proliferation, dendritic cell maturation, and natural killer cell activation. CpG oligos have been applied for antiallergenic or anticancer treatment.

3. Modifications of Nucleic Acid Drugs

Although, the function of nucleic acid drugs is promising, several challenges have been identified, including lack of stability against extracellular and intracellular degradation by nucleases, poor uptake and low potency at target sites of nucleic acid drugs, and off-target effects [12]. Off-target effects are nonspecific suppressive effects of nucleic acid drugs. Although it has been considered that nucleic acid drugs possess high specificity, several nucleic acid drugs can affect gene expression of multiple genes. Furthermore, nucleic acid drugs are quickly degraded by RNase in vivo. In humans, naked nucleic acid drugs preferentially accumulate in the liver and kidneys, which causes the nucleic acid drugs to be rapidly cleared from circulation with poor tissue distribution [13]. The pursuit of clinically viable antisense drugs has led to the development of various types of strategies, such as carriers or chemical modifications. Apart from structural modification of oligonucleotides, different cell-penetrating peptides and ligands conjugated to oligonucleotide-based DDS are normally adopted following the conjugation.

3.1. Structural Modifications of Nucleic Acid Drugs

Important modifications have been implemented to improve the therapeutic potential of nucleic acid medicines. However, the properties of the modifications have also led to some decreased affinity for the target sequence, with associated nonhybridization toxicities such as complement activation, increased coagulation times, or immune activation (Table 2). Another concern relates to the hybridization-dependent toxicity, caused by exaggerated action of the drug or off-target hybridization.

Table 2. Modifications of nucleic acid drugs.

Structural Modifications	Contents	Stability	Cellular Uptake	Gene Silencing Effect	Cytotoxicity	Binding Affinity
Diester modification	Phosphorothioate	superior	superior	inferior	superior	
Ribose modification	2'-O-Me, 2'-O-A, 2'-F	superior		inferior		
Base modification	Adenine methylation and deamination, cytosine methylation, hydroxy methylation and carboxy substitution, Guanine oxidation			superior		
Oligonucleotide analogues replacement	Peptide nucleic acid, locked nucleic acid, morpholino phosphamide	superior		superior		inferior
Conjugation to cell-penetrating peptides	Cysteine, transactivator of transcription peptide, gelatin		superior	superior	inferior	
Aptamer	20–100 nucleotides		superior	superior		

The first of the modifications included phosphorothioate backbone modification, which defined the first-generation nucleic acid drugs [5]. One of the nonbridge oxygen atoms in the diester bond is replaced by sulfur. Chemical modification can help enhance cellular uptake and increase the bioavailability of the modified nucleic acid drugs. Resistance to circumscribed nucleases is also effectively increased. However, although the modified siRNA is found to be significantly stable in the body, it increases the cytotoxicity and decreases the gene silencing effect. Modification of phosphorylated phosphate ester in the phosphorylation location damages RISC activity [14].

The second-generation nucleic acid drugs included the nucleoside analogues containing a modified sugar moiety, such as 2'-O-methyl-modified or 2'-O-methoxyethyl. The 2' modifications inhibit the ability of RNase H to cleave the bound sense RNA strand within the heteroduplex formed between the nucleic acid drugs and the target RNA [15]. The widespread use of thiophosphate modifications results in a certain cytotoxicity, but the 2'-O-methylation improves the siRNA activity and is nontoxic to normal cells [16]. The activity of siRNA depends on the position of the modified parts.

Base modification plays an important role in the function of nucleic acid drugs; for example, it can improve the function of siRNA and increase the ability of the siRNA interaction with the target mRNA. The modification increases the ability of RISC to recognize and cleave the mRNA. The modifications on the base include adenine methylation and deamination, cystosine methylation, hydroxymethylation and carboxyl substitution, and guanine oxidation, etc. [17]. The modified bases are related to the changes of functional groups, which is the basis of triggering the functional changes through the modification of structure of nucleic acid drugs.

Oligonucleotide analogs' replacement includes peptide substitution, and the resulting materials typically include peptide nucleic acid, locked nucleic acid, and morpholino phosphamide. They can reduce the degradation of oligonucleotides by nucleases, and have low toxicity and a slight decrease in affinity compared with unmodified sequences [18]. These nucleotide analogs do not support the cleavage of RNase H-mediated target mRNA in ASPs; thereby, they primarily exhibit their reflective activity by steric hindrance to prevent gene expression during transcription or translation. This method further enhances the binding affinity, nuclease resistance, and targeted effect compared with several other chemical modifications.

3.2. Conjugation of Ligand or Cell-Penetrating Peptides

Cell-penetrating peptides are a class of short peptides that are rich in cations and can efficiently enter cells through penetrating biofilms. Based on these properties, cell-penetrating peptides are used to modified DNA, RNA, and oligonucleotides and are loaded on nanocarriers for therapy. The conjugation of oligonucleotides and cell-penetrating peptides can overcome the deficiencies of cytotoxicity and enhance the efficiency in eukaryotic cells. Complexes formed by cationic cell-penetrating peptides and anionic oligonucleotides which are formed through electrostatic interaction can promote oligonucleotides' entry into cells and initiate RNA interfering, leading to silencing of endogenous genes [19]. Cell-penetrating peptides include cysteine, transactivator of transcription peptide [20], and gelatin [21].

4. Aptamers

Aptamers are synthetic single-stranded oligonucleotides of short length (20–100 nucleotides) whose three-dimensional disposition confers high avidity for their target DNA or RNA. They shows high stability, lack of immunogenicity, flexible structure, and small size, which increases their penetration strength [22]. Aptamer-based targeted delivery of siRNAs using aptamer–siRNA chimeras are becoming a very useful tool for targeting gene-knockdown in cancer therapy [23]. Aptamer–siRNA chimeras bind the aptamer's receptor and upon engagement, the chimera–receptor complex is embedded into an endocytosis vesicle. The chimera reaches the cytoplasm and the duplex siRNA is recognized by Dicer and loaded into Dicer and RNA-induced silencing complex (RISC). Several aptamers have been reported for treatment of prostate, breast, and colon cancer, melanoma, lymphoma, and glioblastoma, for example *PSMA*, *4-1BBm EpCAP*, *CTLA4*, *PDGFRβ*, *HER2*, and *HER3* [23].

5. Drug Delivery Systems of Nucleic Acid Drugs

DDS has been necessary to regulate the drug distribution in the body in terms of quantity and spatiotemporal aspects. Several kinds of DDSs have been developed based on the diameter of medicine, specific antibody for tumor, sustained release, and percutaneous absorption. They are expected to improve the specificity, effects, usability, and economy of drug as well as to suppress the side-effects.

Various carriers of siRNAs have become increasingly available because RNAi can integrate short hairpin RNA into the cell genome, leading to stable siRNA expression and long-term knockdown of a target gene. Nonviral carriers have been increasingly preferred owing to lower toxicity compared with other carrier methods. These carriers typically involve a positively charged vector (cationic cell-penetrating peptides, cationic polymers, and lipids), small molecules (cholesterol, bile acids, lipids, and PEGylated lipids), polymers, antibodies, aptamers, and lipid and polymer-based nanocarriers encapsulating the siRNA [24]. Specific delivery of siRNAs to hepatocytes has been accomplished by conjugation to *N*-acetylgalactosamine in order to target an asialoglycoprotein receptor present in the liver [25].

Different nanocarrier strategies are still needed in practical applications to make them more effective in diagnosing and treating diseases. A combination of chemical modification and a nanoparticle-based DDS is likely to be more effective for oligonucleotide delivery. For example, the siRNA can be modified with the free thiol group of the amino acid cysteine on cell-penetrating peptides, then they are encapsulated into ultrasound-sensitive nanomicrobubbles. When nanomicrobubbles reach the target site, they disintegrate under external ultrasonic irradiation, releasing siRNA to achieve cytoplasmic delivery [26].

Liposomes are widely used as oligonucleotide delivery systems (Table 3). Cationic liposomes include monovalent lipids such as DODMA and DOTAP [27]. Oligonucleotides are negatively charged and easy to encapsulate into cationic liposomes. Neutral liposomes are primarily constructed by neutral lipids, which include PC, PE, cholesterol, and DOPE [28]. Neutral liposomes have good biocompatibility and excellent pharmacokinetic characteristics, but they cannot interact with oligonucleotides to adsorb them and encapsulate them into the liposomes efficiently. Neutral liposomes are adopted to modified cationic liposomes to enhance particle stability. Ionizable liposomes are important for siRNA delivery. They can protonated and deprotonated according to the acidity of the environment [28]. Under hypoxic conditions, tumor tissues are more acidic and pH-responsive liposomes have more positive charges. Cationic liposomes are the most widely used form of liposomes.

Table 3. Drug delivery systems.

	Materials
Liposomes	
Cationic liposome	DOTAP, DODMA, DOGS, DC-Chol
Neutral liposome	PC, Chol, DOPE
Ionizable liposome	DODMA, DODAP
Micelles	
Polymeric micelles	Amphiphilic copolymer, PEG, polyamino acid, polylactic or glycolic acid, polycaprolactone, and short phospholipid chains
Cationic polymer micelles	PEG-PLL-PLLeu, PEI-CG-PEI, PgP
Nanoparticles	
Albumin-based	thiol, arginine-glycine-aspartic acid peptide
Metal-based	gold, silver, magnetic

Polymeric micelles have promising applications in drug delivery including extending the drug cycle time, changing the drug release curve, and easily connecting targeted ligands [29]. Cationic polymer micelles can ensure good oligonucleotide loading capacity through electrostatic adsorption. They show long circulation times, tumor passive targeting by the enhanced permeability and retention effect, and efficient oligonucleotide endosome release by the proton sponge effect [30]. Furthermore, the suitable carrier should can deliver oligonucleotides and chemotherapy drugs together to the tumor tissue and release the two drugs simultaneously, for example polymeric micelles with doxorubicin and siRNA targeting P-glycoprotein [31].

Nanoparticles using albumin, metals, and polymers have been used for drug delivery. Tumor cells can take up human serum albumin through endocytosis; therefore, albumin-based nanoparticles can show high stability without cytotoxicity [32]. Metallic nanometer-sized particles, such as silver, gold, and magnetic metals show the property of the enhanced surface to volume ratio; therefore, they have good applications in oligonucleotide delivery [33].

Another challenge to overcome in the DDS for pancreatic cancer is intratumoral injection [34] or implantation [35,36] of siRNAs in the pancreas (Figure 1). Implantation of Local Drug EluterR

(LODER), can release siRNAs targeting KRAS over months in pancreatic cancer in vivo [36]. LODER is a biodegradable polymeric matrix that shields drugs against enzymatic degradation. EUS have enabled researchers to obtain pancreatic tissue samples and inject medicines into the pancreas repeatedly; therefore, DDS using EUS may improve the effectiveness of siRNA treatment for pancreatic cancer. In an animal model, we have reported that administration of siRNA by intratumoral injection with atelocollagen [37] and intravenous injection [38]. Both settings were effective to reduce targeted mRNA expression in vivo without severe side effects in the short term. Clinical trials are necessary to determine the long-term effects and safety of nucleic acid medicines.

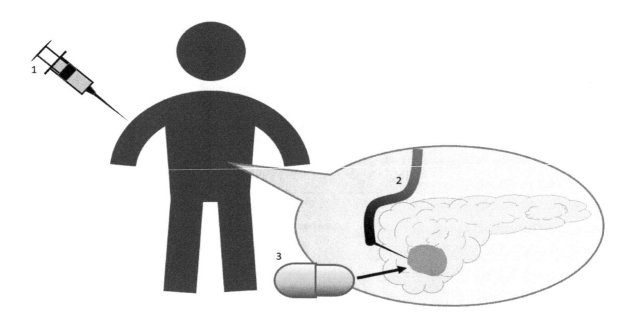

Figure 1. Delivery of nucleic acid medicines. (1) Intravenous injection, (2) intratumoral injection under EUS, and (3) intratumoral implantation.

6. Clinical Trials

6.1. Antisense Oligonucleotide

Eight nucleic acid medicines have been approved by the FDA (Table 4), five of which are ASOs used to treat nervous muscular diseases and familial metabolic diseases.

There have been a lot of reports about ASOs for pancreatic cancer treatment in preclinical studies. KRAS is the most common target because approximately 90% of pancreatic cancer harbor KRAS mutation. AZD-4785, a high-affinity constrained ethyl-containing therapeutic ASO targeting KRAS mRNA, potently depleted KRAS mRNA in KRAS-mutant colon, pancreatic, and lung cancer cell lines, with no feedback activation of MAPK signaling. Significant antitumor activity was obtained in mice bearing KRAS-mutant lung cancer xenografts [39].

ASOs have been tested in more than 1000 clinical trials. Various ASOs have reached clinical trials for the treatment of pancreatic cancer. The targets of these molecules were related to cell proliferation (X-linked inhibitor of apoptosis protein, XIAP [40]; Protein Kinase A, PKA [41]), cell signaling (HRAS) [42], resistance to chemotherapy (heat shock protein 27, Hsp27) [43], or cancer stroma TGFβ2) [44]. However, few ASOs have shown antitumor effects in clinical trials.

Table 4. Food and Drug Administration (FDA)-approved nucleic acid medicines.

Drug	Nucleic Acid	Disease	Modification	Administration	Company
Vitravene [45]	ASO	Cytomegalovirus retinitis	Phosphorothioated	Intravitreous	Isis Pharmaceuticals, Carlsbad, CA
Macugen [46]	Aptamer	Age-related macular degeneration	PEGylation 2'-F 2'-OMe	Intravitreous	Valeant Pharmaceuticals, Laval, Canada
Kynamro [47]	ASO	Homozygous familial hypercholesterolemia	Phosphorothioated 2'-MOE	Subcutaneous	Kastle Therapeutics, Chicago, IL
Exondys 51 [48]	ASO	Duchenne muscular dystrophy	Morpholino nucleic acid	Intravenous	Sarepta Therapeutics, Cambridge, MA
Spinraza [49]	ASO	Myelopathic muscular atrophy	Phosphorothioated 2'-MOE	Intraspinal	Biogen, Cambridge, MA
Heplisav-B [50]	CpG oligo	Hepatitis B	Phosphorothioated	Intramuscular	Dynavax Technologies, Berkeley, CA
Tegsedi [51]	ASO	Hereditary transthyretin-mediated amyloidosis	Phosphorothioated 2'-MOE	Subcutaneous	Akcea Therapeutics, Boston, MA
Onpattro [52]	siRNA	Hereditary transthyretin-mediated amyloidosis	2'-MOE	Intravenous	Alnylam Pharmaceuticals, Cambridge, MA

FDA, Food and Drug Administration; ASO, antisense oligonucleotide; CpG oligo, CpG oligodeoxynucleotide; 2'-MOE, 2'-O-methoxyethyl; 2'-OMe, 2'-O-Methyl; 2'-F, 2'-Fluoro.

ISIS 2503 (ASO targeting *XIAP*) showed evidence of growth inhibition when combined with gemcitabine in locally advanced or metastatic pancreatic cancer in first-line treatment [40]. In that study, 58% of patients who received the combination survived 6 months or longer. Addition of apatorsen, the *Hsp27*-targeting antisense oligonucleotide, to chemotherapy did not improve outcomes in unselected patients with metastatic pancreatic cancer in the first-line setting, although a trend toward prolonged overall survival in patients with high baseline serum Hsp27 suggests that this therapy may warrant further evaluation in this subgroup.

6.2. Clinical Trials for siRNAs

Fourteen years after the first clinical trial using RNAi was entered (2004), the FDA approved the first therapeutic RNAi, ONPATTRO (patisiran), a lipid complex injection for treatment of peripheral nerve disease caused by hereditary transthyretin-mediated amyloidosis in adults [52] (Table 4). However, there is no clinically available therapeutic RNAi for pancreatic cancer.

Some siRNAs have already entered clinical trials for the treatment of locally advanced pancreatic cancer. siRNA targeting mutated *KRAS* is the most common [35,36]. The vast majority of *KRAS* mutations in pancreatic cancer are gain-of-function mutations, most of which occur in codon 12 with substitution of the Glycine for Aspartate (G12D). Golan et al. implanted siRNA targeting *KRAS* (G12D) in the pancreatic tumor using LODER in combination with Gemcitabine treatment [35]. The majority of patients (83%) demonstrated stable disease and 17% of patients showed partial response. Decrease in CA19-9 was observed in 70% of patients. The most frequent adverse events observed were grade 1 or 2 severity (89%); transient abdominal pain, diarrhea, and nausea. They concluded that the combination of mutated *KRAS*-targeting siRNAs and chemotherapy is well tolerated, safe, and demonstrated potential efficacy in pancreatic cancer patients [53].

Nishimura et al. have shown that EUS-guided fine-needle injection (EUS-FNI) of a synthetic double-stranded RNA oligonucleotide directed against *CHST15* (STNM01), an extracellular matrix component, was safe and feasible [34]. There were no adverse effects. STNM01 is also directly injected by endoscopy to treat ulcerative colitis.

Atu027 is a liposomally formulated siRNA with antimetastatic activity, which silences protein kinase N3 (PKN3) expression in the vascular endothelium [54]. PKN3 acts as a Rho effector downstream of PI3K. Combination of Atu027 and gemcitabine for the treatment of advanced pancreatic cancer was safe and well tolerated.

TKM-080301 is a lipid nanoparticle formulation of an siRNA against Polo-like kinase 1 (PLK1), which regulates critical aspects of tumor progression [55]. Preliminary antitumor efficacy for advanced pancreatic cancer has been observed. A potential molecular therapeutic context of increased PLK1 expression with inactivation of p53 or NF1 was observed in a remarkable responder.

However, these data must be interpreted with caution because they are early-phase trials and some are still recruiting patients. The best responses observed so far have been tumor stabilization, with very few complete or partial responses documented. siRNAs were well tolerated but one death and a few grade 3–4 toxic effects due to elevation of liver enzymes were observed [56]. Several trials with different combinations including siRNAs are ongoing, and the combination of several nucleic acid medicines may be explored in the coming years.

7. Conclusions

Despite the challenging obstacles, we hope that nucleic acid drugs will have a significant impact on the treatment of pancreatic cancer. The combination of genetic diagnosis using next-generation sequencing and targeted therapy may provide effective precision medicine for pancreatic cancer patients.

Acknowledgments: We thank Sanae Kushida for preparing the manuscript.

References

1.   Torres, C.; Grippo, P.J. Pancreatic cancer subtypes: A roadmap for precision medicine. *Ann. Med.* **2018**, *50*, 277–287. [CrossRef] [PubMed]

2.   Gleeson, F.C.; Kerr, S.E.; Kipp, B.R.; Voss, J.S.; Minot, D.M.; Tu, Z.J.; Henry, M.R.; Graham, R.P.; Vasmatzis, G.; Cheville, J.C.; et al. Targeted next generation sequencing of endoscopic ultrasound acquired cytology from ampullary and pancreatic adenocarcinoma has the potential to aid patient stratification for optimal therapy selection. *Oncotarget* **2016**, *7*, 54526–54536. [CrossRef] [PubMed]

3.   Barata, P.; Sood, A.K.; Hong, D.S. RNA-targeted therapeutics in cancer clinical trials: Current status and future directions. *Cancer Treat. Rev.* **2016**, *50*, 35–47. [CrossRef] [PubMed]

4.   Jansen, B.; Zangemeister-Wittke, U. Antisense therapy for cancer—The time of truth. *Lancet Oncol.* **2002**, *3*, 672–683. [CrossRef]

5.   Moreno, P.M.; Pego, A.P. Therapeutic antisense oligonucleotides against cancer: Hurdling to the clinic. *Front. Chem.* **2014**, *2*, 87. [CrossRef]

6.   Lee, R.C.; Feinbaum, R.L.; Ambros, V. The C. elegans heterochronic gene lin-4 encodes small RNAs with antisense complementarity to lin-14. *Cell* **1993**, *75*, 843–854. [CrossRef]

7.   Elbashir, S.M.; Harborth, J.; Lendeckel, W.; Yalcin, A.; Weber, K.; Tuschl, T. Duplexes of 21-nucleotide RNAs mediate RNA interference in cultured mammalian cells. *Nature* **2001**, *411*, 494–498. [CrossRef]

8.   Hori, S.I.; Herrera, A.; Rossi, J.J.; Zhou, J. Current Advances in Aptamers for Cancer Diagnosis and Therapy. *Cancers (Basel)* **2018**, *10*, 9. [CrossRef]

9.   Dausse, E.; Da Rocha Gomes, S.; Toulme, J.J. Aptamers: A new class of oligonucleotides in the drug discovery pipeline? *Curr. Opin. Pharmacol.* **2009**, *9*, 602–607. [CrossRef]

10.   Hecker, M.; Wagner, A.H. Transcription factor decoy technology: A therapeutic update. *Biochem. Pharmacol.* **2017**, *144*, 29–34. [CrossRef]

11.   Vollmer, J.; Weeratna, R.D.; Jurk, M.; Davis, H.L.; Schetter, C.; Wullner, M.; Wader, T.; Liu, M.; Kritzler, A.; Krieg, A.M. Impact of modifications of heterocyclic bases in CpG dinucleotides on their immune-modulatory activity. *J. Leukoc. Biol.* **2004**, *76*, 585–593. [CrossRef] [PubMed]

12. Wu, S.Y.; Yang, X.; Gharpure, K.M.; Hatakeyama, H.; Egli, M.; McGuire, M.H.; Nagaraja, A.S.; Miyake, T.M.; Rupaimoole, R.; Pecot, C.V.; et al. 2'-OMe-phosphorodithioate-modified siRNAs show increased loading into the RISC complex and enhanced anti-tumour activity. *Nat. Commun.* **2014**, *5*, 3459. [CrossRef] [PubMed]
13. Geary, R.S.; Norris, D.; Yu, R.; Bennett, C.F. Pharmacokinetics, biodistribution and cell uptake of antisense oligonucleotides. *Adv. Drug Deliv. Rev.* **2015**, *87*, 46–51. [CrossRef] [PubMed]
14. Soutschek, J.; Akinc, A.; Bramlage, B.; Charisse, K.; Constien, R.; Donoghue, M.; Elbashir, S.; Geick, A.; Hadwiger, P.; Harborth, J.; et al. Therapeutic silencing of an endogenous gene by systemic administration of modified siRNAs. *Nature* **2004**, *432*, 173–178. [CrossRef] [PubMed]
15. Yu, R.Z.; Grundy, J.S.; Geary, R.S. Clinical pharmacokinetics of second generation antisense oligonucleotides. *Expert Opin. Drug Metab. Toxicol.* **2013**, *9*, 169–182. [CrossRef] [PubMed]
16. Amarzguioui, M.; Holen, T.; Babaie, E.; Prydz, H. Tolerance for mutations and chemical modifications in a siRNA. *Nucleic Acids Res.* **2003**, *31*, 589–595. [CrossRef] [PubMed]
17. Yi, C.; Pan, T. Cellular dynamics of RNA modification. *Acc. Chem. Res.* **2011**, *44*, 1380–1388. [CrossRef] [PubMed]
18. Fattal, E.; Barratt, G. Nanotechnologies and controlled release systems for the delivery of antisense oligonucleotides and small interfering RNA. *Br. J. Pharmacol.* **2009**, *157*, 179–194. [CrossRef]
19. Simeoni, F.; Morris, M.C.; Heitz, F.; Divita, G. Insight into the mechanism of the peptide-based gene delivery system MPG: Implications for delivery of siRNA into mammalian cells. *Nucleic Acids Res.* **2003**, *31*, 2717–2724. [CrossRef]
20. Xie, X.; Yang, Y.; Lin, W.; Liu, H.; Liu, H.; Yang, Y.; Chen, Y.; Fu, X.; Deng, J. Cell-penetrating peptide-siRNA conjugate loaded YSA-modified nanobubbles for ultrasound triggered siRNA delivery. *Colloids Surf. B Biointerfaces.* **2015**, *136*, 641–650. [CrossRef]
21. Arami, S.; Mahdavi, M.; Rashidi, M.R.; Yekta, R.; Rahnamay, M.; Molavi, L.; Hejazi, M.S.; Samadi, N. Apoptosis induction activity and molecular docking studies of survivin siRNA carried by Fe3O4-PEG-LAC-chitosan-PEI nanoparticles in MCF-7 human breast cancer cells. *J. Pharm. Biomed. Anal.* **2017**, *142*, 145–154. [CrossRef] [PubMed]
22. Keefe, A.D.; Pai, S.; Ellington, A. Aptamers as therapeutics. *Nat. Rev. Drug Discov.* **2010**, *9*, 537–550. [CrossRef] [PubMed]
23. Soldevilla, M.M.; Meraviglia-Crivelli de Caso, D.; Menon, A.P.; Pastor, F. Aptamer-iRNAs as Therapeutics for Cancer Treatment. *Pharmaceuticals (Basel)* **2018**, *11*, 108. [CrossRef] [PubMed]
24. Ramot, Y.; Rotkopf, S.; Gabai, R.M.; Zorde Khvalevsky, E.; Muravnik, S.; Marzoli, G.A.; Domb, A.J.; Shemi, A.; Nyska, A. Preclinical Safety Evaluation in Rats of a Polymeric Matrix Containing an siRNA Drug Used as a Local and Prolonged Delivery System for Pancreatic Cancer Therapy. *Toxicol. Pathol.* **2016**, *44*, 856–865. [CrossRef]
25. Nair, J.K.; Willoughby, J.L.; Chan, A.; Charisse, K.; Alam, M.R.; Wang, Q.; Hoekstra, M.; Kandasamy, P.; Kel'in, A.V.; Milstein, S.; et al. Multivalent N-acetylgalactosamine-conjugated siRNA localizes in hepatocytes and elicits robust RNAi-mediated gene silencing. *J. Am. Chem. Soc.* **2014**, *136*, 16958–16961. [CrossRef]
26. Jing, H.; Cheng, W.; Li, S.; Wu, B.; Leng, X.; Xu, S.; Tian, J. Novel cell-penetrating peptide-loaded nanobubbles synergized with ultrasound irradiation enhance EGFR siRNA delivery for triple negative Breast cancer therapy. *Colloids Surf. B Biointerfaces* **2016**, *146*, 387–395. [CrossRef]
27. Rabbani, P.S.; Zhou, A.; Borab, Z.M.; Frezzo, J.A.; Srivastava, N.; More, H.T.; Rifkin, W.J.; David, J.A.; Berens, S.J.; Chen, R.; et al. Novel lipoproteoplex delivers Keap1 siRNA based gene therapy to accelerate diabetic wound healing. *Biomaterials* **2017**, *132*, 1–15. [CrossRef]
28. Wang, Y.; Miao, L.; Satterlee, A.; Huang, L. Delivery of oligonucleotides with lipid nanoparticles. *Adv. Drug Deliv. Rev.* **2015**, *87*, 68–80. [CrossRef]
29. Amjadi, M.; Mostaghaci, B.; Sitti, M. Recent Advances in Skin Penetration Enhancers for Transdermal Gene and Drug Delivery. *Curr. Gene Ther.* **2017**, *17*, 139–146. [CrossRef]
30. Yin, T.; Wang, L.; Yin, L.; Zhou, J.; Huo, M. Co-delivery of hydrophobic paclitaxel and hydrophilic AURKA specific siRNA by redox-sensitive micelles for effective treatment of breast cancer. *Biomaterials* **2015**, *61*, 10–25. [CrossRef]
31. Shen, J.; Wang, Q.; Hu, Q.; Li, Y.; Tang, G.; Chu, P.K. Restoration of chemosensitivity by multifunctional micelles mediated by P-gp siRNA to reverse MDR. *Biomaterials* **2014**, *35*, 8621–8634. [CrossRef] [PubMed]

32. Ji, S.; Xu, J.; Zhang, B.; Yao, W.; Xu, W.; Wu, W.; Xu, Y.; Wang, H.; Ni, Q.; Hou, H.; et al. RGD-conjugated albumin nanoparticles as a novel delivery vehicle in pancreatic cancer therapy. *Cancer Biol. Ther.* **2012**, *13*, 206–215. [CrossRef] [PubMed]

33. Kahalekar, V.; Gupta, D.T.; Bhatt, P.; Shukla, A.; Bhatia, S. Fully covered self-expanding metallic stent placement for benign refractory esophageal strictures. *Indian J. Gastroenterol.* **2017**, *36*, 197–201. [CrossRef] [PubMed]

34. Nishimura, M.; Matsukawa, M.; Fujii, Y.; Matsuda, Y.; Arai, T.; Ochiai, Y.; Itoi, T.; Yahagi, N. Effects of EUS-guided intratumoral injection of oligonucleotide STNM01 on tumor growth, histology, and overall survival in patients with unresectable pancreatic cancer. *Gastrointest Endosc.* **2018**, *87*, 1126–1131. [CrossRef] [PubMed]

35. Golan, T.; Khvalevsky, E.Z.; Hubert, A.; Gabai, R.M.; Hen, N.; Segal, A.; Domb, A.; Harari, G.; David, E.B.; Raskin, S.; et al. RNAi therapy targeting KRAS in combination with chemotherapy for locally advanced pancreatic cancer patients. *Oncotarget* **2015**, *6*, 24560–24570. [CrossRef] [PubMed]

36. Shemi, A.; Khvalevsky, E.Z.; Gabai, R.M.; Domb, A.; Barenholz, Y. Multistep, effective drug distribution within solid tumors. *Oncotarget* **2015**, *6*, 39564–39577. [CrossRef]

37. Yamahatsu, K.; Matsuda, Y.; Ishiwata, T.; Uchida, E.; Naito, Z. Nestin as a novel therapeutic target for pancreatic cancer via tumor angiogenesis. *Int. J. Oncol.* **2012**, *40*, 1345–1357.

38. Matsuda, Y.; Ishiwata, T.; Yoshimura, H.; Yamashita, S.; Ushijima, T.; Arai, T. Systemic Administration of Small Interfering RNA Targeting Human Nestin Inhibits Pancreatic Cancer Cell Proliferation and Metastasis. *Pancreas* **2016**, *45*, 93–100. [CrossRef]

39. Ross, S.J.; Revenko, A.S.; Hanson, L.L.; Ellston, R.; Staniszewska, A.; Whalley, N.; Pandey, S.K.; Revill, M.; Rooney, C.; Buckett, L.K.; et al. Targeting KRAS-dependent tumors with AZD4785, a high-affinity therapeutic antisense oligonucleotide inhibitor of KRAS. *Sci. Transl. Med.* **2017**, *9*. [CrossRef]

40. Mahadevan, D.; Chalasani, P.; Rensvold, D.; Kurtin, S.; Pretzinger, C.; Jolivet, J.; Ramanathan, R.K.; Von Hoff, D.D.; Weiss, G.J. Phase I trial of AEG35156 an antisense oligonucleotide to XIAP plus gemcitabine in patients with metastatic pancreatic ductal adenocarcinoma. *Am. J. Clin. Oncol.* **2013**, *36*, 239–243. [CrossRef]

41. Goel, S.; Desai, K.; Macapinlac, M.; Wadler, S.; Goldberg, G.; Fields, A.; Einstein, M.; Volterra, F.; Wong, B.; Martin, R.; et al. A phase I safety and dose escalation trial of docetaxel combined with GEM231, a second generation antisense oligonucleotide targeting protein kinase A R1alpha in patients with advanced solid cancers. *Investig. New Drugs* **2006**, *24*, 125–134. [CrossRef] [PubMed]

42. Alberts, S.R.; Schroeder, M.; Erlichman, C.; Steen, P.D.; Foster, N.R.; Moore, D.F., Jr.; Rowland, K.M., Jr.; Nair, S.; Tschetter, L.K.; Fitch, T.R. Gemcitabine and ISIS-2503 for patients with locally advanced or metastatic pancreatic adenocarcinoma: A North Central Cancer Treatment Group phase II trial. *J. Clin. Oncol.* **2004**, *22*, 4944–4950. [CrossRef] [PubMed]

43. Ko, A.H.; Murphy, P.B.; Peyton, J.D.; Shipley, D.L.; Al-Hazzouri, A.; Rodriguez, F.A.; Womack, M.S.; Xiong, H.Q.; Waterhouse, D.M.; Tempero, M.A.; et al. A Randomized, Double-Blinded, Phase II Trial of Gemcitabine and Nab-Paclitaxel Plus Apatorsen or Placebo in Patients with Metastatic Pancreatic Cancer: The RAINIER Trial. *Oncologist* **2017**, *22*, 1427–e129. [CrossRef] [PubMed]

44. Jaschinski, F.; Rothhammer, T.; Jachimczak, P.; Seitz, C.; Schneider, A.; Schlingensiepen, K.H. The antisense oligonucleotide trabedersen (AP 12009) for the targeted inhibition of TGF-beta2. *Curr. Pharm. Biotechnol.* **2011**, *12*, 2203–2213. [CrossRef] [PubMed]

45. Vitravene Study Group. A randomized controlled clinical trial of intravitreous fomivirsen for treatment of newly diagnosed peripheral cytomegalovirus retinitis in patients with AIDS. *Am. J. Ophthalmol.* **2002**, *133*, 467–474.

46. Ng, E.W.; Shima, D.T.; Calias, P.; Cunningham, E.T., Jr.; Guyer, D.R.; Adamis, A.P. Pegaptanib, a targeted anti-VEGF aptamer for ocular vascular disease. *Nat. Rev. Drug Discov.* **2006**, *5*, 123–132. [CrossRef] [PubMed]

47. Raal, F.J.; Santos, R.D.; Blom, D.J.; Marais, A.D.; Charng, M.J.; Cromwell, W.C.; Lachmann, R.H.; Gaudet, D.; Tan, J.L.; Chasan-Taber, S.; et al. Mipomersen, an apolipoprotein B synthesis inhibitor, for lowering of LDL cholesterol concentrations in patients with homozygous familial hypercholesterolaemia: A randomised, double-blind, placebo-controlled trial. *Lancet* **2010**, *375*, 998–1006. [CrossRef]

48. Stein, C.A. Eteplirsen Approved for Duchenne Muscular Dystrophy: The FDA Faces a Difficult Choice. *Mol. Ther.* **2016**, *24*, 1884–1885. [CrossRef]

49. Hua, Y.; Sahashi, K.; Hung, G.; Rigo, F.; Passini, M.A.; Bennett, C.F.; Krainer, A.R. Antisense correction of SMN2 splicing in the CNS rescues necrosis in a type III SMA mouse model. *Genes Dev.* **2010**, *24*, 1634–1644. [CrossRef]

50. Splawn, L.M.; Bailey, C.A.; Medina, J.P.; Cho, J.C. Heplisav-B vaccination for the prevention of hepatitis B virus infection in adults in the United States. *Drugs Today (Barc.)* **2018**, *54*, 399–405. [CrossRef]

51. Gales, L. Tegsedi (Inotersen): An Antisense Oligonucleotide Approved for the Treatment of Adult Patients with Hereditary Transthyretin Amyloidosis. *Pharmaceuticals (Basel)* **2019**, *12*, 78. [CrossRef] [PubMed]

52. Adams, D.; Gonzalez-Duarte, A.; O'Riordan, W.D.; Yang, C.C.; Ueda, M.; Kristen, A.V.; Tournev, I.; Schmidt, H.H.; Coelho, T.; Berk, J.L.; et al. Patisiran, an RNAi Therapeutic, for Hereditary Transthyretin Amyloidosis. *N. Engl. J. Med.* **2019**, *379*, 11–21. [CrossRef] [PubMed]

53. Weng, Y.; Xiao, H.; Zhang, J.; Liang, X.J.; Huang, Y. RNAi therapeutic and its innovative biotechnological evolution. *Biotechnol. Adv.* **2019**, *37*, 801–825. [CrossRef] [PubMed]

54. Suzuki, K.; Yokoyama, J.; Kawauchi, Y.; Honda, Y.; Sato, H.; Aoyagi, Y.; Terai, S.; Okazaki, K.; Suzuki, Y.; Sameshima, Y.; et al. Phase 1 Clinical Study of siRNA Targeting Carbohydrate Sulphotransferase 15 in Crohn's Disease Patients with Active Mucosal Lesions. *J. Crohns Colitis* **2017**, *11*, 221–228. [CrossRef] [PubMed]

55. Schultheis, B.; Strumberg, D.; Kuhlmann, J.; Wolf, M.; Link, K.; Seufferlein, T.; Kaufmann, J.; Gebhardt, F.; Bruyniks, N.; Pelzer, U. A phase Ib/IIa study of combination therapy with gemcitabine and Atu027 in patients with locally advanced or metastatic pancreatic adenocarcinoma. *J. Clin. Oncol.* **2016**, *34*, 385. [CrossRef]

56. Demeure, M.J.; Armaghany, T.; Ejadi, S.; Ramanathan, R.K.; Elfiky, A.; Strosberg, J.R.; Smith, D.C.; Whitsett, T.; Liang, W.S.; Sekar, S.; et al. A phase I/II study of TKM-080301, a PLK1-targeted RNAi in patients with adrenocortical cancer (ACC). *J. Clin. Oncol.* **2016**, *34*, 2547. [CrossRef]

7

The Yin-Yang Regulation of Reactive Oxygen Species and MicroRNAs in Cancer

Kamesh R. Babu [1] and Yvonne Tay [1,2,*]

[1] Cancer Science Institute of Singapore, National University of Singapore, Singapore 117599, Singapore; csirbk@nus.edu.sg

[2] Department of Biochemistry, Yong Loo Lin School of Medicine, National University of Singapore, Singapore 117597, Singapore

* Correspondence: yvonnetay@nus.edu.sg

Abstract: Reactive oxygen species (ROS) are highly reactive oxygen-containing chemical species formed as a by-product of normal aerobic respiration and also from a number of other cellular enzymatic reactions. ROS function as key mediators of cellular signaling pathways involved in proliferation, survival, apoptosis, and immune response. However, elevated and sustained ROS production promotes tumor initiation by inducing DNA damage or mutation and activates oncogenic signaling pathways to promote cancer progression. Recent studies have shown that ROS can facilitate carcinogenesis by controlling microRNA (miRNA) expression through regulating miRNA biogenesis, transcription, and epigenetic modifications. Likewise, miRNAs have been shown to control cellular ROS homeostasis by regulating the expression of proteins involved in ROS production and elimination. In this review, we summarized the significance of ROS in cancer initiation, progression, and the regulatory crosstalk between ROS and miRNAs in cancer.

Keywords: ROS; oxidative stress; antioxidants; miRNA; cancer

1. Introduction

Reactive oxygen species (ROS) are free radicals, ions, or molecules with a single unpaired electron. ROS, including hydrogen peroxide (H_2O_2), hydroxyl radicals (OH^-), nitric oxide (NO), and superoxide radicals (O_2^-) are highly reactive and generated as a byproduct during metabolic processes in various subcellular compartments of a cell [1]. Mitochondria are the main cellular source of ROS. However, ROS are also generated in other cellular organelles including endoplasmic reticulum, lysosomes, and peroxisomes [2]. At lower concentrations, ROS play significant roles in various physiological functions including gene activation, cell growth, proliferation, survival, apoptosis, chemical reaction modulation, blood pressure control, prostaglandin biosynthesis, embryonic development, cognitive function, and immune response [3,4]. However, at higher concentrations, ROS can cause oxidative damage via oxidation of macromolecules such as DNA, RNA, proteins, and lipids that can contribute to the pathogenesis of various diseases including cancer [5–9]. Elevated ROS production is associated with tumorigenesis and suggested to be a hallmark of cancer. Nevertheless, the molecular mechanisms responsible for sustained high ROS levels in cancer is not well understood.

MicroRNAs (miRNAs) are a class of small non-coding RNAs that are approximately 22 nucleotides long and regulate gene expression at the post-transcriptional level [10]. They regulate gene expression by binding to the target messenger RNA (mRNA) transcript which activates either degradation or translation suppression based on the extent of basepairing. However, several studies reported that miRNAs can also target and regulate the stability of non-coding RNAs. Studies have demonstrated that the deregulation of miRNA expression is associated with cancer development, and miRNAs may function as potential oncogenes or tumor suppressors [11]. Surprisingly, studies show the existence of a regulatory connection between ROS and miRNA. For example, H_2O_2 treatment has been shown to dysregulate the expression of certain miRNAs in vascular smooth muscle cells and macrophage cells [12,13]. Another study has shown that miR-30e regulates oxidative stress and ROS levels by targeting SNAI1 mRNA in human umbilical endothelial vein cells [14]. These findings suggest that ROS and miRNAs may co-regulate each other in cancer to maintain cellular ROS levels that support cancer development. In this review, we discuss the significance of ROS in cancer development, as well as the crosstalk between ROS and miRNAs in the regulation of redox homeostasis and cancer progression.

2. Significance of ROS in Cancer Development

ROS are required by cells to carry out physiological cellular functions and this is also true in the case of cancer cells. However, cancer cells show elevated levels of ROS when compared to normal cells, which is mainly due to persistent and high metabolic rate in mitochondria, endoplasmic reticulum (ER). and cell membranes. In this section, we discuss how ROS play a significant role in the whole process of cancer development, including initiation, promotion, and progression.

2.1. ROS in Cancer Initiation

ROS are potent mutagens that can stimulate cancer initiation. High levels of ROS oxidize DNA bases resulting in DNA lesions including base damage, strand breaks, and mutations, which are usually repaired by the endogenous DNA repair enzymes of the base excision repair, nuclear excision repair, or mismatch repair pathways [15]. Cells unable to repair DNA lesions undergo apoptosis to prevent the passage of DNA mutations to progeny cells. However, under certain conditions, cells harboring DNA lesions evade apoptosis, which eventually leads to cancer. In a similar fashion to DNA, RNA also undergoes oxidation under oxidative stress that results in strand breaks and oxidative base modifications. Oxidized mRNA can cause several defects during protein translation, which include synthesis of truncated, mutated, or non-functional proteins, ribosome stalling, and ribosome dysfunction [6]. Oxidized RNA can promote the pathogenesis of chronic degenerative diseases including cancer [7]. For example, oxidation of tumor suppressor mRNAs results in the synthesis of mutated or truncated proteins that lack proper function, and this may lead to carcinogenesis. It is important to note that RNA oxidation is not limited only to mRNA as all RNA species including non-coding RNAs are subjected to oxidative damage. Since several studies have shown the significant participation of non-coding RNAs, including miRNAs and long non-coding RNAs (lncRNAs) in cancer development [16], oxidative modification of non-coding RNAs may also promote cancer initiation. ROS-induced mutation or modification is not only restricted to nucleotides, but even protein molecules are also susceptible to such modifications. Oxidation of proteins by ROS results in amino acid modification, protein carbonylation, nitration of tyrosine and phenylalanine residues, protein degradation, or formation of cross-linked proteins or glycated proteins [17,18]. Oxidized amino acid residues can affect their protein activity. For example, oxidation of DNA polymerase affects its fidelity during replication/synthesis, transcription, or DNA repair activity, which is closely associated with cancer initiation [19]. Finally, ROS can also damage polyunsaturated or polydesaturated fatty acids by the process of lipid peroxidation which generates various toxic molecules including malondialdehyde, 2-alkenals, 4-hydroxynonenal (HNE), and lipoperoxyl radical (LOO^-) [9,20]. The LOO^- reacts with the lipids to generate lipid peroxides, which are unstable and can produce new peroxyl and alkoxy radicals. These radicals may further increase the oxidation of macromolecules. Furthermore, HNE is a chemically reactive molecule that can react with macromolecules and form covalent modifications, which has been proposed as the mechanism to induce carcinogenesis [20]. These studies indicate that higher levels of ROS are detrimental to cells and can increase the risk of developing cancer.

2.2. ROS in Cancer Cell Proliferation

ROS function as secondary messengers in cellular signaling and activate ROS-sensitive signaling pathways by regulating protein activity through the reversible oxidation of target proteins. Redox-sensitive signaling pathways, including the mitogen-activated protein kinase (MAPK)/extracellular signal-regulated kinase (ERK), phosphoinositide-3 kinase (PI3K)/protein kinase B (AKT), and nuclear factor κ-B (NF-κB) signaling pathways, are constantly upregulated in various cancer subtypes, where they play a pivotal role in cell proliferation, growth, protein synthesis, glucose metabolism, cell survival, and inflammation [21]. Activation of MAPK/ERK signaling has been shown to increase anchorage-independent growth, cell survival, and motility of many cancer subtypes including breast cancer, leukemia, melanoma, and ovarian cancer. Studies have shown that high ROS levels in cancer cells can elevate MAPK/ERK signaling and can increase cancer cell proliferation [19]. Analogously, high levels of ROS, either produced endogenously or added exogenously, have shown to increase the activation PI3K/AKT signaling pathway in breast and ovarian cancer. Furthermore, studies show that elevated ROS levels can activate the transcription factor NF-κB. Oxidative stress-induced through the exogenous treatment of sodium arsenite, rotenone, H_2O_2, or through inhibition of endogenous antioxidants elevated the NF-κB activation and increased cancer cell proliferation [19,22]. Moreover, ROS play a significant role in the cell cycle by regulating mRNA levels of cyclins that promote G1 to S phase transition, which include cyclin B2, cyclin D3, cyclin E1, and cyclin E3 [23]. In breast cancer cells, ROS generated by sodium arsenite treatment promote S phase transition and aberrant cell proliferation [24], whereas reduction of ROS levels through antioxidant N-acetyl cysteine (NAC) treatment reduces cyclin D1 levels and slowed the G1 to S phase transition in the non-cancerous human breast epithelial cells [25]. All these studies suggest that besides being a highly reactive mutagen, ROS can also function as a secondary messenger that mediate physiological signaling pathways involved in cell proliferation, thus higher ROS production in cancer cells favor cancer progression through elevated and sustained activation of these pathways.

2.3. ROS in Cancer Metastasis

Metastasis is a multistep process that involves the spread of cancer cells from its original site to distal parts of the body, the process comprises migration, invasion, intravasation into the blood, anchorage-independent survival in the blood, and extravasation into distal organs [26]. Several studies show that ROS levels are increased in cells that undergo metastasis, and they play a significant role in the cancer cell metastasis. A study has shown that endogenous ROS levels are increased in circulating melanoma cells and metastasis nodules of xenografted mice compared to primary subcutaneous tumors [27]. Importantly, cancer cells treated with H_2O_2 have shown high metastasis upon injected intravenously into mice. Likewise, a sub-population of the breast cancer cells that has elevated intracellular ROS levels compared to the parental cells exhibits high motility and metastasized to distant organs including lung, liver, and spleen [19]. It is noteworthy that the levels and activity of endogenous antioxidants are decreased in metastatic cancer cells. For example, the levels and catalytic activity of manganese-dependent superoxide dismutase (MnSOD) are lower in highly invasive pancreatic cancer cells and metastatic breast cancer cells [28,29]. Cancer cells go through epithelial to mesenchymal transition (EMT) before migrating to distant sites of the body. During the EMT process, expression of matrix metalloproteinases (MMPs) is increased to mediate degradation and reorganization of extracellular matrix and their elevated activation is associated with tumor growth, angiogenesis, invasion, and metastasis [30]. ROS play a significant role in the EMT process in which they regulate the expression of MMPs and their inhibitors tissue inhibitor of metalloproteinases (TIMP) [31]. A study has shown that treatment of MMP-3, a stromal protease whose expression is upregulated in mammary tumors, has increased cellular ROS and induced EMT in murine mammary epithelial cells. In contrast, scavenging cellular ROS through NAC treatment abrogated MMP-3-induced EMT, suggesting that high levels of cellular ROS can lead to malignant transformation [32]. Moreover, ROS also facilitate metastasis by increasing vascular permeability through various mechanisms. Oxidative stress in endothelial cells mediate Rac-1-induced loss of cell–cell adhesion and loosens the endothelium integrity, which favors the cancer cell intravasation [33]. ROS regulate the expression of IL-8 and intracellular adhesion protein 1 (ICAM-1) via NF-κB activation. Both IL-8 and ICAM-1 regulate transendothelial migration of tumor cells [34,35]. Furthermore, ROS induce actin reorganization in vascular endothelial cells through p38-mediated phosphorylation of the heat shock protein Hsp27, which may contribute to promote invasive processes [36]. Taken altogether, these studies suggest that ROS has a versatile role in the pathogenesis of cancer, therefore it would be interesting to identify further novel roles of ROS in other physiological processes that could possibly support the process of cancer development.

2.4. ROS in Cancer Stem Cells

Cancer stem cells (CSCs) are a subset of tumorigenic cells that possess similar characteristics as normal stem cells, in particular the capabilities of self-renewal or differentiation. Interestingly, CSCs have been shown to have a high capacity to grow into tumors. Similar to cancer cells, ROS also play an important role in CSCs. However, in contrast to cancer cells in which ROS levels are elevated, CSCs exhibit lower levels of ROS. This is similar to the levels found in normal stem cells [37]. The lower cellular ROS levels in CSCs are associated with increased expression of ROS scavenging systems and are essential for the maintenance of self-renewal and stemness. A study has shown that pharmacological depletion of ROS scavengers in breast CSCs reduces their clonogenicity and results in radiosensitization [38]. Conversely, ovarian CSCs exhibit higher mitochondrial ROS production, and inhibition of the mitochondrial respiratory chain in CSCs results in apoptosis [39]. Furthermore, a study has shown that the population of hematopoietic stem cells (HSCs) with higher ROS levels possess higher myeloid differentiation potential compared to the HSC fraction with lower ROS levels [40]. These findings suggest that ROS levels in CSCs are crucial for their survival and differentiation. Nevertheless, the effects of ROS and the regulation of ROS levels in CSCs have not been studied extensively. Future investigations may unravel the molecular mechanisms behind the regulation of redox homeostasis in CSCs.

3. ROS Regulate MiRNA Expression

Accumulating studies show functional regulatory links between ROS and miRNAs in carcinogenesis. ROS also contribute to cancer development by regulating the expression of miRNAs that target genes responsible for enhancing or suppressing carcinogenesis. In this section, we discuss how the ROS affect the miRNA expression in cancer via different mechanisms including alteration of epigenetic signatures, transcription, and biogenesis.

3.1. Regulation of MiRNA Expression via Epigenetic Modifications

Dysregulated miRNA expression in cancer is associated with altered DNA methylation and histone modifications such as acetylation, methylation, and phosphorylation. ROS can regulate miRNA expression by altering the epigenetic signatures including DNA methylation or histone modifications (Figure 1a). For example, ROS inhibit the expression of miR-199a and miR-125b in ovarian cancer cells via increasing promoter methylation of the miR-199a and miR-125b genes, which is mediated by the DNA methyltransferase 1 (DNMT1) [41]. Interestingly, overexpression of miR-199a and miR-125b in ovarian cancer cells decreased the expression of the hypoxia-inducible factor 1-alpha (HIF-1α) and vascular endothelial growth factor, which suppressed tumor-induced angiogenesis [42]. Histone modifications play an important role in chromatin remodeling in order to regulate gene transcription. Histone acetylation is a type of histone modification in which the lysine residues of histone are acetylated to relax the chromatin structure for gene transcription. In contrast, deacetylation of lysine residues catalyzed by the histone deacetylases (HDACs) causes chromatin condensation and transcriptional gene silencing [43]. ROS can regulate the activity of HDACs. For example, the Cys667 and Cys669 amino acid residues of HDAC4 are oxidized to form an intramolecular dis-sulfide bond, which promotes its nuclear export [44]. Cancer cells promote nuclear translocation of HDAC4 by increasing endogenous antioxidants, which decreases miR-206 expression through deacetylation of its promoter and promotes cancer progression [45,46]. Furthermore, oxidative stress-induced by glucose depletion increases the expression of miR-466h-5p by inhibiting HDAC2 activity, which results in increased apoptosis due to the fact that miR-466h-5p directly targets and downregulates many anti-apoptotic proteins including BCL212, DAD1, BIRC6, STAT5A, and SMO [47,48]. These findings suggest that ROS can affect the epigenetic status of miRNA genes thereby regulating its expression in cancer. It is important to note that ROS-mediated regulation of DNMT1 and HDACs in cancer may change its global epigenetic signature, therefore the expression of other genes including oncogenes and tumor suppressors can also be activated or silenced.

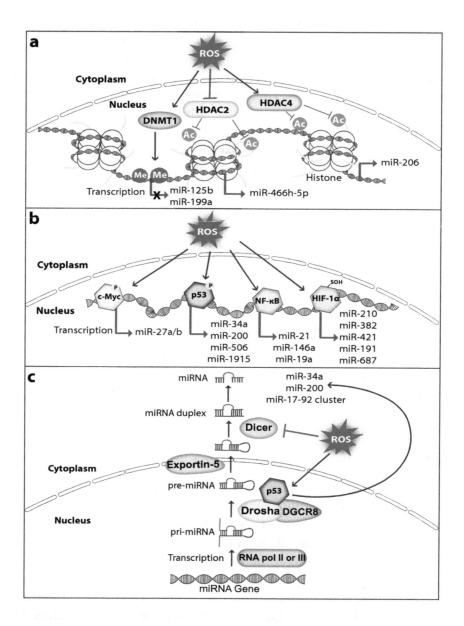

Figure 1. ROS regulate miRNA expression and processing in cancer. (**a**) ROS induce epigenetic modifications to regulate miRNA expression. ROS affect DNMT1 (DNA methyltransferase 1), HDAC2 (Histone deacetylase 2), or HDAC4 (Histone deacetylase 4) to either inhibit or activate miRNA expression. (**b**) ROS can induce miRNA transcription through activating transcription factors c-Myc, p53, NF-κB (nuclear factor κ-B), or HIF-1α (hypoxia-inducible factor 1-alpha). (**c**) ROS affect miRNA biogenesis and maturation through regulating the activity and expression of miRNA processing enzymes Drosha and Dicer, respectively. Me, methyl group; Ac, acetyl group; P, phosphoryl group (The black arrow indicates upmodulation, the red T arrow indicates inhibition, the green arrow indicates transcription activation, the green arrow with red cross indicates transcription inhibition).

3.2. Regulation of MiRNA Expression via Transcription Factors

ROS can also control miRNA expression by regulating the ROS-responsive transcription factors that activate miRNA transcription (Figure 1b). ROS regulate the activation of transcription factors through direct or indirect mechanisms. The activated transcription factor binds to the target miRNA promoter and upregulates miRNA transcription.

3.2.1. C-Myc

C-Myc is a well-studied transcription factor and characterized as an oncoprotein whose expression is elevated in a wide range of tumors. It promotes tumorigenesis by activating the transcription of several oncogenes including the miR-17-92 cluster, or by inhibiting the transcription of tumor-suppressors including let-7a which functions as a negative regulator of CSC features by regulating PTEN and Lin28b expression in pancreatic and prostate cancer [49]. c-Myc is a redox-sensitive transcription factor. Under oxidative stress, ROS cause ERK-dependent phosphorylation at the Ser62 amino acid residue of c-Myc which enhances the c-Myc recruitment to the promoter of gamma-glutamyl-cysteine synthetase, the rate-limiting enzyme catalyzing glutathione (GSH) synthesis. The c-Myc phosphorylation-dependent activation of GSH promotes the survival of cancer cells under oxidative-stress conditions [50]. Lithocholic acid (LCA)-induced ROS increased c-Myc expression in the human hepatocellular carcinoma (HCC) cells and in mouse liver. Importantly, LCA mediated c-Myc overexpression activates the expression of miR-27a/b that promotes HCC proliferation [51]. miR-27a/b directly targets and suppresses the expression of nuclear factor-erythroid 2-related factor 2 (NRF2) and prohibitin 1 (PHB1), a mitochondrial chaperone function as a tumor suppressor in liver cancer [52], whereas knockdown of c-Myc or miR-27a/b in Huh-7 cells rescued the LCA-mediated suppression of NRF2 and PHB1. This suggests that the interplay of ROS, c-Myc, and miR-27 has a significant role in HCC progression.

3.2.2. P53

The tumor suppressor protein p53 maintains genome integrity by inducing antiproliferative programs such as cell cycle arrest, senescence, and apoptosis through differential activation of key effector genes including the tumor suppressor miR-34a [53,54]. p53 is an oxidative stress-responsive transcription factor whose expression can be induced by ROS to protect genome stability via selectively activating its target genes [55]. Furthermore, the transcriptional activity of p53 is affected by oxidative stress, as the endogenous antioxidants thioredoxin (TRX) and GSH modify the cysteine amino acids of p53, which affects p53 activity including DNA binding capacity, activation of target gene transcription, and apoptosis induction [56–58]. A study has shown that H_2O_2 treatment phosphorylates the Ser33 amino acid residue of p53 in hepatic cells, which promotes miR-200 transcription and cell death [59]. Interestingly, p53 knockdown reversed the H_2O_2 mediated miR-200 expression [60], confirming that miR-200 expression under oxidative stress is p53-dependent. Importantly, miR-200 has shown to function as a tumor suppressor by inhibiting the CSC self-renewal potential and EMT process in various cancer subtypes including bladder cancer, gastric cancer, ovarian cancer, pancreatic cancer, and prostate cancer [49,61,62]. Furthermore, ROS mediated p53 activation also upregulates the expression of miR-506, which inhibited the growth of lung tumor in-vitro and in-vivo [63]. In addition, expression of miR-34a-5p and miR-1915 is regulated by p53 in HCC cells during oxidative stress [64]. Moreover, miR-34 inhibits pancreatic CSC proliferation, self-renewal, and induces apoptosis and cell cycle arrest [49]. Altogether, these studies strongly suggest that p53 mediates anticancer roles through promoting the expression of tumor suppressor miRNAs in a redox-dependent fashion.

3.2.3. NFκB

NF-κB is an inducible transcription factor that plays a pivotal role in DNA transcription, cytokine production, cell proliferation, survival, differentiation, cell cycle regulation, and especially in inflammation [65]. The activity of NF-κB is inhibited by its inhibitor IκB which sequesters NF-κB in the cytosol to prevent its translocation to the nucleus. The canonical NF-κB activation is mediated through the degradation of IκB, induced via site-specific phosphorylation by NF-κB-inducing kinase (NIK) and IκB kinase (IKK) protein complex, consisting of IKKα, IKKβ, and NF-κB essential modulator. ROS activate the NF-κB pathway by activating NIK through oxidative inhibition of regulatory phosphatases, and through tyrosine phosphorylation of IκBα [22]. NFκB mediates transcription of several miRNAs including let-7, miR-21, and miR-146 [66]. miR-21 is a well-studied oncomiR which mediates pro-survival and anti-proliferative effects through directly targeting and suppressing the expression of tumor suppressors such as PTEN, PDCD4, IGFBP3, and MKK3 [67–70]. Overexpression of miR-21 is associated with the progression of many cancer types and considered as a biomarker and target for cancer treatment [71]. Interestingly, miR-21 is elevated in breast CSC subpopulations and regulates the EMT phenotype [49]. ROS-induced miR-21 expression has been shown to contribute to the invasion and metastasis of prostate cancer [72]. NFκB activates miR-21 transcription by directly binding to the promoter of the miR-21 gene [73]. Likewise, ROS-activated NFκB can also upregulate miR-146a transcription, which suppresses the progression of acute myeloid leukemia (AML) [74]. In contrast, berberine-treatment-induced oxidative stress, suppressed miR-21 expression by inhibiting the nuclear translocation of NFκB in human multiple myeloma cells, which induces apoptosis [75]. In addition, oxidative stress deactivated NFκB activity that downregulated miR-19a transcription and activated apoptosis of the pheochromocytoma cells [76]. These findings suggest that the transcription factor NFκB can be either activated or inhibited under oxidative stress.

3.2.4. HIF-1α

HIF-1α is a subunit of heterodimeric transcription factor hypoxia-inducible factor 1, which regulates the expression of genes involved in the process of angiogenesis and erythropoiesis, which is important for blood vessel formation and the survival of cells under hypoxic condition [77,78]. Under hypoxia, HIF-1α activates the transcription of certain miRNAs called hypoxamiRs, which function as key regulators of the cell against decreased oxygen tension [79]. miR-210 is one such miRNA whose transcription is activated through direct binding of HIF-1α to the hypoxia-responsive element located within its promoter. Interestingly, miR-210 can negatively regulate HIF-1α expression by directly targeting its mRNA forming a negative-feedback loop, and disruption of this loop has been implicated in autoimmune diseases and tumor initiation [79,80]. Studies have shown that miR-210 promotes CSC proliferation, migration, metastasis, and self-renewal [49]. Furthermore, HIF-1α activates the transcription of many other miRNAs including miR-382, miR-421, miR-191, and miR-687 that promote migration, angiogenesis, metastasis, tumor growth, or drug resistance in cancer [81–84]. ROS regulate HIF-1α directly by oxidizing the Cys533 amino acid residue of HIF-1α, which increases the HIF-1α protein stability under oxidative stress [85]. In addition, ROS can activate HIF-1α indirectly through downregulating SIRT1 deacetylase, which results in acetylation at the Lys647 amino acid residue of HIF-1α [86]. This strongly suggests that ROS may regulate the expression of a broad range of miRNA genes in cancer by regulating the redox-sensitive HIF-1α transcription factor.

3.3. Regulation of MiRNA Processing

ROS can also affect miRNA expression by regulating proteins involved in miRNA processing. Generally, miRNAs are transcribed as primary miRNA (pri-miRNA) transcripts by RNA polymerase II or RNA polymerase III. Pri-miRNAs are then processed into premature miRNA (pre-miRNA) transcripts that are approximately 60–70 nucleotide long by the RNA-specific RNAse III type ribonuclease Drosha and DGCR8 protein complex. The pre-miRNA hairpins are then exported to the cytoplasm by the Exportin-5 and are processed into mature miRNA duplex by the ribonuclease Dicer [87] (Figure 1c).

Interestingly, p53 regulates the processing of pri-miRNA to pre-miRNA by interacting with the Drosha processing complex via the association with DEAD-box RNA helicase p68 (DDX5), thus indirectly inducing the transcription of miR-34a, miR-200c, and miR-17-92 cluster [88]. A study has demonstrated that H_2O_2 treatment in endothelial cells decreased the expression of Dicer which in turn downregulated the majority of miRNAs that are normally expressed in cerebromicrovascular endothelial cells [89]. Strikingly, ROS production is also regulated by cellular Dicer levels. A study has shown that dicer knockdown downregulated miRNA expression and decreased the production of ROS in human microvascular endothelial cells [90]. Although this has been investigated in non-cancerous cell models, it would be interesting to analyze whether this phenomenon also exists in cancer cells. Furthermore, NFκB can also regulate miRNA expression indirectly by expressing proteins involved in miRNA processing. A study has shown that NFκB activates the transcription of the miRNA processing inhibitor Lin28, which decreased the let-7 levels rapidly leading to Src-induced cellular transformation [91]. Moreover, ROS not only affect miRNA expression but also modify miRNAs directly through oxidation. A study has shown that upon oxidative modification, miR-184 can target the 3'UTR of antiapoptotic proteins BCL-XL and BCL-W, which are non-native targets of miR-184. Oxidized miR-184 induces apoptosis through downregulating the expression of BCL-XL and BCL-W in the rat heart cell line H9c2 [92]. Altogether, these studies indicate that ROS promote cancer progression through controlling miRNA expression, and the mechanisms involved in the ROS-mediated miRNA expression are not limited. Therefore, more novel mechanisms involved in ROS-dependent miRNA regulation continue to be unraveled in future studies.

4. MiRNAs Regulate ROS Homeostasis

MiRNAs can affect cellular redox homeostasis by regulating the expression of endogenous ROS producers and antioxidants. They usually manipulate ROS levels by directly targeting the genes involved in ROS production or elimination processes (Figure 2). In this section, we discuss how miRNAs control cellular ROS levels in cancer by targeting genes involved in redox homeostasis.

4.1. Regulation of ROS Producer

Studies have shown that miRNAs can affect the expression and function of endogenous ROS producers through functional interactions, thereby controlling cellular ROS production in cancer cells. The membrane-bound enzyme NADPH oxidases (NOXs) produce O_2^- through catalyzing the reduction of O_2 by transferring an electron from NADPH [93]. The tumor suppressor miR-34a regulates NOX2, the catalytic subunit of NADPH oxidase and overexpression of miR-34a in glioma cells induced apoptosis through NOX2 mediated ROS production [94]. Proline oxidase (POX) is a p53-activated ROS producer whose expression is decreased in human cancer tissues including renal cancer. POX is a direct target of miR-23b, and knockdown of miR-23b promotes ROS production and apoptosis thereby inhibiting kidney tumor growth [95]. Knockdown of dicer in mouse endothelial cells increased the activity of miR-21a-3p targeting NOX4 3'UTR, which resulted in decreased cellular ROS production and endothelial cell tumor formation [96]. These findings indicate that ROS can act as a double-edged sword, thus both overproduction or inhibition of ROS can have a significant effect on cancer progression.

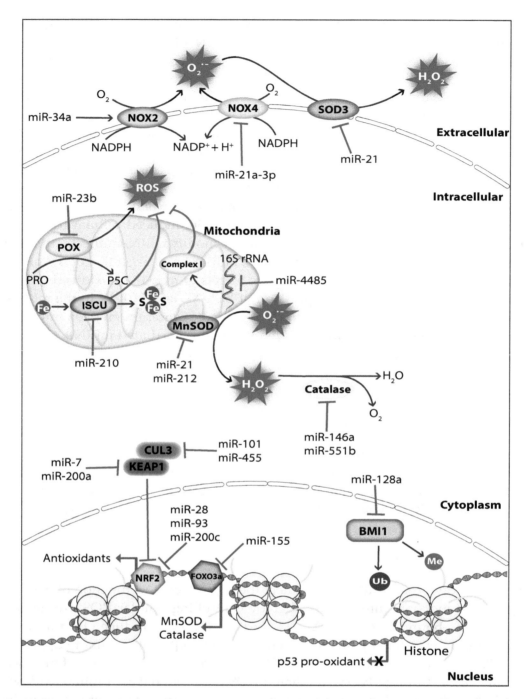

Figure 2. MiRNAs regulate ROS levels in cancer. MiRNAs regulate ROS levels in cancer by inhibiting the expression of ROS producers or antioxidants. MiRNAs decrease ROS levels through inhibiting ROS producers NOX2 (NADPH oxidase 2), NOX4 (NADPH oxidase 4), POX (Proline oxidase), or indirectly by inhibiting the polycomb complex protein BMI1 which repress p53 pro-oxidant expression. ROS levels are elevated by miRNAs through direct or indirect inhibition of antioxidants including catalase, SOD3 (Superoxide dismutase 3), MnSOD (Manganese-dependent superoxide dismutase), and proteins involved in mitochondrial function including mitochondrial complex I (NADH Coenzyme Q reductase) and ISCU (Iron-sulfur cluster assembly enzyme). BMI1, B lymphoma Mo-MLV insertion region 1 homolog; CUL3, cullin-3, Fe, iron; FOXO3a, forkhead box O3; KEAP1, Kelch-like ECH-associated protein 1, NRF2, nuclear factor-erythroid 2-related factor 2; H_2O_2, hydrogen peroxide; Me, methyl group; NADPH, nicotinamide adenine dinucleotide phosphate; O_2^-, superoxide; PRO, proline; P5C, 1-pyrroline-5-carboxylate; S, sulfur; Ub, ubiquitin (The black arrow indicates upmodulation, the red T arrow indicates inhibition, the green arrow indicates transcription activation, the green arrow with red cross indicates transcription inhibition).

4.2. Regulation of Mitochondrial Functions

Mitochondria are the major site for ROS production, and redox homeostasis in mitochondria is crucial for normal cellular processes. MiRNAs have been shown to affect the ROS production of mitochondria in cancer cells by regulating genes associated with mitochondrial function. The hypoxia-induced miR-210 promotes ROS production by repressing the iron-sulfur cluster assembly enzyme (ISCU) which is essential for the assembly of iron-sulfur (Fe-S) cluster and mitochondria respiratory activity [97]. However, a study suggests that miR-210 mediated ROS accumulation may be due to the repression of other gene targets since the ISCU knockdown in colon cancer cells does not increase ROS levels significantly [98]. A study has shown that miR-128a promotes intracellular ROS levels and cellular senescence in medulloblastoma cells by directly targeting the polycomb complex protein BMI-1 which is involved in the maintenance of mitochondrial activities and redox homeostasis [99]. Surprisingly, a study demonstrated that miRNAs regulate ROS production by targeting non-coding RNAs, during cellular stress miR-4485 translocates to mitochondria and directly targets mitochondrial 16S ribosomal RNA (rRNA), thus modulates mitochondrial function and subsequent ROS accumulation (Figure 2). Importantly, miR-4485 levels are decreased in human breast cancer tissues and overexpression of miR-4485 suppressed breast cancer tumorigenesis in-vitro and in-vivo [100]. These findings strongly suggest that hindrance in mitochondrial metabolism can promote carcinogenesis through ROS accumulation.

4.3. Regulation of Antioxidants

Antioxidant enzymes and non-enzymatic antioxidants mediate the detoxification of ROS to protect cells from oxidative damage. Superoxide dismutase (SOD) is an antioxidant metalloenzyme expressed in both eukaryotes and prokaryotes, which utilizes the metal ions including copper, iron, manganese, and zinc as cofactors to catalyze the dismutation of O_2^- into molecular oxygen (O_2) and H_2O_2. Similarly, catalase is an antioxidant enzyme located mostly in the cytosol and peroxisomes scavenge ROS through catalyzing the conversion of H_2O_2 into water (H_2O) and O_2 [101]. Several studies have shown that miRNAs can upregulate cellular ROS levels in cancer cells by inhibiting antioxidants including SOD and catalase. The oncomiR miR-21 promotes tumorigenesis through increasing cellular ROS levels by directly targeting the SOD3 or by targeting TNFα that results in MnSOD downregulation (Figure 2) [102]. Furthermore, the miR-212 which is downregulated in human colorectal cancer (CRC) can regulate MnSOD by directly targeting its mRNA, and overexpression of miR-212 inhibited metastasis of CRC cells by suppressing MnSOD expression [103]. In cancer cells, catalase expression is regulated by miR-551b and miR-146a, and inhibition of catalase by these miRNAs promotes ROS accumulation [104,105]. Interestingly, miRNAs can also control the expression of antioxidants indirectly through targeting transcription factors that promote the transcription of antioxidants. For example, K-Ras-induced miR-155 increases ROS levels by directly targeting FOXO3a, a transcription factor that activates the transcription of antioxidants MnSOD and catalase (Figure 2) [106]. These findings suggest that the endogenous expression of endogenous antioxidants is crucial for the prevention of cellular ROS accumulation, which is manipulated by miRNAs in cancer cells to support cancer progression.

4.4. Regulation of NRF2/KEAP1 System

Cellular redox homeostasis is controlled by the nuclear factor-erythroid 2-related factor 2 (NRF2)/ Kelch-like ECH-associated protein 1 (KEAP1) system. NRF2 is a transcriptional factor, which activates the transcription of genes that encode antioxidant enzymes and non-enzymatic antioxidants in response to oxidative stress. Under normal conditions, NRF2 is inactivated by the KEAP1-cullin3 (CUL3) complex, which sequesters NRF2 in the cytoplasm and promotes NRF2 degradation through ubiquitination. During oxidative stress, NRF2 is dissociated from the KEAP1-CUL3 complex caused by the rapid oxidation on cys151 residue of KEAP1 [107]. In cancer, miRNAs can affect the cellular redox homeostasis by targeting genes involved in the NRF2/KEAP1 regulatory system. Overexpression of miR-200c in lung cancer cells increases ROS levels through suppressing the expression of proteins involved in oxidative stress defense including peroxiredoxin 2, NRF2, and Sestrin 1 [108]. A study has shown that miR-28 decreases NRF2 expression by directly targeting its 3'UTR, which increased the colony formation capacity in breast cancer cells [109]. Similarly, miR-93 regulates NRF2 and is associated with breast cancer development [110]. Moreover, a bioinformatic prediction showed that about 85 miRNAs may negatively regulate NRF2 expression by directly targeting its mRNA [111]. miRNAs also regulate NRF2 activity indirectly through targeting its inhibitors KEAP1 and CUL3 (Figure 2). miR-7 and miR-200a target KEAP1 mRNA and decrease its protein expression thereby mediating NRF2 nuclear localization and target gene transcription in neuroblastoma and breast cancer cells, respectively [112,113]. Likewise, miR-101 and miR-455 target CUL3 mRNA, which promotes NRF2 nuclear localization that leads to angiogenesis and oxidative stress protection, respectively [114,115]. Altogether, these studies strongly suggest that cancer cells manipulate ROS levels by controlling miRNA expression to support their survival and promotion.

5. The Interplay of ROS and MiRNAs in Cancer

Oxidative stress induces DNA damage or mutation that may affect the expression and function of genes associated with the damaged genomic loci, and can eventually cause cancer initiation. ROS may also affect miRNA expression and function directly by causing oxidative damage-induced mutation on miRNA genes and mature miRNA sequences, or indirectly by altering its epigenetic signature or biogenesis pathway. Deregulated miRNA expression caused through genomic deletion, epigenetic silencing, or overexpression can contribute to cancer initiation and progression by controlling oncogenes and tumor suppressor genes. Therefore, ROS can regulate miRNA-mediated carcinogenesis. Elevated ROS production is observed in various cancer types and high cellular ROS can activate oncogenic signaling pathways that support cancer progression. MiRNAs are able to control the cellular ROS levels by targeting genes involved in ROS production and elimination, thus miRNA can control ROS-mediated carcinogenesis. These facts suggest that ROS and miRNAs can function synergistically in the process of cancer development (Figure 3). ROS upregulate the expression of the oncomiR miR-21 and miR-146a through activating NFκB, and these miRNAs can increase cellular ROS levels by downregulating endogenous antioxidants [73,102,105]. Similarly, the miR-210 expression is upregulated through ROS-mediated activation of HIF-1α, and the miR-210 has been shown to increase ROS production by negatively regulating ISCU [97,116]. Interestingly, ROS can also upregulate the expression of the tumor suppressor miR-34 through p53 activation, whereas the miR-34 has been shown to increase ROS production by upregulating the expression of NOX2 [64,94]. These studies strongly suggest that ROS and miRNAs crosstalk in cancer cells to orchestrate the ROS production to activate and promote cancer development. Furthermore, it is of importance to investigate whether the mRNA of genes involved in miRNA expression, ROS production, and detoxification would function as potential competing endogenous RNAs (ceRNAs) which can co-regulate each other's expression by competing for binding to shared miRNAs [117]. For example, miR-210 can directly target the mRNA of HIF-1α and ISCU, suggesting that HIF-1α and ISCU could function as potential ceRNAs [80,97]. Likewise, miR-21 has been shown to downregulate the expression of antioxidants SOD3 and MnSOD [102]. However, miR-21 targets only the mRNA of SOD3 but not the MnSOD. Therefore, it would be interesting to investigate whether the MnSOD mRNA encompasses a binding site for miR-21 or any other miRNA that can target SOD3. Nevertheless, more studies should be done in this perspective to unravel the complete regulatory network between miRNA and ROS in cancer.

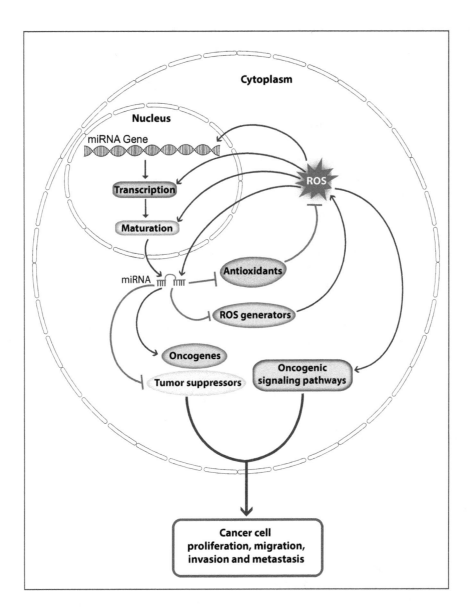

Figure 3. Schematic model illustrates the interplay of ROS and miRNAs in cancer progression. ROS regulate miRNA expression and function by altering miRNA transcription, maturation, or sequences. Dysregulated miRNA expression promotes cancer progression by regulating oncogenes and tumor suppressors. MiRNAs control cellular ROS levels by regulating endogenous antioxidants and ROS generators, which favor cancer development through activating oncogenic signaling pathways (The black arrow indicates upmodulation, the red T arrow indicates inhibition).

6. Challenges in Using Antioxidants for Anti-Cancer Therapy

Since ROS mediate cellular damage and oncogenic mutations, usage of dietary supplement with antioxidants was proposed to prevent or treat cancer. Dietary supplement containing antioxidants such as selenium, vitamin E, and β-carotene was tested to reduce the occurrence of cancer in individuals with a history of cancer. This resulted in a significant decrease in total cancer occurrence and overall mortality. Conversely, studies also show that nutritional supplements of antioxidants may promote cancer incidence and mortality [118–120]. Moreover, the usage of antioxidants as additional therapy in cancer treatment failed to show beneficiary effect, supplementing breast and colorectal cancer patients with ascorbate/vitamin C does not improve overall or progression-free survival [121]. Even though antioxidants are often ineffective for cancer prevention/treatment in humans with a high risk of cancer, it was shown that antioxidant treatment might suppress cancer risk in mice with certain genetic modifications. NAC treatment reduces ROS generation, DNA damage, and cancer occurrence in mice deficient of ATM and p53 [122,123]. However, this was not consistent since another study demonstrated that treating mouse models of lung cancer with antioxidants NAC and vitamin E promotes tumor progression and decreases mouse survival [124]. The underlying cause of antioxidants promoting cancer progression may be due to the fact that cancer cells are more susceptible to oxidative stress when compared to normal cells. Therefore, cancer cells depend on endogenous antioxidants including GSH, TRX, NRF2, thioredoxin-like 2, SOD, MnSOD, and glutamate-cysteine ligase to protect them from oxidative stress during cancer development [125–130]. In cancer cells, several oncogenes increase NRF2 transcription to promote ROS detoxification and tumorigenesis, whereas the deletion of NRF2 promotes DNA damage and suppresses tumorigenesis in pancreatic cancer cells [131]. In some cancers, ROS levels are suppressed by continuous activation of NRF2 achieved via mutations in NRF2 or its inhibitors KEAP1 that prevents NRF2 translocation from nucleus to cytoplasm [132]. Oxidative stress also limits metastasis by melanoma cells, whereas antioxidant treatment in a mouse model of malignant melanoma promotes the distant metastasis without affecting the growth of primary subcutaneous tumors [133]. Furthermore, cancer cells manage ROS levels by increasing NAPDH generation through accelerating multiple metabolic pathways including the pentose phosphate, folate, and malic enzyme pathway [134–136]. These studies suggest that antioxidant treatments are beneficial for cancer progression instead of being detrimental to cancer cells. Importantly, the inconsistent outcome from the clinical trials and experimental mouse models suggest that the application of antioxidants for anti-cancer therapy may not be a promising approach.

On the other hand, miRNAs are suggested as promising therapeutic agents for cancer treatment. In recent years, several studies proposed many novel miRNA-based cancer therapies that have significantly improved the survival of cancer patients [137]. The application of miRNAs as therapeutic agents has many potential advantages. Basically, miRNAs are highly conserved endogenous small RNA molecules with known sequences which may simplify the process of designing therapeutic agents with less off-target effects. A single miRNA can potentially regulate multiple target genes associated with single or multiple pathways which could be a very efficient way to treat multi-pathway diseases including cancer. For anti-cancer therapy, two miRNA-based strategies are applied. MiRNA replacement therapy is applied to either induce apoptosis or suppress the proliferation of cancer cells by using exogenous tumor suppressor miRNA mimics. MiRNA reduction therapy is applied to inhibit the function of oncogenic miRNAs by using antagomiRs or locked-nucleic acids antisense oligonucleotides (LNAs) [138,139]. To date, there is no miRNA-based drug available for cancer treatment. However, some miRNA drug candidates have entered into the early phase of human clinical trials. These include MesomiR-1, the miRNA mimic of tumor-suppressing miR-16 for treating lung cancer; MRX34, the miRNA mimic of tumor-suppressing miR-34 for treating liver cancer, lymphoma and melanoma; and MRG106, the LNA-modified anti-miR of miR-155 for treating T-cell lymphoma [140]. Although miRNA-based therapy has made progress, still there are some challenges ahead to become an efficient therapeutic approach. The adverse effect is one of the major challenges encountered by this therapy. For example, MRX34 has been withdrawn from entering phase 2 trials due to the serious immune response observed in some patients during phase 1 trials [140]. There are limitations in the efficiency of in-vivo delivery of miRNA mimics and antagomiRs as the oligonucleotides are degraded by the endonucleases in the blood. Understanding the regulatory network of miRNA and ROS production in cancer would further help to develop an alternative effective therapeutic approach to treat cancer. One such approach would be aggravating oxidative stress in cancer cells through miRNA-based therapy that either enhance ROS production or inhibit endogenous antioxidant system.

7. Concluding Remarks

ROS function as a mediator of cellular signaling pathways involved in proliferation, growth, survival, and apoptosis, and the redox homeostasis is actively maintained by endogenous antioxidant systems. Cancer cells manipulate the cellular ROS levels to favor their proliferation, survival, and metastasis. ROS levels are regulated via fine-tuning the expression of ROS producers and scavengers by miRNAs. On the other hand, ROS regulate miRNA expression by altering the activity of proteins involved in miRNA transcription and maturation. The regulatory network of ROS and miRNAs is orchestrated in cancer to promote cancer progression and to cope with oxidative stress. Identification of regulatory crosstalk between miRNA and redox signaling opens up new horizons for using miRNAs as potential therapeutic targets in cancer treatment. However, further understanding of the miRNA-ROS regulatory network is needed for the application of miRNAs to augment ROS-mediated cancer cell death.

Author Contributions: K.R.B. wrote the manuscript. K.R.B. and Y.T. edited and revised the manuscript.

Acknowledgments: We apologize to all colleagues whose work could not be cited due to space constraints. We thank Yvonne Tay lab members (Cancer Science Institute of Singapore) for critical reading of the manuscript. Yvonne Tay was supported by a Singapore National Research Foundation Fellowship and a National University of Singapore President's Assistant Professorship.

References

1. Snezhkina, A.V.; Kudryavtseva, A.V.; Kardymon, O.L.; Savvateeva, M.V.; Melnikova, N.V.; Krasnov, G.S.; Dmitriev, A.A. ROS Generation and antioxidant defense systems in normal and malignant cells. *Oxidative Med. Cell. Longev.* **2019**, *2019*, 6175804. [CrossRef] [PubMed]

2. Di Meo, S.; Reed, T.T.; Venditti, P.; Victor, V.M. Role of ROS and RNS sources in physiological and pathological conditions. *Oxidative Med. Cell. Longev.* **2016**, *2016*, 1245049. [CrossRef] [PubMed]

3. Brieger, K.; Schiavone, S.; Miller, F.J., Jr.; Krause, K.H. Reactive oxygen species: From health to disease. *Swiss Med Wkly.* **2012**, *142*, 13659. [CrossRef] [PubMed]

4. Schieber, M.; Chandel, N.S. ROS function in redox signaling and oxidative stress. *Curr. Biol.* **2014**, *24*, 453–462. [CrossRef] [PubMed]

5. Jena, N.R. DNA damage by reactive species: Mechanisms, mutation and repair. *J. Biosci.* **2012**, *37*, 503–517. [CrossRef] [PubMed]

6. Poulsen, H.E.; Specht, E.; Broedbaek, K.; Henriksen, T.; Ellervik, C.; Mandrup-Poulsen, T.; Tonnesen, M.; Nielsen, P.E.; Andersen, H.U.; Weimann, A. RNA modifications by oxidation: A novel disease mechanism? *Free. Radic. Biol. Med.* **2012**, *52*, 1353–1361. [CrossRef] [PubMed]

7. Fimognari, C. Role of oxidative RNA damage in chronic-degenerative diseases. *Oxidative Med. Cell. Longev.* **2015**, *2015*, 358713. [CrossRef]

8. Ma, Y.; Zhang, L.; Rong, S.; Qu, H.; Zhang, Y.; Chang, D.; Pan, H.; Wang, W. Relation between gastric cancer and protein oxidation, DNA damage, and lipid peroxidation. *Oxidative Med. Cell. Longev.* **2013**, *2013*, 543760. [CrossRef]

9. Barrera, G. Oxidative stress and lipid peroxidation products in cancer progression and therapy. *ISRN Oncol.* **2012**, *2012*, 137289. [CrossRef]

10. Bartel, D.P. Metazoan MicroRNAs. *Cell* **2018**, *173*, 20–51. [CrossRef]

11. Peng, Y.; Croce, C.M. The role of MicroRNAs in human cancer. *Signal Transduct. Target. Ther.* **2016**, *1*, 15004. [CrossRef] [PubMed]

12. Thulasingam, S.; Massilamany, C.; Gangaplara, A.; Dai, H.; Yarbaeva, S.; Subramaniam, S.; Riethoven, J.J.; Eudy, J.; Lou, M.; Reddy, J. MiR-27b, an oxidative stress-responsive microRNA modulates nuclear factor-kB pathway in RAW 264.7 cells. *Mol. Cell. Biochem.* **2011**, *352*, 181–188. [CrossRef] [PubMed]

13. Peng, J.; He, X.; Zhang, L.; Liu, P. MicroRNA26a protects vascular smooth muscle cells against H_2O_2 induced injury through activation of the PTEN/AKT/mTOR pathway. *Int. J. Mol. Med.* **2018**, *42*, 1367–1378. [CrossRef] [PubMed]

14. Cheng, Y.; Zhou, M.; Zhou, W. MicroRNA-30e regulates TGF-beta-mediated NADPH oxidase 4-dependent oxidative stress by Snai1 in atherosclerosis. *Int. J. Mol. Med.* **2019**, *43*, 1806–1816. [CrossRef]

15. Van Houten, B.; Santa-Gonzalez, G.A.; Camargo, M. DNA repair after oxidative stress: Current challenges. *Curr. Opin. Toxicol.* **2018**, *7*, 9–16. [CrossRef]

16. Anastasiadou, E.; Jacob, L.S.; Slack, F.J. Non-coding RNA networks in cancer. *Nat. Rev. Cancer* **2018**, *18*, 5–18. [CrossRef]

17. Davies, M.J. Protein oxidation and peroxidation. *Biochem. J.* **2016**, *473*, 805–825. [CrossRef]

18. Gangwar, A.; Paul, S.; Ahmad, Y.; Bhargava, K. Competing trends of ROS and RNS-mediated protein modifications during hypoxia as an alternate mechanism of NO benefits. *Biochimie* **2018**, *148*, 127–138. [CrossRef]

19. Liou, G.Y.; Storz, P. Reactive oxygen species in cancer. *Free Radic. Res.* **2010**, *44*, 479–496. [CrossRef]

20. Zhong, H.; Yin, H. Role of lipid peroxidation derived 4-hydroxynonenal (4-HNE) in cancer: Focusing on mitochondria. *Redox Biol.* **2015**, *4*, 193–199. [CrossRef]

21. Liao, Z.; Chua, D.; Tan, N.S. Reactive oxygen species: A volatile driver of field cancerization and metastasis. *Mol. Cancer* **2019**, *18*, 65. [CrossRef] [PubMed]

22. Lingappan, K. NF-kappaB in oxidative stress. *Curr. Opin. Toxicol.* **2018**, *7*, 81–86. [CrossRef] [PubMed]

23. Verbon, E.H.; Post, J.A.; Boonstra, J. The influence of reactive oxygen species on cell cycle progression in mammalian cells. *Gene* **2012**, *511*, 1–6. [CrossRef] [PubMed]

24. Ruiz-Ramos, R.; Lopez-Carrillo, L.; Rios-Perez, A.D.; De Vizcaya-Ruiz, A.; Cebrian, M.E. Sodium arsenite induces ROS generation, DNA oxidative damage, HO-1 and c-Myc proteins, NF-kappaB activation and cell proliferation in human breast cancer MCF-7 cells. *Mutat. Res.* **2009**, *674*, 109–115. [CrossRef] [PubMed]

25. Menon, S.G.; Coleman, M.C.; Walsh, S.A.; Spitz, D.R.; Goswami, P.C. Differential susceptibility of nonmalignant human breast epithelial cells and breast cancer cells to thiol antioxidant-induced G(1)-delay. *Antioxid. Redox Signal.* **2005**, *7*, 711–718. [CrossRef] [PubMed]

26. Vanharanta, S.; Massague, J. Origins of metastatic traits. *Cancer cell* **2013**, *24*, 410–421. [CrossRef]

27. Piskounova, E.; Agathocleous, M.; Murphy, M.M.; Hu, Z.; Huddlestun, S.E.; Zhao, Z.; Leitch, A.M.; Johnson, T.M.; DeBerardinis, R.J.; Morrison, S.J. Oxidative stress inhibits distant metastasis by human melanoma cells. *Nature* **2015**, *527*, 186–191. [CrossRef]

28. Hitchler, M.J.; Oberley, L.W.; Domann, F.E. Epigenetic silencing of SOD2 by histone modifications in human breast cancer cells. *Free Radic. Biol. Med.* **2008**, *45*, 1573–1580. [CrossRef]

29. Lewis, A.; Du, J.; Liu, J.; Ritchie, J.M.; Oberley, L.W.; Cullen, J.J. Metastatic progression of pancreatic cancer: Changes in antioxidant enzymes and cell growth. *Clin. Exp. Metastasis* **2005**, *22*, 523–532. [CrossRef]

30. Roy, R.; Morad, G.; Jedinak, A.; Moses, M.A. Metalloproteinases and their roles in human cancer. *Anat. Rec. (Hoboken)* **2019**. [CrossRef]

31. Kang, K.A.; Ryu, Y.S.; Piao, M.J.; Shilnikova, K.; Kang, H.K.; Yi, J.M.; Boulanger, M.; Paolillo, R.; Bossis, G.; Yoon, S.Y.; et al. DUOX2-mediated production of reactive oxygen species induces epithelial mesenchymal transition in 5-fluorouracil resistant human colon cancer cells. *Redox. Biol.* **2018**, *17*, 224–235. [CrossRef] [PubMed]

32. Radisky, D.C.; Levy, D.D.; Littlepage, L.E.; Liu, H.; Nelson, C.M.; Fata, J.E.; Leake, D.; Godden, E.L.; Albertson, D.G.; Nieto, M.A.; et al. Rac1b and reactive oxygen species mediate MMP-3-induced EMT and genomic instability. *Nature* **2005**, *436*, 123–127. [CrossRef] [PubMed]

33. van Wetering, S.; van Buul, J.D.; Quik, S.; Mul, F.P.; Anthony, E.C.; ten Klooster, J.P.; Collard, J.G.; Hordijk, P.L. Reactive oxygen species mediate Rac-induced loss of cell-cell adhesion in primary human endothelial cells. *J. Cell Sci.* **2002**, *115*, 1837–1846. [PubMed]

34. Kuai, W.X.; Wang, Q.; Yang, X.Z.; Zhao, Y.; Yu, R.; Tang, X.J. Interleukin-8 associates with adhesion, migration, invasion and chemosensitivity of human gastric cancer cells. *World J. Gastroenterol.* **2012**, *18*, 979–985. [CrossRef]

35. Ghislin, S.; Obino, D.; Middendorp, S.; Boggetto, N.; Alcaide-Loridan, C.; Deshayes, F. LFA-1 and ICAM-1 expression induced during melanoma-endothelial cell co-culture favors the transendothelial migration of melanoma cell lines in vitro. *BMC Cancer* **2012**, *12*, 455. [CrossRef] [PubMed]

36. Sawada, J.; Li, F.; Komatsu, M. R-Ras Inhibits VEGF-Induced p38MAPK Activation and HSP27 Phosphorylation in Endothelial Cells. *J. Vasc. Res.* **2015**, *52*, 347–359. [CrossRef]

37. Shi, X.; Zhang, Y.; Zheng, J.; Pan, J. Reactive oxygen species in cancer stem cells. *Antioxid. Redox Signal.* **2012**, *16*, 1215–1228. [CrossRef]

38. Diehn, M.; Cho, R.W.; Lobo, N.A.; Kalisky, T.; Dorie, M.J.; Kulp, A.N.; Qian, D.; Lam, J.S.; Ailles, L.E.; Wong, M.; et al. Association of reactive oxygen species levels and radioresistance in cancer stem cells. *Nature* **2009**, *458*, 780–783. [CrossRef]

39. Pasto, A.; Bellio, C.; Pilotto, G.; Ciminale, V.; Silic-Benussi, M.; Guzzo, G.; Rasola, A.; Frasson, C.; Nardo, G.; Zulato, E.; et al. Cancer stem cells from epithelial ovarian cancer patients privilege oxidative phosphorylation, and resist glucose deprivation. *Oncotarget* **2014**, *5*, 4305–4319. [CrossRef]

40. Jang, Y.Y.; Sharkis, S.J. A low level of reactive oxygen species selects for primitive hematopoietic stem cells that may reside in the low-oxygenic niche. *Blood* **2007**, *110*, 3056–3063. [CrossRef]

41. He, J.; Xu, Q.; Jing, Y.; Agani, F.; Qian, X.; Carpenter, R.; Li, Q.; Wang, X.R.; Peiper, S.S.; Lu, Z.; et al. Reactive oxygen species regulate ERBB2 and ERBB3 expression via miR-199a/125b and DNA methylation. *EMBO Rep.* **2012**, *13*, 1116–1122. [CrossRef] [PubMed]

42. He, J.; Jing, Y.; Li, W.; Qian, X.; Xu, Q.; Li, F.S.; Liu, L.Z.; Jiang, B.H.; Jiang, Y. Roles and mechanism of miR-199a and miR-125b in tumor angiogenesis. *PLoS ONE* **2013**, *8*, e56647. [CrossRef] [PubMed]

43. Li, Y.; Seto, E. HDACs and HDAC inhibitors in cancer development and therapy. *Cold Spring Harb. Perspect. Med.* **2016**, *6*. [CrossRef] [PubMed]

44. Ago, T.; Liu, T.; Zhai, P.; Chen, W.; Li, H.; Molkentin, J.D.; Vatner, S.F.; Sadoshima, J. A redox-dependent pathway for regulating class II HDACs and cardiac hypertrophy. *Cell* **2008**, *133*, 978–993. [CrossRef] [PubMed]

45. Ciesla, M.; Marona, P.; Kozakowska, M.; Jez, M.; Seczynska, M.; Loboda, A.; Bukowska-Strakova, K.; Szade, A.; Walawender, M.; Kusior, M.; et al. Heme oxygenase-1 controls an HDAC4-miR-206 pathway of oxidative stress in rhabdomyosarcoma. *Cancer Res.* **2016**, *76*, 5707–5718. [CrossRef]

46. Singh, A.; Happel, C.; Manna, S.K.; Acquaah-Mensah, G.; Carrerero, J.; Kumar, S.; Nasipuri, P.; Krausz, K.W.; Wakabayashi, N.; Dewi, R.; et al. Transcription factor NRF2 regulates miR-1 and miR-206 to drive tumorigenesis. *J. Clin. Investig.* **2013**, *123*, 2921–2934. [CrossRef]

47. Druz, A.; Betenbaugh, M.; Shiloach, J. Glucose depletion activates mmu-miR-466h-5p expression through oxidative stress and inhibition of histone deacetylation. *Nucleic Acids Res.* **2012**, *40*, 7291–7302. [CrossRef]

48. Druz, A.; Chu, C.; Majors, B.; Santuary, R.; Betenbaugh, M.; Shiloach, J. A novel microRNA mmu-miR-466h affects apoptosis regulation in mammalian cells. *Biotechnol. Bioeng.* **2011**, *108*, 1651–1661. [CrossRef]

49. Dando, I.; Cordani, M.; Dalla Pozza, E.; Biondani, G.; Donadelli, M.; Palmieri, M. Antioxidant Mechanisms and ROS-Related MicroRNAs in Cancer Stem Cells. *Oxidative Med. Cell. Longev.* **2015**, *2015*, 425708. [CrossRef]

50. Benassi, B.; Fanciulli, M.; Fiorentino, F.; Porrello, A.; Chiorino, G.; Loda, M.; Zupi, G.; Biroccio, A. c-Myc phosphorylation is required for cellular response to oxidative stress. *Mol. cell* **2006**, *21*, 509–519. [CrossRef]

51. Huang, S.; He, X.; Ding, J.; Liang, L.; Zhao, Y.; Zhang, Z.; Yao, X.; Pan, Z.; Zhang, P.; Li, J.; et al. Upregulation of miR-23a approximately 27a approximately 24 decreases transforming growth factor-beta-induced tumor-suppressive activities in human hepatocellular carcinoma cells. *Int. J. Cancer* **2008**, *123*, 972–978. [CrossRef] [PubMed]

52. Yang, H.; Li, T.W.; Zhou, Y.; Peng, H.; Liu, T.; Zandi, E.; Martinez-Chantar, M.L.; Mato, J.M.; Lu, S.C. Activation of a novel c-Myc-miR27-prohibitin 1 circuitry in cholestatic liver injury inhibits glutathione synthesis in mice. *Antioxid. Redox Signal.* **2015**, *22*, 259–274. [CrossRef] [PubMed]

53. Yeo, C.Q.X.; Alexander, I.; Lin, Z.; Lim, S.; Aning, O.A.; Kumar, R.; Sangthongpitag, K.; Pendharkar, V.; Ho, V.H.B.; Cheok, C.F. p53 Maintains genomic stability by preventing interference between transcription and replication. *Cell Rep.* **2016**, *15*, 132–146. [CrossRef] [PubMed]

54. Navarro, F.; Lieberman, J. miR-34 and p53: New insights into a complex functional relationship. *PLoS ONE* **2015**, *10*, e0132767. [CrossRef]

55. Chen, Y.; Liu, K.; Shi, Y.; Shao, C. The tango of ROS and p53 in tissue stem cells. *Cell Death Differ.* **2018**, *25*, 639–641. [CrossRef]

56. Peuget, S.; Bonacci, T.; Soubeyran, P.; Iovanna, J.; Dusetti, N.J. Oxidative stress-induced p53 activity is enhanced by a redox-sensitive TP53INP1 SUMOylation. *Cell Death Differ.* **2014**, *21*, 1107–1118. [CrossRef]

57. Maillet, A.; Pervaiz, S. Redox regulation of p53, redox effectors regulated by p53: A subtle balance. *Antioxid. Redox Signal.* **2012**, *16*, 1285–1294. [CrossRef]

58. Haffo, L.; Lu, J.; Bykov, V.J.N.; Martin, S.S.; Ren, X.; Coppo, L.; Wiman, K.G.; Holmgren, A. Inhibition of the glutaredoxin and thioredoxin systems and ribonucleotide reductase by mutant p53-targeting compound APR-246. *Sci. Rep.* **2018**, *8*, 12671. [CrossRef]

59. Xiao, Y.; Yan, W.; Lu, L.; Wang, Y.; Lu, W.; Cao, Y.; Cai, W. p38/p53/miR-200a-3p feedback loop promotes oxidative stress-mediated liver cell death. *Cell Cycle* **2015**, *14*, 1548–1558. [CrossRef]

60. Magenta, A.; Cencioni, C.; Fasanaro, P.; Zaccagnini, G.; Greco, S.; Sarra-Ferraris, G.; Antonini, A.; Martelli, F.; Capogrossi, M.C. miR-200c is upregulated by oxidative stress and induces endothelial cell apoptosis and senescence via ZEB1 inhibition. *Cell Death Differ.* **2011**, *18*, 1628–1639. [CrossRef]

61. Gibbons, D.L.; Lin, W.; Creighton, C.J.; Rizvi, Z.H.; Gregory, P.A.; Goodall, G.J.; Thilaganathan, N.; Du, L.; Zhang, Y.; Pertsemlidis, A.; et al. Contextual extracellular cues promote tumor cell EMT and metastasis by regulating miR-200 family expression. *Genes Dev.* **2009**, *23*, 2140–2151. [CrossRef] [PubMed]

62. Feng, X.; Wang, Z.; Fillmore, R.; Xi, Y. MiR-200, a new star miRNA in human cancer. *Cancer Lett.* **2014**, *344*, 166–173. [CrossRef] [PubMed]

63. Yin, M.; Ren, X.; Zhang, X.; Luo, Y.; Wang, G.; Huang, K.; Feng, S.; Bao, X.; Huang, K.; He, X.; et al. Selective killing of lung cancer cells by miRNA-506 molecule through inhibiting NF-kappaB p65 to evoke reactive oxygen species generation and p53 activation. *Oncogene* **2015**, *34*, 691–703. [CrossRef]

64. Wan, Y.; Cui, R.; Gu, J.; Zhang, X.; Xiang, X.; Liu, C.; Qu, K.; Lin, T. Identification of Four Oxidative Stress-Responsive MicroRNAs, miR-34a-5p, miR-1915-3p, miR-638, and miR-150-3p, in Hepatocellular Carcinoma. *Oxidative Med. Cell. Longev.* **2017**, *2017*, 5189138. [CrossRef] [PubMed]

65. Liu, T.; Zhang, L.; Joo, D.; Sun, S.C. NF-kappaB signaling in inflammation. *Signal Transduct. Target. Ther.* **2017**, *2*. [CrossRef] [PubMed]

66. Markopoulos, G.S.; Roupakia, E.; Tokamani, M.; Alabasi, G.; Sandaltzopoulos, R.; Marcu, K.B.; Kolettas, E. Roles of NF-kappaB Signaling in the regulation of miRNAs impacting on inflammation in cancer. *Biomedicines* **2018**, *6*, 40. [CrossRef] [PubMed]

67. Li, X.; Xin, S.; He, Z.; Che, X.; Wang, J.; Xiao, X.; Chen, J.; Song, X. MicroRNA-21 (miR-21) post-transcriptionally downregulates tumor suppressor PDCD4 and promotes cell transformation, proliferation, and metastasis in renal cell carcinoma. *Cell. Physiol. Biochem.* **2014**, *33*, 1631–1642. [CrossRef]

68. Li, Z.; Deng, X.; Kang, Z.; Wang, Y.; Xia, T.; Ding, N.; Yin, Y. Elevation of miR-21, through targeting MKK3, may be involved in ischemia pretreatment protection from ischemia-reperfusion induced kidney injury. *J. Nephrol.* **2016**, *29*, 27–36. [CrossRef]

69. Yang, C.H.; Yue, J.; Pfeffer, S.R.; Fan, M.; Paulus, E.; Hosni-Ahmed, A.; Sims, M.; Qayyum, S.; Davidoff, A.M.; Handorf, C.R.; et al. MicroRNA-21 promotes glioblastoma tumorigenesis by down-regulating insulin-like growth factor-binding protein-3 (IGFBP3). *J. Biol. Chem.* **2014**, *289*, 25079–25087. [CrossRef]

70. Peralta-Zaragoza, O.; Deas, J.; Meneses-Acosta, A.; De la, O.G.F.; Fernandez-Tilapa, G.; Gomez-Ceron, C.; Benitez-Boijseauneau, O.; Burguete-Garcia, A.; Torres-Poveda, K.; Bermudez-Morales, V.H.; et al. Relevance of miR-21 in regulation of tumor suppressor gene PTEN in human cervical cancer cells. *BMC Cancer* **2016**, *16*, 215. [CrossRef]

71. Feng, Y.H.; Tsao, C.J. Emerging role of microRNA-21 in cancer. *Biomed. Rep.* **2016**, *5*, 395–402. [CrossRef] [PubMed]

72. Jajoo, S.; Mukherjea, D.; Kaur, T.; Sheehan, K.E.; Sheth, S.; Borse, V.; Rybak, L.P.; Ramkumar, V. Essential role of NADPH oxidase-dependent reactive oxygen species generation in regulating microRNA-21 expression and function in prostate cancer. *Antioxid. Redox Signal.* **2013**, *19*, 1863–1876. [CrossRef] [PubMed]

73. Ling, M.; Li, Y.; Xu, Y.; Pang, Y.; Shen, L.; Jiang, R.; Zhao, Y.; Yang, X.; Zhang, J.; Zhou, J.; et al. Regulation of miRNA-21 by reactive oxygen species-activated ERK/NF-kappaB in arsenite-induced cell transformation. *Free Radic. Biol. Med.* **2012**, *52*, 1508–1518. [CrossRef] [PubMed]

74. Mohr, S.; Doebele, C.; Comoglio, F.; Berg, T.; Beck, J.; Bohnenberger, H.; Alexe, G.; Corso, J.; Strobel, P.; Wachter, A.; et al. Hoxa9 and Meis1 cooperatively induce addiction to syk signaling by suppressing miR-146a in acute myeloid leukemia. *Cancer Cell* **2017**, *31*, 549–562. [CrossRef]

75. Hu, H.Y.; Li, K.P.; Wang, X.J.; Liu, Y.; Lu, Z.G.; Dong, R.H.; Guo, H.B.; Zhang, M.X. Set9, NF-kappaB, and microRNA-21 mediate berberine-induced apoptosis of human multiple myeloma cells. *Acta Pharmacol. Sin.* **2013**, *34*, 157–166. [CrossRef]

76. Hong, J.; Wang, Y.; Hu, B.C.; Xu, L.; Liu, J.Q.; Chen, M.H.; Wang, J.Z.; Han, F.; Zheng, Y.; Chen, X.; et al. Transcriptional downregulation of microRNA-19a by ROS production and NF-kappaB deactivation governs resistance to oxidative stress-initiated apoptosis. *Oncotarget* **2017**, *8*, 70967–70981. [CrossRef]

77. Gerri, C.; Marin-Juez, R.; Marass, M.; Marks, A.; Maischein, H.M.; Stainier, D.Y.R. Hif-1alpha regulates macrophage-endothelial interactions during blood vessel development in zebrafish. *Nat. Commun.* **2017**, *8*, 15492. [CrossRef]

78. Koh, M.Y.; Lemos, R. Jr.; Liu, X.; Powis, G. The hypoxia-associated factor switches cells from HIF-1alpha- to HIF-2alpha-dependent signaling promoting stem cell characteristics, aggressive tumor growth and invasion. *Cancer Res.* **2011**, *71*, 4015–4027. [CrossRef]

79. Serocki, M.; Bartoszewska, S.; Janaszak-Jasiecka, A.; Ochocka, R.J.; Collawn, J.F.; Bartoszewski, R. miRNAs regulate the HIF switch during hypoxia: A novel therapeutic target. *Angiogenesis* **2018**, *21*, 183–202. [CrossRef]

80. Wang, H.; Flach, H.; Onizawa, M.; Wei, L.; McManus, M.T.; Weiss, A. Negative regulation of Hif1a expression and TH17 differentiation by the hypoxia-regulated microRNA miR-210. *Nat. Immunol.* **2014**, *15*, 393–401. [CrossRef]

81. Seok, J.K.; Lee, S.H.; Kim, M.J.; Lee, Y.M. MicroRNA-382 induced by HIF-1alpha is an angiogenic miR targeting the tumor suppressor phosphatase and tensin homolog. *Nucleic Acids Res.* **2014**, *42*, 8062–8072. [CrossRef] [PubMed]

82. Ge, X.; Liu, X.; Lin, F.; Li, P.; Liu, K.; Geng, R.; Dai, C.; Lin, Y.; Tang, W.; Wu, Z.; et al. MicroRNA-421 regulated by HIF-1alpha promotes metastasis, inhibits apoptosis, and induces cisplatin resistance by targeting E-cadherin and caspase-3 in gastric cancer. *Oncotarget* **2016**, *7*, 24466–24482. [CrossRef] [PubMed]

83. Nagpal, N.; Ahmad, H.M.; Chameettachal, S.; Sundar, D.; Ghosh, S.; Kulshreshtha, R. HIF-inducible miR-191 promotes migration in breast cancer through complex regulation of TGFbeta-signaling in hypoxic microenvironment. *Sci. Rep.* **2015**, *5*, 9650. [CrossRef] [PubMed]

84. Bhatt, K.; Wei, Q.; Pabla, N.; Dong, G.; Mi, Q.S.; Liang, M.; Mei, C.; Dong, Z. MicroRNA-687 Induced by hypoxia-inducible factor-1 targets phosphatase and tensin homolog in renal ischemia-reperfusion injury. *JASN* **2015**, *26*, 1588–1596. [CrossRef] [PubMed]

85. Li, F.; Sonveaux, P.; Rabbani, Z.N.; Liu, S.; Yan, B.; Huang, Q.; Vujaskovic, Z.; Dewhirst, M.W.; Li, C.Y. Regulation of HIF-1alpha stability through S-nitrosylation. *Mol. Cell* **2007**, *26*, 63–74. [CrossRef]

86. Lim, J.H.; Lee, Y.M.; Chun, Y.S.; Chen, J.; Kim, J.E.; Park, J.W. Sirtuin 1 modulates cellular responses to hypoxia by deacetylating hypoxia-inducible factor 1alpha. *Mol. Cell* **2010**, *38*, 864–878. [CrossRef]

87. Macfarlane, L.A.; Murphy, P.R. MicroRNA: Biogenesis, function and role in cancer. *Curr. Genom.* **2010**, *11*, 537–561. [CrossRef]

88. Abdi, J.; Rastgoo, N.; Li, L.; Chen, W.; Chang, H. Role of tumor suppressor p53 and micro-RNA interplay in multiple myeloma pathogenesis. *J. Hematol. Oncol.* **2017**, *10*, 169. [CrossRef]

89. Ungvari, Z.; Tucsek, Z.; Sosnowska, D.; Toth, P.; Gautam, T.; Podlutsky, A.; Csiszar, A.; Losonczy, G.; Valcarcel-Ares, M.N.; Sonntag, W.E.; et al. Aging-induced dysregulation of dicer1-dependent microRNA expression impairs angiogenic capacity of rat cerebromicrovascular endothelial cells. *J. Gerontol. Ser. A Biol. Sci. Med Sci.* **2013**, *68*, 877–891. [CrossRef]

90. Shilo, S.; Roy, S.; Khanna, S.; Sen, C.K. Evidence for the involvement of miRNA in redox regulated angiogenic response of human microvascular endothelial cells. *Arterioscler. Thromb. Vasc. Biol.* **2008**, *28*, 471–477. [CrossRef]

91. Iliopoulos, D.; Hirsch, H.A.; Struhl, K. An epigenetic switch involving NF-kappaB, Lin28, Let-7 MicroRNA, and IL6 links inflammation to cell transformation. *Cell* **2009**, *139*, 693–706. [CrossRef] [PubMed]

92. Wang, J.X.; Gao, J.; Ding, S.L.; Wang, K.; Jiao, J.Q.; Wang, Y.; Sun, T.; Zhou, L.Y.; Long, B.; Zhang, X.J.; et al. Oxidative modification of miR-184 enables it to target Bcl-xL and Bcl-w. *Mol. Cell* **2015**, *59*, 50–61. [CrossRef] [PubMed]

93. Panday, A.; Sahoo, M.K.; Osorio, D.; Batra, S. NADPH oxidases: An overview from structure to innate immunity-associated pathologies. *Cell. Mol. Immunol.* **2015**, *12*, 5–23. [CrossRef] [PubMed]

94. Li, S.Z.; Hu, Y.Y.; Zhao, J.; Zhao, Y.B.; Sun, J.D.; Yang, Y.F.; Ji, C.C.; Liu, Z.B.; Cao, W.D.; Qu, Y.; et al. MicroRNA-34a induces apoptosis in the human glioma cell line, A172, through enhanced ROS production and NOX2 expression. *Biochem. Biophys. Res. Commun.* **2014**, *444*, 6–12. [CrossRef]

95.   Liu, W.; Zabirnyk, O.; Wang, H.; Shiao, Y.H.; Nickerson, M.L.; Khalil, S.; Anderson, L.M.; Perantoni, A.O.; Phang, J.M. miR-23b targets proline oxidase, a novel tumor suppressor protein in renal cancer. *Oncogene* **2010**, *29*, 4914–4924. [CrossRef]

96.   Gordillo, G.M.; Biswas, A.; Khanna, S.; Pan, X.; Sinha, M.; Roy, S.; Sen, C.K. Dicer knockdown inhibits endothelial cell tumor growth via microRNA 21a-3p targeting of Nox-4. *J. Biol. Chem.* **2014**, *289*, 9027–9038. [CrossRef]

97.   Kim, J.H.; Park, S.G.; Song, S.Y.; Kim, J.K.; Sung, J.H. Reactive oxygen species-responsive miR-210 regulates proliferation and migration of adipose-derived stem cells via PTPN2. *Cell Death Dis.* **2013**, *4*, e588. [CrossRef]

98.   Chen, Z.; Li, Y.; Zhang, H.; Huang, P.; Luthra, R. Hypoxia-regulated microRNA-210 modulates mitochondrial function and decreases ISCU and COX10 expression. *Oncogene* **2010**, *29*, 4362–4368. [CrossRef]

99.   Venkataraman, S.; Alimova, I.; Fan, R.; Harris, P.; Foreman, N.; Vibhakar, R. MicroRNA 128a increases intracellular ROS level by targeting Bmi-1 and inhibits medulloblastoma cancer cell growth by promoting senescence. *PLoS ONE* **2010**, *5*, e10748. [CrossRef]

100. Sripada, L.; Singh, K.; Lipatova, A.V.; Singh, A.; Prajapati, P.; Tomar, D.; Bhatelia, K.; Roy, M.; Singh, R.; Godbole, M.M.; et al. hsa-miR-4485 regulates mitochondrial functions and inhibits the tumorigenicity of breast cancer cells. *J. Mol. Med.* **2017**, *95*, 641–651. [CrossRef]

101. Glorieux, C.; Calderon, P.B. Catalase, a remarkable enzyme: Targeting the oldest antioxidant enzyme to find a new cancer treatment approach. *Biol. Chem.* **2017**, *398*, 1095–1108. [CrossRef] [PubMed]

102. Zhang, X.; Ng, W.L.; Wang, P.; Tian, L.; Werner, E.; Wang, H.; Doetsch, P.; Wang, Y. MicroRNA-21 modulates the levels of reactive oxygen species by targeting SOD3 and TNFalpha. *Cancer Res.* **2012**, *72*, 4707–4713. [CrossRef] [PubMed]

103. Meng, X.; Wu, J.; Pan, C.; Wang, H.; Ying, X.; Zhou, Y.; Yu, H.; Zuo, Y.; Pan, Z.; Liu, R.Y.; et al. Genetic and epigenetic down-regulation of microRNA-212 promotes colorectal tumor metastasis via dysregulation of MnSOD. *Gastroenterology* **2013**, *145*, 426–436. [CrossRef] [PubMed]

104. Xu, X.; Wells, A.; Padilla, M.T.; Kato, K.; Kim, K.C.; Lin, Y. A signaling pathway consisting of miR-551b, catalase and MUC1 contributes to acquired apoptosis resistance and chemoresistance. *Carcinogenesis* **2014**, *35*, 2457–2466. [CrossRef] [PubMed]

105. Wang, Q.; Chen, W.; Bai, L.; Chen, W.; Padilla, M.T.; Lin, A.S.; Shi, S.; Wang, X.; Lin, Y. Receptor-interacting protein 1 increases chemoresistance by maintaining inhibitor of apoptosis protein levels and reducing reactive oxygen species through a microRNA-146a-mediated catalase pathway. *J. Biol. Chem.* **2014**, *289*, 5654–5663. [CrossRef] [PubMed]

106. Wang, P.; Zhu, C.F.; Ma, M.Z.; Chen, G.; Song, M.; Zeng, Z.L.; Lu, W.H.; Yang, J.; Wen, S.; Chiao, P.J.; et al. Micro-RNA-155 is induced by K-Ras oncogenic signal and promotes ROS stress in pancreatic cancer. *Oncotarget* **2015**, *6*, 21148–21158. [CrossRef] [PubMed]

107. Ma, Q. Role of nrf2 in oxidative stress and toxicity. *Annu. Rev. Pharmacol. Toxicol.* **2013**, *53*, 401–426. [CrossRef]

108. Cortez, M.A.; Valdecanas, D.; Zhang, X.; Zhan, Y.; Bhardwaj, V.; Calin, G.A.; Komaki, R.; Giri, D.K.; Quini, C.C.; Wolfe, T.; et al. Therapeutic delivery of miR-200c enhances radiosensitivity in lung cancer. *J. Am. Soc. Gene Ther.* **2014**, *22*, 1494–1503. [CrossRef]

109. Yang, M.; Yao, Y.; Eades, G.; Zhang, Y.; Zhou, Q. MiR-28 regulates Nrf2 expression through a Keap1-independent mechanism. *Breast Cancer Res. Treat.* **2011**, *129*, 983–991. [CrossRef]

110. Singh, B.; Ronghe, A.M.; Chatterjee, A.; Bhat, N.K.; Bhat, H.K. MicroRNA-93 regulates NRF2 expression and is associated with breast carcinogenesis. *Carcinogenesis* **2013**, *34*, 1165–1172. [CrossRef]

111. Papp, D.; Lenti, K.; Modos, D.; Fazekas, D.; Dul, Z.; Turei, D.; Foldvari-Nagy, L.; Nussinov, R.; Csermely, P.; Korcsmaros, T. The NRF2-related interactome and regulome contain multifunctional proteins and fine-tuned autoregulatory loops. *FEBS Lett.* **2012**, *586*, 1795–1802. [CrossRef] [PubMed]

112. Kabaria, S.; Choi, D.C.; Chaudhuri, A.D.; Jain, M.R.; Li, H.; Junn, E. MicroRNA-7 activates Nrf2 pathway by targeting Keap1 expression. *Free Radic. Biol. Med.* **2015**, *89*, 548–556. [CrossRef] [PubMed]

113. Eades, G.; Yang, M.; Yao, Y.; Zhang, Y.; Zhou, Q. miR-200a regulates Nrf2 activation by targeting Keap1 mRNA in breast cancer cells. *J. Biol. Chem.* **2011**, *286*, 40725–40733. [CrossRef] [PubMed]

114. Kim, J.H.; Lee, K.S.; Lee, D.K.; Kim, J.; Kwak, S.N.; Ha, K.S.; Choe, J.; Won, M.H.; Cho, B.R.; Jeoung, D.; et al. Hypoxia-responsive microRNA-101 promotes angiogenesis via heme oxygenase-1/vascular endothelial growth factor axis by targeting cullin 3. *Antioxid. Redox Signal.* **2014**, *21*, 2469–2482. [CrossRef] [PubMed]

115. Xu, D.; Zhu, H.; Wang, C.; Zhu, X.; Liu, G.; Chen, C.; Cui, Z. microRNA-455 targets cullin 3 to activate Nrf2 signaling and protect human osteoblasts from hydrogen peroxide. *Oncotarget* **2017**, *8*, 59225–59234. [CrossRef] [PubMed]

116. Li, L.; Huang, K.; You, Y.; Fu, X.; Hu, L.; Song, L.; Meng, Y. Hypoxia-induced miR-210 in epithelial ovarian cancer enhances cancer cell viability via promoting proliferation and inhibiting apoptosis. *Int. J. Oncol.* **2014**, *44*, 2111–2120. [CrossRef] [PubMed]

117. Tay, Y.; Rinn, J.; Pandolfi, P.P. The multilayered complexity of ceRNA crosstalk and competition. *Nature* **2014**, *505*, 344–352. [CrossRef]

118. Vinceti, M.; Filippini, T.; Del Giovane, C.; Dennert, G.; Zwahlen, M.; Brinkman, M.; Zeegers, M.P.; Horneber, M.; D'Amico, R.; Crespi, C.M. Selenium for preventing cancer. *Cochrane Database Syst. Rev.* **2018**, *1*, CD005195. [CrossRef]

119. Harvie, M. Nutritional supplements and cancer: Potential benefits and proven harms. In *American Society of Clinical Oncology Educational Book*; American Society of Clinical Oncology: Alexandria, VA, USA, 2014. [CrossRef]

120. Druesne-Pecollo, N.; Latino-Martel, P.; Norat, T.; Barrandon, E.; Bertrais, S.; Galan, P.; Hercberg, S. Beta-carotene supplementation and cancer risk: A systematic review and metaanalysis of randomized controlled trials. *Int. J. Cancer* **2010**, *127*, 172–184. [CrossRef]

121. Jacobs, C.; Hutton, B.; Ng, T.; Shorr, R.; Clemons, M. Is there a role for oral or intravenous ascorbate (vitamin C) in treating patients with cancer? A systematic review. *Oncologist* **2015**, *20*, 210–223. [CrossRef] [PubMed]

122. Sablina, A.A.; Budanov, A.V.; Ilyinskaya, G.V.; Agapova, L.S.; Kravchenko, J.E.; Chumakov, P.M. The antioxidant function of the p53 tumor suppressor. *Nature Med.* **2005**, *11*, 1306–1313. [CrossRef] [PubMed]

123. Schubert, R.; Erker, L.; Barlow, C.; Yakushiji, H.; Larson, D.; Russo, A.; Mitchell, J.B.; Wynshaw-Boris, A. Cancer chemoprevention by the antioxidant tempol in Atm-deficient mice. *Hum. Mol. Genet.* **2004**, *13*, 1793–1802. [CrossRef] [PubMed]

124. Sayin, V.I.; Ibrahim, M.X.; Larsson, E.; Nilsson, J.A.; Lindahl, P.; Bergo, M.O. Antioxidants accelerate lung cancer progression in mice. *Sci. Transl. Med.* **2014**, *6*, 221ra215. [CrossRef] [PubMed]

125. Harris, I.S.; Treloar, A.E.; Inoue, S.; Sasaki, M.; Gorrini, C.; Lee, K.C.; Yung, K.Y.; Brenner, D.; Knobbe-Thomsen, C.B.; Cox, M.A.; et al. Glutathione and thioredoxin antioxidant pathways synergize to drive cancer initiation and progression. *Cancer Cell* **2015**, *27*, 211–222. [CrossRef]

126. Wang, H.; Liu, X.; Long, M.; Huang, Y.; Zhang, L.; Zhang, R.; Zheng, Y.; Liao, X.; Wang, Y.; Liao, Q.; et al. NRF2 activation by antioxidant antidiabetic agents accelerates tumor metastasis. *Sci. Transl. Med.* **2016**, *8*, 334ra351. [CrossRef]

127. Qu, Y.; Wang, J.; Ray, P.S.; Guo, H.; Huang, J.; Shin-Sim, M.; Bukoye, B.A.; Liu, B.; Lee, A.V.; Lin, X.; et al. Thioredoxin-like 2 regulates human cancer cell growth and metastasis via redox homeostasis and NF-kappaB signaling. *J. Clin. Investig.* **2011**, *121*, 212–225. [CrossRef]

128. Kamarajugadda, S.; Cai, Q.; Chen, H.; Nayak, S.; Zhu, J.; He, M.; Jin, Y.; Zhang, Y.; Ai, L.; Martin, S.S.; et al. Manganese superoxide dismutase promotes anoikis resistance and tumor metastasis. *Cell Death Dis.* **2013**, *4*, e504. [CrossRef]

129. Glasauer, A.; Sena, L.A.; Diebold, L.P.; Mazar, A.P.; Chandel, N.S. Targeting SOD1 reduces experimental non-small-cell lung cancer. *J. Clin. Investig.* **2014**, *124*, 117–128. [CrossRef]

130. Nguyen, A.; Loo, J.M.; Mital, R.; Weinberg, E.M.; Man, F.Y.; Zeng, Z.; Paty, P.B.; Saltz, L.; Janjigian, Y.Y.; de Stanchina, E.; et al. PKLR promotes colorectal cancer liver colonization through induction of glutathione synthesis. *J. Clin. Investig.* **2016**, *126*, 681–694. [CrossRef]

131. DeNicola, G.M.; Karreth, F.A.; Humpton, T.J.; Gopinathan, A.; Wei, C.; Frese, K.; Mangal, D.; Yu, K.H.; Yeo, C.J.; Calhoun, E.S.; et al. Oncogene-induced Nrf2 transcription promotes ROS detoxification and tumorigenesis. *Nature* **2011**, *475*, 106–109. [CrossRef] [PubMed]

132. Kitamura, H.; Motohashi, H. NRF2 addiction in cancer cells. *Cancer Sci.* **2018**, *109*, 900–911. [CrossRef] [PubMed]

133. Le Gal, K.; Ibrahim, M.X.; Wiel, C.; Sayin, V.I.; Akula, M.K.; Karlsson, C.; Dalin, M.G.; Akyurek, L.M.; Lindahl, P.; Nilsson, J.; et al. Antioxidants can increase melanoma metastasis in mice. *Sci. Transl. Med.* **2015**, *7*, 308re308. [CrossRef] [PubMed]

134. Patra, K.C.; Hay, N. The pentose phosphate pathway and cancer. *Trends Biochem. Sci.* **2014**, *39*, 347–354. [CrossRef]

135. Fan, J.; Ye, J.; Kamphorst, J.J.; Shlomi, T.; Thompson, C.B.; Rabinowitz, J.D. Quantitative flux analysis reveals folate-dependent NADPH production. *Nature* **2014**, *510*, 298–302. [CrossRef]

136. Ren, J.G.; Seth, P.; Clish, C.B.; Lorkiewicz, P.K.; Higashi, R.M.; Lane, A.N.; Fan, T.W.; Sukhatme, V.P. Knockdown of malic enzyme 2 suppresses lung tumor growth, induces differentiation and impacts PI3K/AKT signaling. *Sci. Rep.* **2014**, *4*, 5414. [CrossRef]

137. Mollaei, H.; Safaralizadeh, R.; Rostami, Z. MicroRNA replacement therapy in cancer. *J. Cell. Physiol.* **2019**, *234*, 12369–12384. [CrossRef]

138. Bader, A.G.; Brown, D.; Winkler, M. The promise of microRNA replacement therapy. *Cancer Res.* **2010**, *70*, 7027–7030. [CrossRef]

139. Ling, H.; Fabbri, M.; Calin, G.A. MicroRNAs and other non-coding RNAs as targets for anticancer drug development. *Nature reviews. Drug Discov.* **2013**, *12*, 847–865. [CrossRef]

140. Hanna, J.; Hossain, G.S.; Kocerha, J. The Potential for microRNA Therapeutics and Clinical Research. *Front. Genet.* **2019**, *10*, 478. [CrossRef]

miR526b and miR655 Induce Oxidative Stress in Breast Cancer

Bonita Shin [†], Riley Feser [†], Braydon Nault [‡], Stephanie Hunter [‡], Sujit Maiti,
Kingsley Chukwunonso Ugwuagbo and Mousumi Majumder *

Department of Biology, Brandon University, 3rd Floor, John R. Brodie Science Centre, 270—18th Street, Brandon, MB R7A6A9, Canada
* Correspondence: majumderm@brandonu.ca
† These authors contributed equally to this work.
‡ These authors contributed equally to this work.

Abstract: In eukaryotes, overproduction of reactive oxygen species (ROS) causes oxidative stress, which contributes to chronic inflammation and cancer. MicroRNAs (miRNAs) are small, endogenously produced RNAs that play a major role in cancer progression. We established that overexpression of miR526b/miR655 promotes aggressive breast cancer phenotypes. Here, we investigated the roles of miR526b/miR655 in oxidative stress in breast cancer using in vitro and in silico assays. miRNA-overexpression in MCF7 cells directly enhances ROS and superoxide (SO) production, detected with fluorescence assays. We found that cell-free conditioned media contain extracellular miR526b/miR655 and treatment with these miRNA-conditioned media causes overproduction of ROS/SO in MCF7 and primary cells (HUVECs). Thioredoxin Reductase 1 (TXNRD1) is an oxidoreductase that maintains ROS/SO concentration. Overexpression of *TXNRD1* is associated with breast cancer progression. We observed that miR526b/miR655 overexpression upregulates *TXNRD1* expression in MCF7 cells, and treatment with miRNA-conditioned media upregulates *TXNRD1* in both MCF7 and HUVECs. Bioinformatic analysis identifies two negative regulators of TXNRD1, TCF21 and PBRM1, as direct targets of miR526b/miR655. We validated that *TCF21* and *PBRM1* were significantly downregulated with miRNA upregulation, establishing a link between miR526b/miR655 and TXNRD1. Finally, treatments with oxidative stress inducers such as H_2O_2 or miRNA-conditioned media showed an upregulation of miR526b/miR655 expression in MCF7 cells, indicating that oxidative stress also induces miRNA overexpression. This study establishes the dynamic functions of miR526b/miR655 in oxidative stress induction in breast cancer.

Keywords: MicroRNA (miRNA); miR526b; miR655; oxidative stress; reactive oxygen species (ROS); superoxide (SO); Thioredoxin Reductase 1 (TXNRD1); breast cancer

1. Introduction

Breast cancer is the most common cancer affecting women and is responsible for the highest number of cancer-related deaths among women worldwide [1]. Breast cancer progression follows a complex multistep process, which depends on multiple exogenous and endogenous factors. The production of reactive oxygen species (ROS) such as superoxide (SO) leads to the induction of oxidative stress, which has been largely associated with breast cancer [2]. Oxidative stress is the result of cellular inability to neutralize and eliminate excess ROS, which is frequently associated with cancer development and progression. Under normal physiological conditions, cells endogenously produce ROS such as H_2O_2, $ONOO^-$, OH^-, $HClO^-$, NO^-, ROO^-, and SO, during metabolism, respiration, and biosynthesis of macromolecules. Thus, cell metabolites are great resources for understanding oxidative stress. Excessive ROS production can induce inflammation, regulate the cell cycle, and stimulate intracellular transduction pathways, which leads to the promotion of cancer [3]. Specifically, SO production is the consequence of oxygen (O_2) acting as the final electron acceptor in the electron transport chain, and has been shown to regulate signaling cascades that lead to cell survival and proliferation [4]. Within the cell, there is a homeostatic balance of various protective molecules and ROS. However, in cancer, tumor cells demonstrate deviations in oxidative metabolism and signaling pathways as a result of the constitutive activation of growth signaling pathways, leading to increased levels of ROS and induction of oxidative stress [5].

A high concentration of ROS is a signature feature of the tumor microenvironment. Cells have a natural defense mechanism to reduce damage caused by oxidative stress. Antioxidants, which are stable molecules that donate electrons to neutralize free radicals, belong to this natural defense mechanism of the cell [6]. Cellular detoxification pathways are regulated by enzymes that eliminate ROS, which include SO dismutase, catalase, glutathione peroxidase, cysteine, and thioredoxin (TXN). Specifically, TXN is a ubiquitous antioxidant protein that is responsible for the regulation of dithiol/disulfide balance [7,8]. TXN is active when it is in its reduced form. When active, it will participate in a reaction catalyzed by peroxiredoxin to neutralize H_2O_2 and peroxynitrate, both of which are products of oxidative stress activity [9]. TXNRD1 is responsible for the conversion of TXN into its active state (Figure 1). Malfunctions in antioxidant pathways can lead to increased oxidative stress and consequential damage to the cells. High expression of *TXNRD1* is associated with increased oxidative stress and correlates with poor prognosis in breast cancer [10]. In cancers, excessive production of ROS can cause mutations in the DNA, overexpression of tumor-promoting microRNAs (miRNAs, miRs), release of inflammatory molecules, and inactivation of oxidoreductive enzymes; making antioxidant pathways dysfunctional. Overexpression of oncogenic miRNAs leads to the regulation and promotion of tumor growth; however, the regulation of oxidative stress in cancer by miRNAs remains unclear.

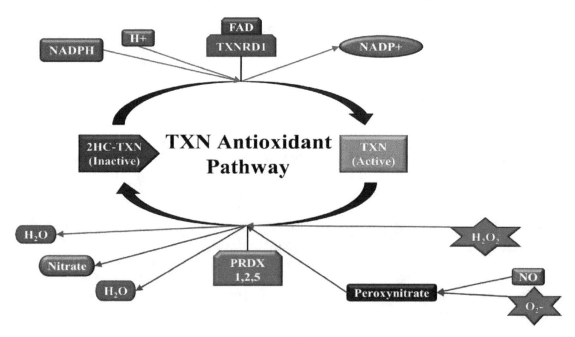

Figure 1. Thioredoxin (TXN) is a main constituent in an antioxidant pathway that neutralizes Hydrogen Peroxide (H_2O_2) and superoxide (O_2-), to prevent oxidative damage. TXN exists in active (reduced) and inactive (oxidized) states. Thioredoxin Reductase 1 (TXNRD1) is responsible for reducing 2HC-TXN (TXN with attached double hydrocarbon) into its active form. Therefore, in the presence of more ROS, an increased expression of *TXNRD1* occurs to protect the cells from oxidative damage.

miRNAs are small, endogenously produced RNAs which regulate gene expression at the post-transcriptional level [11]. Release of circulating miRNAs in the tumor microenvironment can regulate tumor growth and metastasis. Previously, miR526b and miR655 have been established as oncogenic and tumor-promoting miRNAs in human breast cancer [12–14]. The roles of miR526b and miR655 have been implicated in many hallmarks of cancer, including: Driving primary tumor growth, induction of stem-like cell (SLC) phenotypes, epithelial-to-mesenchymal transition (EMT), invasion and migration, distant metastasis. We have shown that cell metabolites and cell-free conditioned media of these two miRNA-high cells induce tumor-associated angiogenesis and lymphangiogenesis in breast cancer [15]. It has also been shown that cellular stress and ROS production can also induce oncogenic miRNA expression in tumors, and it is well-established that both ROS and miRNA expression signatures are associated with tumor development, progression, metastasis, and therapeutic response [16]. Thus, we wanted to investigate the relationship between ROS and miR526b/miR655 in breast cancer.

In this study, we investigate the roles of oncogenic miR526b and miR655 in oxidative stress in breast cancer. First, we show that both miR526b/miR655 directly and indirectly regulate oxidative stress. Next, we use the expression of *TXNRD1* as a molecular marker of oxidative stress to further validate the link between miRNA and ROS production. Moreover, we identify a positive feedback loop between oxidative stress and miRNA expression in breast cancer, showing that while the upregulation of miR526b and miR655 led to the induction of ROS production, the induction of oxidative stress also further upregulated miR526b and miR655 expression in breast tumor cells. Hence, we establish the dynamic roles of miR526b and miR655 in oxidative stress in breast cancer.

2. Results

To test the effects of miR526b and miR655 in oxidative stress in breast cancer, we used an estrogen receptor (ER)-positive, poorly metastatic breast cancer cell line, MCF7, and highly aggressive, miR526b/miR655-overexpressing MCF7-miR526b and MCF7-miR655 cell lines. We also used a primary endothelial cell line, human umbilical vein endothelial cells (HUVEC), to test the indirect or paracrine effects of miR526b and miR655 on oxidative stress induction. Finally, we used a breast epithelial cell line MCF10A and breast cancer cell lines T47D, MCF7, SKBR3, MCF7-COX2, Hs578T, and MDA-MB-231 to measure *TXNRD1* expression.

2.1. miR526b and miR655 Directly Induce Oxidative Stress by Overproduction of ROS and SO

2.1.1. Fluorescence Microplate Assay

Previously, studies have used a total ROS detection kit for the measurement of ROS and SO in triple negative breast cancer cell lines, colon cancer cells, colorectal cancer cell lines, and in hepatocellular carcinoma cells [17–21]. We used the same ROS-ID Total ROS/SO detection kit (Enzo Life Sciences, Farmingdale, NY, USA) to measure fluorescence due to ROS/SO production following manufacturer's protocol. Microplate readings were carried out at 1 and 21 h following Pyocyanin (ROS inducer) treatment and addition of non-fluorescent, cell-permeable ROS detection dyes. We monitored cellular morphology at various time points from 1–24 h after the addition of the ROS inducer in MCF7 cells (data not shown). With minimum dosage of ROS inducer, we observed oxidative stress in the cell within an hour, and after 21 h a decrease in cell viability was recorded due to the toxicity of the ROS inducer. Therefore, fluorescence was measured at two different timepoints; at 1 and 21 h. Fluorescence emissions were captured using two different filters to detect green (Fluorescein) and red (Rhodamine) emissions. ROS/SO production was calculated by subtracting the negative control emissions (basal emissions) from the test group emissions (with treatment) (Figure S1A). Overall, we found that ROS and SO production was greater in miRNA-high cells compared to MCF7 cells. Specifically, ROS production was found to be statistically significant at 21 h for MCF7-miR655 (Figure 2A). Similarly, SO production was found to be significantly greater in the MCF7-miR655 cell line compared to MCF7 at both 1 h and 21 h. ROS and SO production was not statistically significant for MCF7-miR526b compared to MCF7 (Figure 2A,B).

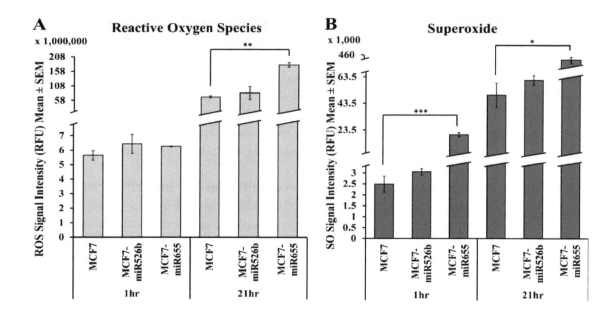

Figure 2. Fluorescence microplate assays to quantify ROS (Green) and SO (Red) production by MCF7, MCF7-miR526b, and MCF7-miR655 cell lines. (**A**) Quantitative data represents the ROS signal intensity in MCF7, MCF7-miR526b, and MCF7-miR655 cell lines at 1 and 21 h. (**B**) Quantitative data represents SO signal intensity in MCF7, MCF7-miR526b, and MCF7-miR655 cell lines at the 1 and 21 h. Data presented as the mean ± SEM of triplicate replicates; * $p < 0.05$, ** $p < 0.01$, *** $p < 0.001$.

2.1.2. Fluorescence Microscopy Assay

Fluorescence microscopy assays were conducted to measure the difference in cellular fluorescence expression with individual fluorescent cell quantification, determining the fraction of cells producing ROS and SO. Using the green (Fluorescein) and red (Rhodamine) fluorescence filter sets, photos of the fluorescent cells were captured with an inverted fluorescence microscope 1 h after the detection dyes were added. We also captured bright field images of cells without using fluorescence filters to quantify total number of viable cells (Figure S2M,R,W). Results show that wells containing MCF7-miR526b (Figure 3D,E) or MCF7-miR655 cell lines (Figure 3G,H) had more fluorescing cells than MCF7 (Figure 3A,B) under both red and green filters. Similarly, quantifications show significantly higher green (Figure 3J) and red (Figure 3K) cells in both MCF7-miR526b and MCF7-MCF7-miR655-high cells compared to MCF7 cells. Furthermore, we measured the ratio of cells positive for both ROS and SO production using merged channels and found that miRNA-high cell lines (Figure 3F,I) had a significantly higher ratio of fluorescing cells under both filters compared to MCF7 cells (Figure 3C,L).

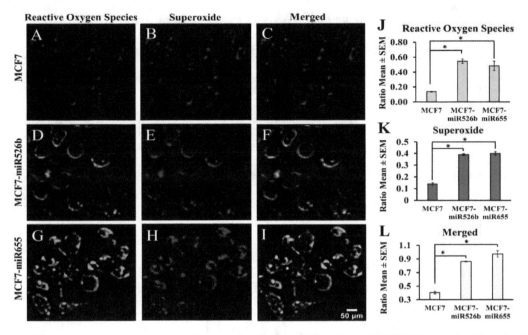

Figure 3. Fluorescence microscopy images and quantification of ROS/SO production in MCF7, MCF7-miR526b, and MCF7-miR655 cell lines. (**A–C**) Images of fluorescent MCF7 cells with green, red, or merged filters. (**D–F**) Images of fluorescent MCF7-miR526b cells with green, red, or merged filters. (**G–I**) Images of fluorescent MCF7-miR655 cells with green, red, or merged filters. Scale bar: 50 μm. (**J**) Quantification of ratios of cells positive for ROS detection. (**K**) Quantification ratios of cells positive for SO detection. (**L**) Quantification ratios of cells showing both ROS and SO production. Quantitative data presented as the mean ± SEM of triplicate replicates. Quantifications presented in ratios of fluorescence-positive cells to the total number of cells; * $p < 0.05$.

2.2. Cell-Free Conditioned Media from miR526b/miR655-High Cells Indirectly Induce Production of ROS and SO

The tumor microenvironment is very heterogeneous, containing tumor cells, endothelial cells, macrophages, miRNAs, cell metabolites, inflammatory molecules, growth factors, and also ROS. In the following assays, we first tested the paracrine effect of miRNA in oxidative stress. To test the paracrine effect of miRNA, we used the cell-free conditioned media from miR526b/miR655-high cells as an ROS inducer using MCF7 (tumor model) and HUVEC (primary endothelial model) cell lines. Next, we quantified pri-miR526b and pri-miR655 in the conditioned media to investigate if the indirect induction of oxidative stress in breast cancer is due to the presence of miR526b and miR655 in the cell secretions, and to justify our use of conditioned media as an ROS inducer.

2.2.1. Fluorescence Microplate Assay with MCF7 Cells

MCF7 cells were grown and then treated with basal media or cell-free conditioned media (containing cell metabolites and secretory proteins) collected from MCF7-miR526b and MCF7-miR655 cells for 24 h. Then we added the ROS inducer as described earlier and fluorescence data were collected at 1 and 21 h. These two time points were selected to remain consistent with our previous experiments that used the ROS inducer. MCF7 cells treated with miRNA-conditioned media showed significantly higher ROS production than the basal media treated MCF7 control group at both 1 and 21 h. Specifically, the change is extremely significant for MCF7 cells treated with MCF7-miR526b conditioned media at 1 h, and with MCF7-miR655 conditioned media at 21 h (Figure 4A). Similarly, MCF7 cells treated with MCF7-miR655 cell-free conditioned media had significantly higher SO production than the basal media treated cells at 1 h (Figure 4B). At 21 h, MCF7 cells treated with MCF7-miR526b cell-free conditioned media had a significantly higher SO production than MCF7 cells treated with basal media. While MCF7 cells treated with MCF7-miR655 cell-free conditioned media did show slightly higher SO production than MCF7 treated with basal media, this was not statistically significant (Figure 4B).

MCF7 Cells in Various Conditioned Media

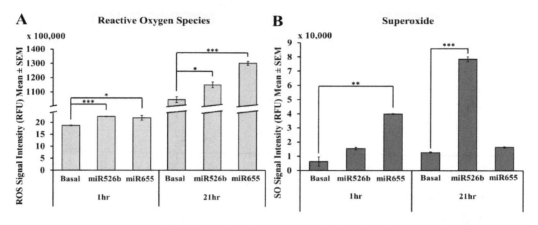

HUVECs in Various Conditioned Media

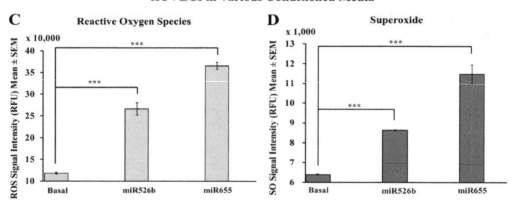

Figure 4. Fluorescence microplate assay with MCF7 and HUVEC cells cultured in miRNA conditioned media. (**A**) MCF7 cells treated with MCF7-miR526b or MCF7-miR655 conditioned media show an overproduction of ROS as compared to basal media treated cells at both 1 and 21 h. (**B**) MCF7 cells treated with MCF7-miR655 conditioned media show a significant overproduction of SO at 1 h, and MCF7 cells treated with MCF7-miR526b conditioned media show a significant overproduction of SO at 21 h compared to MCF7 cells treated with basal media. (**C**) HUVECs treated with MCF7-miR526b or MCF7-miR655 conditioned media show overproduction of ROS compared to HUVECs treated with basal media after 30 min. (**D**) HUVECs treated with MCF7-miR526b or MCF7-miR655 conditioned media show a significant overproduction of SO as compared to non-treated MCF7 cells after 30 min. Data presented as the mean ± SEM of triplicate replicates; * $p < 0.05$, ** $p < 0.01$, *** $p < 0.001$.

2.2.2. Fluorescence Microplate Assay with HUVECs

Previously, we have shown that cell-free conditioned media from miRNA-high cells induce angiogenic potential in HUVECs [15]. Here, we tested if cell-free conditioned media containing all secretory proteins and metabolites from miR526b/miR655-high cells can induce oxidative stress in HUVECs. HUVECs treated with MCF7-miR526b or MCF7-miR655 conditioned media for 12–18 h had significantly higher ROS/SO production compared to HUVECs treated with basal media (Figure 4C,D). It should be noted that HUVECs are very sensitive to changes in growth conditions and treatments, as they can only survive for 12–18 h without native growth condition. Thus, HUVECs were treated with conditioned media from miRNA-high cells for 12 h. We found that HUVECs were extremely stressed, observing cell death after an hour following the addition of the ROS inducer (Figure S4). Therefore, the microplate assay was done only 30 min after ROS inducer was added.

2.2.3. Fluorescence Microscopy Assay with MCF7 Cells in miRNA- Conditioned Media

In this experiment, cell-free conditioned media was used as an inducer of oxidative stress. MCF7 cells were grown and treated with basal media or cell-free conditioned media from MCF7-miR526b or MCF7-miR655 cells for 12–18 h. No other ROS inducer was added, only cell-permeable dyes from the ROS detection kit were added to detect cell-free conditioned media-induced oxidative stress. Images were captured after 1 h, and the number of fluorescent cells were measured with ImageJ as mentioned above. Results show that MCF7 cells treated with MCF7-miR526b (Figure 5D–F) or MCF7-miR655 conditioned media (Figure 5G–I) had more fluorescing cells than basal media treated MCF7 cells (Figure 5A–C) for both Fluorescein and Rhodamine filters. Quantification of MCF7 cells treated with MCF7-miR526b or MCF7-miR655 conditioned media show a significant increase in ROS production (Figure 5J) and SO production (Figure 5K). The ratio of cells positive for both ROS and SO production was also significantly higher in cells treated with MCF7-miR526b or MCF7-miR655 conditioned media than those treated with basal media (Figure 5L).

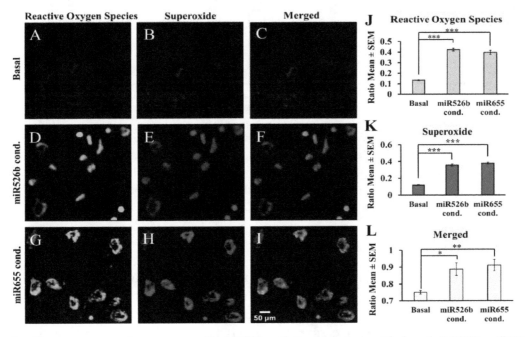

Figure 5. Fluorescence microscopy with MCF7 cell line treated with basal, MCF7-miR526b, or MCF7-miR655 cell-free conditioned media to quantify ROS/SO producing cells. MCF7 treated with basal media under the Rhodamine filter was used as a threshold to quantify ROS positive cells. (**A–C**) Images of MCF7 cells treated with basal media in green, red, or merged filters. (**D–F**) Images of MCF7 cells treated with cell-free conditioned media from MCF7-miR526b cells in green, red, or merged filters. (**G–I**) Images of MCF7 cells treated with cell-free conditioned media from MCF7-miR655 cells in green, red, or merged filters. Scale bar: 50 µm. (**J**) Quantification of cells positive for ROS detection presented as ratios. (**K**) Quantification of cells positive for SO detection presented as ratios. (**L**) Ratio of cells showing both ROS and SO production. Quantitative data presented as the mean ± SEM of quadruplicate replicates; * $p < 0.05$, ** $p < 0.01$, *** $p < 0.001$.

2.2.4. miRNA-High Cells Release miR526b and miR655 in Cell-Free Conditioned Media

To test if the indirect induction of oxidative stress with conditioned media is due to the presence of miRNA itself, we measured pri-miR526b and pri-miR655 expression in MCF7, MCF7-miR526, and MCF7-miR655 cell-free conditioned media. We found that both pri-miRNAs' expressions were significantly higher in MCF7-miR526b conditioned media compared to MCF7 conditioned media (Figure 6). The expression of pri-miR526b was significantly higher and the expression of pri-miR655 was marginally higher in MCF7-miR655 conditioned media. It should be noted that in the MCF7-miR526b conditioned media, the overall expression of pri-miR526b was higher than pri-miR655, while in

MCF7-miR655 conditioned media, the overall expression of pri-miR655 was higher than pri-miR526b (Figure 6). This result confirms that due to the release of miRNA in the conditioned media of serum starved cells, extracellular miR526b and miR655 act as an ROS inducer, therefore indirectly inducing oxidative stress in nearby cells.

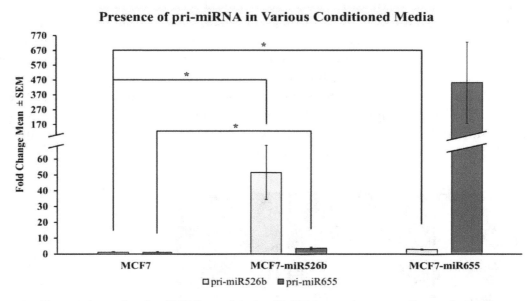

Figure 6. Expression of pri-miR526b and pri-miR655 in various conditioned media measured using qRT-PCR. MCF7-miR526b conditioned media show a significantly higher expression of both pri-miRNAs with prominent change in pri-miR526b expression compared to MCF7 conditioned media. MCF7-miR655 conditioned media show a significantly higher expression of miR526b, and very high expression of pri-miR655, which was not significant. Data is presented as the mean ± SEM of duplicate replicates; * $p < 0.05$.

2.3. TXNRD1 is a Marker for Oxidative Stress

TXN is an antioxidant protein that is responsible for neutralizing ROS within the cell [8]. TXNRD1 is the enzyme responsible for reducing TXN into its active form. Previous analyses of *TXNRD1* expression have shown that *TXNRD1* is upregulated in pancreatic, colon, lung, prostate, and breast cancers, and is associated with poor cancer prognosis [10]. To further investigate the direct and indirect roles of miR526b and miR655 in the induction of oxidative stress, *TXNRD1* was validated as a marker of oxidative stress using various breast cancer cell lines and its expression was measured in miR526b/miR655-high cell lines. Furthermore, bioinformatic analysis was done to investigate the regulation of *TXNRD1* by miR526b and miR655, which showed that miR526b and miR655 target two transcription factors that regulate *TXNRD1* expression. The expression of these transcription factors was then measured in miR526b/miR655-high cell lines. Moreover, with the success of using miRNA-conditioned media as an ROS inducer in our previous assays, we tested to see if cell-free conditioned media from miR526b/miR655-high cells regulate *TXNRD1* expression in both tumor and endothelial cells.

2.3.1. Highly Metastatic Breast Cancer Cell Lines Show Upregulation of TXNRD1

MCF10A, T47D, MCF7, SKBR3, MCF7-COX2, Hs578T, and MDA-MB-231 cell lines were used to quantify the expression of *TXNRD1* using qRT-PCR. Since MCF10A is a breast epithelial cell line, gene expression changes for all breast cancer cell lines were measured and compared to MCF10A. Results show that *TXNRD1* was significantly downregulated in the poorly metastatic MCF7 and SKBR3 cell lines, while the T47D cell line showed no change in expression (Figure 7A). *TXNRD1* was significantly upregulated in all highly metastatic cell lines, MCF7-COX2, Hs578T, and MDA-MB-231; with maximum upregulation seen in MDA-MB-231 (Figure 7A). We have previously found that these aggressive breast cancer cell lines (MCF7-COX2, Hs578T, and MDA-MB-231) show overexpression of both miR526b and miR655; while poorly metastatic cells (MCF7, T47D) show low expression of both miRNAs [12,13]. These observations validate the use of *TXNRD1* as a marker of oxidative stress in breast cancer, and show a link between *TXNRD1*, miR526b, and miR655 expression.

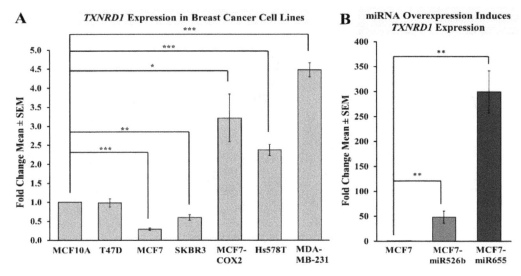

Figure 7. Expression of oxidative stress marker *TXNRD1* in various cell lines measured using qRT-PCR. (**A**) Breast cancer cell lines with various degrees of metastatic potential show a difference in *TXNRD1* expression. The more metastatic cell lines including MCF7-COX2, Hs578T, and MDA-MB-231 show the greatest fold change of *TXNRD1* expression and MCF7 cells showing lowest *TXNRD1* expression compared to the breast epithelial MCF10A cell line. (**B**) Expression of *TXNRD1* is quantified in MCF7 cells, MCF7-miR526b, and MCF7-miR655 cell lines, showing how these oncogenic miRNAs impact the expression of this oxidative stress marker. Large fold change increases are seen in both miRNA cell lines. Data is presented as the mean ± SEM of triplicate replicates; * $p < 0.05$, ** $p < 0.01$, *** $p < 0.001$.

2.3.2. miRNA Overexpression Directly Upregulates TXNRD1 Expression

To establish the direct role of miRNA in oxidative stress, total RNA extraction followed by qRT-PCR was carried out with MCF7, MCF7-miR526b, and MCF7-miR655 cell lines to quantify the expression of *TXNRD1*. Results show that *TXNRD1* was significantly upregulated in both MCF7-miR526b and MCF7-miR655 cell lines compared to MCF7, with greater fold change in *TXNRD1* expression measured in the MCF7-miR655 cell line (Figure 7B).

2.3.3. Bioinformatic Analysis to Identify a Link between miRNAs and TXNRD1

Since we observed that miRNA overexpression results in the upregulation of *TXNRD1* in breast cancer, we further wanted to investigate this mechanism in silico. Thus, we conducted bioinformatic analysis to investigate how miR526b and miR655 regulate *TXNRD1* expression. Both miRNA target gene lists were extracted from the miRBase database, using TargetScan analysis tool which can predict miRNA target genes in mammalian mRNA pool [22–26]. By virtue, miRNAs bind to target genes, degrading the corresponding mRNA at the post-transcriptional level, and thus block the protein expression of the target. We found that *TXNRD1* is not a direct target of miR526b and miR655, so we instead attempted to identify transcription factors (TFs) that regulate *TXNRD1* and are also targets of miR526b and miR655. In miR526b/miR655 overexpressing cells, we observed that *TXNRD1* expression is high, which indicates that these miRNAs might be targeting negative regulators of *TXNRD1*. To identify these TFs, we used Enrichr, a tool that consists of both a validated user-submitted gene list and a search engine for further analysis [27]. By comparing miRNA target genes and *TXNRD1* regulatory TFs, we identified eight TFs as direct targets of miR526b (blue down arrows in the yellow circle) and eleven TFs as direct targets of miR655 (blue down arrows in the pink circle) (Figure 8A). Finally, we identified two TFs (PBRM1 and TCF21) as common targets of both miRNAs (blue down arrows in the green circle), which negatively regulate *TXNRD1* (Figure 8A). Both *PBRM1* and *TCF21* have been shown to have tumor suppressor-like functions in breast cancer [28,29]. Therefore, we hypothesize that when miR526b and miR655 are upregulated, their targets *PBRM1* and *TCF21* are downregulated, leading to the upregulation of *TXNRD1*. This result justifies the abundance of *TXNRD1* in miRNA-high cells.

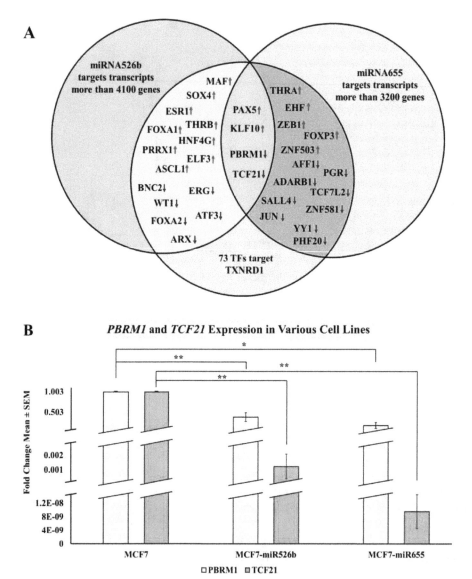

Figure 8. (A) Overlap of TFs regulating *TXNRD1*, and miR526b, miR655 target genes. The purple area represents the list of 4133 miR526b target genes, the brown area represents the list of 3264 miR655 target genes, and the blue area represents all 155 TFs regulating *TXNRD1*. The yellow area shows the miR526b target genes which are also TFs regulating *TXNRD1*. The pink area indicates miR655 targets which are also TFs regulating *TXNRD1*. The green center represents the overlap of all three criteria, which shows the four TFs of *TXNRD1* which are common targets of both miRNAs. The red up arrow symbolizes that the TF upregulates *TXNRD1* expression and the blue down arrow signifies that the TFs downregulates *TXNRD1* expression. Because we observed that *TXNRD1* is upregulated in miRNA-high cells, we considered both miRNAs targeting the two TFs, *PBRM1*, and *TCF21*, which are the negative regulators of *TXNRD1*. (B) *PBRM1* and *TCF21* expression in MCF7, MCF7-miR526b, and MCF7-miR655 cell lines. miR526b/miR655-high cell lines show significantly lower expression of both *PBRM1* and *TCF21*. This indicates both miRNAs target the negative regulator of *TXNRD1*. Data presented as the mean of quadruplicate replicates; $* p < 0.01$, $** p < 0.001$.

2.3.4. miRNA Overexpression Indirectly Upregulates TXNRD1 by Targeting Negative Regulator of the Gene

The expression of *PBRM1* and *TCF21* was measured in MCF7, MCF7-miR526b, and MCF7-miR655 cell lines to further confirm that miR526b and miR655 target these TFs to regulate the expression of *TXNRD1*. Results show that both *PBRM1* and *TCF21* are significantly downregulated in both miR526b/miR655-high cell lines as compared to MCF7 cells, validating the in silico analysis (Figure 8B).

2.3.5. MCF7 Cells Treated with miR526b and miR655-High Cell-Free Conditioned Media Show Upregulation of TXNRD1

MCF7 cells were treated with basal media or miR526b/miR655-high cell-free conditioned media for 12–18 h as mentioned before. RNA extraction and gene expression assays were carried out to quantify the expression of *TXNRD1*. It was found that *TXNRD1* expression in MCF7-miR655 conditioned media-treated cells was significantly higher compared to the basal media treated MCF7 cells. MCF7 cells treated with MCF7-miR526b conditioned media had marginally higher, but statistically non-significant *TXNRD1* expression compared to basal media treated MCF7 cells (Figure 9A). These results piqued our interest in the paracrine effect of miRNA-overexpressing cells. In the tumor microenvironment, oxidative stress in neighboring normal, immune, and endothelial cells would also be increased. Thus, we wanted to investigate this principle in non-cancerous cells, using a primary endothelial cell line (HUVECs).

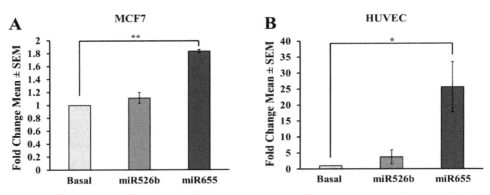

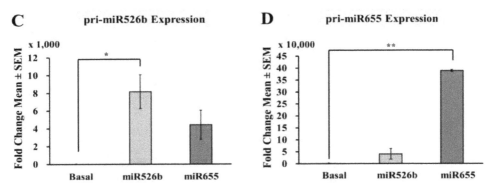

Figure 9. Indirect effects of miRNA overexpression on *TXNRD1*, pri-miR526b, and pri-miR655 expression. (**A**) *TXNRD1* expression in MCF7 cells treated with MCF7-miR526b or MCF7-miR655 conditioned media compared to non-treated MCF7 cells. (**B**) *TXNRD1* expression in HUVEC cells treated with MCF7-miR526b or MCF7-miR655 conditioned media compared to non-treated MCF7 cells. (**C**) pri-miR526b expression in MCF7 cells treated with MCF7-miR526b or MCF7-miR655 conditioned media compared to non-treated MCF7 cells. (**D**) pri-miR655 expression in MCF7 cells treated with MCF7-miR526b or MCF7-miR655 conditioned media compared to non-treated MCF7 cells. Data presented as the mean ± SEM of triplicate replicates; * $p < 0.05$, ** $p < 0.001$.

2.3.6. HUVECs Treated with Cell-Free miR526b and miR655 Conditioned Media Show Upregulation of TXNRD1

HUVECs were treated with basal media, MCF7-miR526b conditioned media, or MCF7-miR655 conditioned media for 12 h. Using qRT-PCR, quantification for the expression of *TXNRD1* in treated and non-treated HUVECs was performed. Results show that HUVECs treated with MCF7-miR526b conditioned media containing secretory proteins and metabolites show a marginal upregulation of *TXNRD1* compared to HUVECs treated with basal media, but was not statistically significant (Figure 9B). However, HUVECs treated with MCF7-miR655 conditioned media show a significant overexpression of *TXNRD1* when compared to HUVECs treated with basal media (Figure 9B).

2.4. Cell-Free miRNA Conditioned Media Indirectly Induces miRNA Overexpression in MCF7 Cells

Since we have shown that cell-free conditioned media from miR526b/miR655-high cell lines induces ROS production and *TXNRD1* expression in MCF7 cells, we wanted to test if cell-metabolites and secretory proteins could also induce oncogenic miRNA upregulation in poorly metastatic MCF7 cells. MCF7 cells were treated with serum-free basal media, MCF7-miR526b conditioned media, or MCF7-miR655 conditioned media for 21 h. RNA was extracted and reverse transcribed into cDNA to quantify pri-miR526b and pri-miR655 expressions. After relative gene expression analysis, the results showed that all miRNA conditioned media-treated MCF7 cells had a significant increase in expression of pri-miR526b and pri-miR655 compared to basal control MCF7 cells (Figure 9C,D). These results establish the dynamic roles of miR526b/miR655 and ROS in the tumor microenvironment where a complex interplay between the tumor cell, tumor cell secretions, and endothelial cells is ongoing, thus promoting tumor growth.

2.5. Induction of Oxidative Stress Upregulates miR526b and miR655 Expression in MCF7 Cells

Finally, we wanted to investigate if miR526b and miR655 are key responders to oxidative stress. Thus, we induced oxidative stress in MCF7 cells using a chemical inducer, H_2O_2, and measured its effects on miR526b and miR655 expression. MCF7 cells were grown until 80% confluent, and then treated with either 25 μM or 50 μM of H_2O_2 for 24 h. Following treatment, RNA was extracted and reverse transcribed into cDNA. qRT-PCR was then carried out to quantify the expression of pri-miR526b and pri-miR655 in the H_2O_2-treated and non-treated MCF7 cells. Results show a significant dose-dependent increase in the expression of pri-miR655 in H_2O_2-treated MCF7 cells at both 25 μM and 50 μM concentrations (Figure 10). Expression of pri-miR526b following treatment with 50 μM of H_2O_2 showed marginal upregulation; however, this was not statistically significant (Figure 10). These results support the notion that these two miRNAs are immediate responders to oxidative stress in breast cancer.

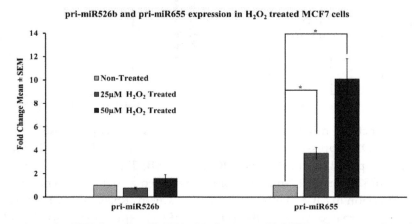

Figure 10. Expression of miR526b and miR655 in H_2O_2 treated MCF7 cells. pri-miR526b and pri-miR655 expression quantified in MCF7 cells, MCF7 cells treated with 25 μM H_2O_2, or 50 μM H_2O_2 using qRT-PCR. Data presented as the mean ± SEM of triplicate replicates; * $p < 0.05$.

3. Discussion

Previously, we have established the roles of oncogenic miR526b and miR655 in breast cancer disease progression, angiogenesis, cancer stem cell regulation, and metastasis [12–15]. We have also previously shown that overexpression of miR526b and miR655 is associated with poor breast cancer patient survival and found that miRNA expression was elevated in advanced grades of breast cancer [12,13], suggesting these two miRNAs are oncogenic and metastasis-promoting miRNAs. Here, we tested the potential roles of these miRNAs in the induction of oxidative stress and the effects of this potential regulation within the tumor microenvironment. ROS including SO, free radicals, and charged ions are the byproducts of cellular metabolism. Under normal physiological conditions, cells keep a balance of ROS production and neutralization to maintain tissue homeostasis [4,5]. However, overproduction of ROS induces oxidative stress, which is associated with cancer development and progression. Production of ROS causes DNA mutation, oncogenic miRNA expression, protein malfunction, apoptosis, and the induction of oxidative stress, which has been identified as a major cause of breast cancer [30]. Superoxide (SO) serves as a growth-stimulating molecule that regulates signaling cascades, which leads to cell survival and proliferation [4]. Moreover, it has been shown that ER-positive breast cancer tumor samples exhibit higher SO levels compared to matched normal tissues, and that SO levels are higher in the blood of breast cancer patients [31,32]. In this study we used the ER-positive MCF7 breast cancer cell line as an in vitro tumor model to establish the link between miRNA and ROS/SO production in breast cancer. The link between various miRNAs and oxidative stress has also been previously reported, such as the expression of miR155 shown to regulate oxidative stress in endothelial cells, and the in vitro induction of oxidative stress being shown to regulate the expression of miR146a and miR34a [33,34]. Oxidative stress is the result of excess ROS, which is due to an imbalance between the generation of ROS and the cell's ability to neutralize and eliminate them. Previous studies have linked the roles of miRNA with ROS production; for example, Zhang et al. showed that miR21 modulates oxidative stress by measuring ROS production in cells through ROS detection [35].

We wanted to investigate if miRNA expression can regulate ROS production and induce oxidative stress, while oxidative stress can also regulate miRNA expression in breast cancer. In this study, we used cell-permeable dyes which interact with cellular ROS and SO to detect and quantify ROS/SO production in cells using fluorescence assays. First, we measured and compared ROS/SO production in MCF7, MCF7-miR526b, and MCF7-miR655 cell lines, to test for the direct regulation of oxidative stress in breast cancer cells by these miRNAs. We have shown that MCF7-miR526b and MCF7-miR655 cell lines, especially the MCF7-miR655 cell line, have higher production of ROS/SO than MCF7 cells, showing that miR526b and miR655 have a role in the endogenous or "direct" induction of oxidative stress.

Breast tumors consist of heterogeneous cells and interactions between tumor cells and cells within the tumor microenvironment to promote tumor sustenance and metastasis. Specifically, cell metabolites and secretions from tumor cells into the tumor microenvironment function to communicate between tumor cells with nearby non-tumor cells, which can regulate many different pathways and networks to promote tumor metastasis [5,36]. Therefore, we investigated the paracrine or "indirect" induction of oxidative stress by treating MCF7 cells and HUVECs with tumor cell metabolites and secretions from miR526b/miR655-high cell lines. We observed that cell-free conditioned media collected from miR526b/miR655-high cells induced ROS/SO production in both MCF7 cells and HUVECs, which suggests that miR526b and miR655 secretory proteins and metabolites indirectly induce oxidative stress in the tumor microenvironment.

The roles of extracellular or cell-free miR526b and miR655 in the complexity of breast tumor metastasis has not been well investigated. Although in recent years many reports studied the detection of miRNAs in the blood of cancer patients, it was only recently shown by a group that extracellular miRNAs can be found in the media of *Drosophila* cell lines growth in petri dish [37]; giving an excellent model to test cell-free miRNA in vitro. We previously have shown that miR526b/miR655 cell-free conditioned media contain stimulatory proteins which induce angiogenesis in the tumor microenvironment [15]. However, we never measured the presence of miR526b/miR655 themselves in the cell-free conditioned media. Here, for the first time, we showed that cell-free supernatant (cell-free conditioned media) collected from MCF7, MCF7-miR526b, and MCF7-miR655 serum-starved cells media contain miR5256b/miR655. Moreover, we found that both MCF7-miR526b and MCF7-miR655 conditioned media had a higher expression of both pri-miR526b and pri-miR655 compared to the MCF7 conditioned media. These results prove that cell secretions from miR526b/miR655-high cell lines also contain miRNAs and indirectly play a role in oxidative stress induction in the tumor microenvironment.

Since ROS activates signaling cascades that promote cell survival and tumor growth, it is expected that highly metastatic and aggressive breast cancer cell lines will be under higher oxidative stress than poorly metastatic breast cancer cell lines [38]. Here, we observed that a key regulatory protein of oxidative stress, TXNRD1, is upregulated in highly metastatic and aggressive breast cancer cell lines, which is supported by other studies showing a link between oxidative stress and breast cancer [10,39]. Next, it was found that miRNA overexpression induced *TXNRD1* expression in MCF7-miR526b and MCF7-miR655 cell lines. These results led us to investigate potential targets of miR526b and miR655 to explain the upregulation of *TXNRD1* in miRNA-overexpressing cell lines. It was found that two transcription factors, PBRM1 (polybromo 1) and TCF21 (Transcription Factor 21), which are negative regulators of *TXNRD1*, are both targets of miR526b and miR655. *PBRM1* has been described as a tumor suppressor gene that is responsible for the control of the cell cycle [28]. Low *PBRM1* expression has been shown to predict poor prognosis in breast cancer and mutations in *PBRM1* have been reported in many tumor types such as renal cell carcinoma, biliary carcinoma, gallbladder carcinoma, and intrahepatic cholangiocarcinoma [28,40]. *TCF21* has also been reported as a tumor suppressor gene in gastric cancer, colorectal cancer, head and neck carcinomas, and breast cancer [29,41–43]. Following the Bioinformatics analysis, we validated this observation by measuring the expression of these two TFs in MCF7, MCF7-miR526b, and MCF7-miR655 cell lines. Our results showed that *PBRM1* and *TCF21* are indeed downregulated in miR526b/miR655-high cell lines, proving that miR526b and miR655 upregulate *TXNRD1* by targeting these two negative regulators of *TXNRD1*.

In the tumor microenvironment, dynamics between tumor cell secretion of inflammatory molecules and growth factors, communication with endothelial cells, and activation of immune cells are well established [15,44]. We have previously shown that treatment of HUVECs with MCF7-miR526b or MCF7-miR655 conditioned media induced cancer related phenotypes, such as angiogenesis and lymphangiogenesis via paracrine regulation [15]. In addition, here we showed that even cell-free conditioned media contain miR526b/miR655. To further investigate the roles of miR526b and miR655 in the indirect induction of oxidative stress, *TXNRD1* expression was quantified and compared in MCF7 cells and HUVECs treated with MCF7-miR526b or MCF7-miR655 cell-free conditioned media. Here, we have found that in both MCF7 and HUVECs treated with MCF7-miR526b or MCF7-miR655 cell-conditioned media, there is an upregulation of *TXNRD1*, which supports our findings of miR526b and miR655 indirectly regulating the production of ROS and induction of oxidative stress.

While ROS production is a component of the cell's physiological process, high concentrations of ROS are detrimental for the cell, which induces apoptosis. However, epigenetic changes, such as miRNA overexpression by tumor cells, protect cellular death and promote cell proliferation. It has been shown that the induction of oxidative stress can alter the expression of specific miRNAs by inhibiting or inducing their expression [16]. Similarly, we have shown that conditioned media collected from MCF7-miR526b and MCF7-miR655 cell lines induce ROS production in MCF7 cells, thus miR526b and miR655 are involved in the regulation of oxidative stress both directly, and indirectly.

Next, we tested to see if miR526b and miR655 are immediate responders to cellular oxidative stress. In this study, we have shown that cell-free conditioned media collected from MCF7-miR526b and MCF7-miR655 cell lines induce oxidative stress; thus, we again used conditioned media from miRNA-overexpressing cells to induce oxidative stress in MCF7 cells and examined miR526b and miR655 expression. We observed that MCF7 cells treated with MCF7-miR526b or MCF7-miR655 conditioned media had increased expressions of both pri-miR526b and pri-miR655. Interestingly, we observed that MCF7 cells treated with miR526b conditioned media and metabolites showed a higher expression of pri-miR526b than pri-miR655, and MCF7 cells treated with conditioned media from MCF7-miR655 showed a higher expression of pri-miR655 than pri-miR526b. It has previously been shown by other groups that H_2O_2 treatment induces oxidative stress in MCF7 cells [45,46]. Therefore, to further validate that miR526b and miR655 are immediate responders to cellular oxidative stress, we tested the effects of H_2O_2 treatment on MCF7 cells. Interestingly, H_2O_2 treatment significantly increased the expression of pri-miR655 in MCF7 cells, and marginally increased pri-miR526b expression in a dose dependent manner. Taken together, these results suggest that a positive feedback loop exists between oxidative stress and miRNA in breast cancer, which is driven by miRNA-high cell line secretions.

Interestingly, we noticed a common trend in which miR655 appeared to have a stronger role in both the direct and indirect induction of oxidative stress than miR526b. MCF7-miR655 was shown to have the greatest expression of *TXNRD1*, and the greatest production of ROS/SO as compared to MCF7 cells. Furthermore, MCF7 and HUVECs treated with cell-free conditioned media from MCF7-miR655 showed the greatest expression of *TXNRD1* and the greatest amount of ROS/SO production. This shows that while miR526b still appeared to be involved in oxidative stress and the *TXNRD1* pathway, miR655 has a stronger role in oxidative stress pathways in breast cancer. Differential roles of miRNAs in regulating oxidative stress may be due to various targets of miRNAs (Figure 8A). In the future, it would be interesting to investigate the signaling pathways involved in miR526b and miR655's regulation of oxidative stress in breast cancer.

In this study, we identified the novel roles of miR526b and miR655 in oxidative stress in breast cancer. Specifically, this is the first time that miR526b and miR655 has been linked to oxidative stress, as we show that miR526b and miR655 regulate ROS production, as well as show greater expression of miRNAs during cellular oxidative stress. Furthermore, we suggest a positive feedback loop exists between miR526b/miR655 and oxidative stress in breast cancer. Here we also show that miR526b and miR655 are present in the extracellular tumor microenvironment, which suggests that these cell free miRNAs might also be regulating extracellular signaling and regulating oxidative stress hence promoting tumor growth and metastasis. These discoveries add to the accumulation of evidence that miR526b and miR655 are strong candidates for potential biomarkers in breast cancer. Future studies require a complete analysis of miRNA cell metabolites and cell secretome to discover new functions of miR526b and miR655. This will allow us to discover complex mechanisms behind oxidative stress induction in breast cancer and the possibility of these miRNAs as therapeutic targets to abrogate oxidative stress.

4. Materials and Methods

We conducted all experiments at Brandon University, following the regulations of Brandon University Research Ethics (#21986, approved on April 21, 2017) and Biohazard Committee (#2017-BIO-02, approved on September 13, 2017). An overview of the methods workflow is presented in Figure 11.

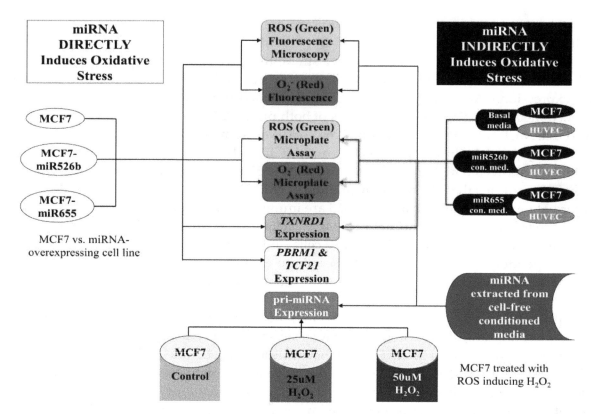

Figure 11. Outline of the in vitro approaches taken in establishing the direct and indirect induction of oxidative stress by miR526b and miR655, as well as the effect of ROS induction on the regulation of miRNA.

4.1. Cell Culture

All human breast cancer cell lines MCF7, SKBR3, T47D, MDAMB231, and Hs578T were purchased from American Type Culture Collection (ATCC, Rockville, MD, USA). All breast cancer cell lines were grown in Dulbecco's Modified Eagle Medium (DMEM) (Gibco, Mississauga, ON, Canada) supplemented with 10% fetal bovine serum (FBS) and 1% Penstrep as described before following manufacturer protocols [12,13,47]. Stable miRNA-overexpressing MCF7-miR526b and MCF7-miR655 cell lines were established as previously described [12,13]. MCF7-miR526b and MCF7-miR655 cells were grown in Roswell Park Memorial Institute (RPMI) 1640 medium (Gibco, Mississauga, ON, Canada) supplemented with 10% FBS and 1% Penstrep. Furthermore, MCF7-miR526b and MCF7-miR655 cell lines were sustained with Geneticin (Gibco, Mississauga, ON, CAN) at 40 mg/mL. An immortalized non-tumorigenic mammary epithelial cell line MCF10A was cultured and maintained by Ling Liu at the University of Western Ontario in Professor Peeyush K Lala's laboratory as described earlier [47] and they kindly shared an aliquot of MCF10A cDNA.

HUVECs were purchased from Life Technologies (NY, USA) and grown in Medium 200 (Gibco, Mississauga, ON, Canada), supplemented with Low Serum Growth Supplement Kit containing 2% FBS, hydrocortisone (1 µg/mL), human epidermal growth factor (10ng/mL), basic fibroblast growth factor (3 ng/mL), and heparin (10 µg/mL). All cell lines were maintained in a humidified incubator at 37 °C with 5% CO_2.

4.2. Collection of Conditioned Media

MCF7, MCF7-miR526b, and MCF7-miR655 cell lines were grown in complete RPMI 1640 until 90% confluent. Cells were then washed with phosphate buffered saline (PBS) to remove any trace of the complete media. The cells were then starved with basal RPMI 1640 medium (serum-free) for 12–16 h prior to collection of media, and then centrifuged. Cell-free supernatant was then collected for assays testing the indirect induction of oxidative stress by miR526b and miR655. We hypothesized that these cell supernatants contain cell metabolites and secretory proteins with unknown function.

4.3. RNA Extraction and Quantitative Real-Time PCR

Total RNA was extracted from all cell lines using the miRNeasy Mini Kit (Qiagen, Toronto, ON, Canada) and reverse transcribed using the microRNA and mRNA cDNA Reverse Transcription Kit (Applied Biosystems, Waltham, MA, USA). For conditioned media miRNA extraction, MCF7, MCF7-miR526b, and MCF7-miR655 conditioned media were centrifuged at 3000 RPM for 5 min, and the supernatants were collected for RNA extraction following the miRNeasy Mini Kit protocol (Qiagen, Toronto, ON, Canada). The TaqMan miRNA or Gene Expression Assays was used for qRT-PCR. The expressions of two endogenous control genes, *Beta-actin* (Hs01060665_g1) and *RPL5* (Hs03044958_g1), were quantified using qRT-PCR and were used to normalize the expression of *TXNRD1* (Hs00917067), *PBRM1* (Hs01015916_m1), *TCF21* (Hs00162646_m1), pri-miR526b (Hs03296227), and pri-miR655 (Hs03304873) markers using relative analysis. Gene expression was measured using CT values from each curve, which are obtained from the point at which each curve reaches the threshold. To determine the relative levels of gene expression, the comparative threshold cycle method (ΔCt) was used [15,47].

4.4. Fluorescence Microplate Assay

MCF7, MCF7-miR526b, and MCF7-miR655 cells were seeded in a 96-well plate as shown in Figure S1A and were grown until 70% confluent. Total ROS and SO levels were detected using the ROS-ID Total ROS/SO detection kit (Enzo Life Sciences, Farmingdale, NY, USA) according to manufacturer's instructions. Negative controls and test groups were prepared for each cell line. The negative controls were treated with 5 μM of *N*-acetyl-L-cysteine (ROS inhibitor) for 30 min to eliminate all ROS present in the cells. Following this, 200 μM of Pyocyanin (ROS inducer) was added to induce ROS production in all wells. The test groups were treated with only the ROS inducer. Detection reagents from the ROS-ID kit were used to measure ROS/SO production. Microplate readings were done at 1 and 21 h following the addition of detection dyes, using the standard Fluorescein filter (Ex/Em: 485/535 nm) and Rhodamine filter (Ex/Em: 550/625 nm). Data was collected using the SoftMax Pro 6 Microplate Data Acquisition and Analysis software (Molecular Devices, San Jose, CA, USA). Concentrations of the ROS inhibitor, inducer, and detection reagents were determined based on a known standard curve. For normalization, negative control emissions were subtracted from the test group emissions to show the total production ROS in each cell line (Figure S1A).

Two more plate-reading experiments were done using MCF7 (Figure S1B) or HUVECs (Figure S1C) treated with basal media (no serum added) or MCF7-miR526b/miR655 conditioned media. MCF7 cells/HUVECs were seeded as shown in Figure S1B/C, and when 70% confluent, they were washed with PBS to remove traces of the serum and growth factors. They were then treated with basal media or MCF7-miR526b/miR655 supernatant for 12–18 h. The assay was then performed as described above.

4.5. Fluorescence Microscopy Assay

We used the same ROS/SO detection kit to determine the number of cells producing ROS and SO following the manufacturer's protocol. Test groups and negative controls were prepared for the MCF7, MCF7-miR526b, and MCF7-miR655 cell lines and seeded as described above. When 70% confluent, the cells were washed PBS and treated as described above. The assay was performed on the NIS Elements Advanced Research software (Nikon, Melville, NY, USA), using a Nikon Ds-Ri1 microscopy camera. The fluorescent cells in each experiment were quantified using the ImageJ software (National Institute of Health, Bethesda, MD, USA). Fluorescent images were converted to 8-bit and adjustments were made. Particle analysis was then done on ImageJ to quantify the number of fluorescing cells (Figure S2B,D,F,H,J,L,N,S,X) and (Figure S3B,D,F,H,J,L,N,S,X) For each condition, the negative control was used as a threshold for quantification (Figures S2 and S3). Negative control quantifications were subtracted from test group quantifications, and then divided by the total number of cells to present the total ROS/SO production in each cell line as ratios.

A second experiment was conducted using the same ROS-ID kit, following the same protocol as described above. MCF7 cells were seeded in a 96-well plate as shown in Figure S1B/C, and once they have reached 70% confluency, washed with PBS and treated with miRNA-conditioned media. The assay was performed and fluorescent cells were quantified and presented using the same methods outlined above. The assay was then performed as described above.

4.6. Bioinformatics Analysis

A total of 4133 target transcript genes for human miR526b (hsa-miR-526b) and 3264 target transcription genes for human miR655 (hsa-miR-655) were found using TargetScan (analysis tool which can predict miRNA target genes in mammalian mRNA pool) and miRBase database [22–26]. Finding the TFs of the *TXNRD1* gene allows us to distinguish what up/down regulates *TXNRD1* expression within the human system.

We used the Enrichr (a tool that consists of both a validated user-submitted gene list and a search engine for further analysis) and found 155 TFs perturbations followed by gene expression [27]. These 155 TFs upregulate or downregulate *TXNRD1* gene expression. We then compared the two data sets to find common genes between miR526b/miR655 target genes and *TXNRD1* regulatory genes (TFs). We observed 19 genes that are common between miR526b targets and *TXNRD1* regulators, of which 8 down-regulated the *TXNRD1* gene expression. We then compared the gene list to find common genes between miR655 targets and *TXNRD1* regulators, and observed 18 genes that are common in both gene sets, of which 11 down-regulated the *TXNRD1* gene. Finally, we found two TFs as common targets of both miRNAs that are negative regulators of *TXNRD1*. To determine the target a nominal $p < 0.05$ was used and the p value was calculated with Fisher exact test, which is a proportion test that assumes a binomial distribution and independence for the probability of any gene belonging to any set.

4.7. Treatment of MCF7 Cells with H_2O_2

MCF7 cells were grown and maintained until 90% confluent. H2O2 at a concentration of either 25 µM or 50 µM was added to confluent MCF7 cells for 24 h. H_2O_2 was used instead of pyocyanin to test the effects of a different ROS inducer. These concentrations of H_2O_2 have been previously reported to induce oxidative stress in the MCF7 cell line [45,46]. Following the addition of H_2O_2 for 24 h, MCF7 cells were collected for RNA extraction, carried out with miRNeasy Mini Kit (Qiagen, Toronto, ON, Canada) and reverse transcribed using the TaqMan microRNA and mRNA cDNA Reverse Transcription Kit (Applied Biosystems, Waltham, MA, USA). qRT-PCR was carried out as mentioned above to measure pri-miR526b and pri-miR655 expression, and were normalized to *Beta-actin* and *RPL5*.

4.8. Statistical Analysis

Statistical calculations were performed using GraphPad Prism software version 8 (https://www. graphpad.com/quickcalcs/ttest1/?Format=SEM). All parametric data were analyzed with one-way ANOVA followed by Tukey–Kramer or Dunnett post-hoc comparisons. Student's t-test was used when comparing two datasets. Statistically relevant differences between means were accepted at $p < 0.05$. Fisher exact test was performed for miRNA database and target TFs analysis followed by false positive rate (FDR) correction to identify significant changes in target gene expression ($p < 0.05$).

Author Contributions: Concept, project design, and supervision: M.M.; Experiments: B.S., R.F., B.N., S.H., S.M., K.C.U.; Data Analysis: B.S., R.F., B.N., S.M.; Figures and Image Data Processing: B.S., R.F., B.N., S.M.; Manuscript writing: B.S., R.F., S.H., B.N., and M.M.

Acknowledgments: The authors of this article would like to thank Danielle Laroque at Brandon University for her help with cell number quantification with ImageJ. We sincerely thank Peeyush K Lala at the University of Western Ontario to share MCF10A cDNA with us. Thanks to Bernadette Ardelli at Brandon University to give us access to Fluorescence Microscope, Microplate Reader, and the Rotor Gene PCR machine in her laboratory. We also want to thank Vincent Chen at Brandon University to help us with the analysis of cell metabolites.

Abbreviations

ROS	Reactive Oxygen Species
SO	Superoxide
miRNA	microRNA
TXNRD1	Thioredoxin Reductase 1
TXN	Thioredoxin
TF	Transcription Factor
ER	Estrogen Receptor

References

1. Breast Cancer—Early Diagnosis and Screening. Available online: https://www.who.int/cancer/prevention/diagnosis-screening/breast-cancer/en/ (accessed on 19 July 2019).

2. Hecht, F.; Pessoa, C.F.; Gentile, L.B.; Rosenthal, D.; Carvalho, D.P.; Fortunato, R.S. The role of oxidative stress on breast cancer development and therapy. *Tumor Biol.* **2016**, *37*, 4281–4291. [CrossRef] [PubMed]

3. Federico, A.; Morgillo, F.; Tuccillo, C.; Ciardiello, F.; Loguercio, C. Chronic inflammation and oxidative stress in human carcinogenesis. *Int. J. Cancer* **2007**, *121*, 2381–2386. [CrossRef] [PubMed]

4. Buetler, T.M.; Krauskopf, A.; Ruegg, U.T. Role of superoxide as a signaling molecule. *Physiology* **2004**, *19*, 120–123. [CrossRef]

5. Calaf, G.M.; Urzua, U.; Termini, L.; Aguayo, F. Oxidative stress in female cancers. *Oncotarget* **2018**, *9*, 23824. [CrossRef] [PubMed]

6. Lobo, V.; Patil, A.; Phatak, A.; Chandra, N. Free radicals, antioxidants and functional foods: Impact on human health. *Pharmacogn. Rev.* **2010**, *4*, 118–126. [CrossRef]

7. Arnér, E.S.; Holmgren, A. Physiological functions of thioredoxin and thioredoxin reductase. *Eur. J. Biochem.* **2000**, *267*, 6102–6109. [CrossRef] [PubMed]

8. Lu, J.; Vlamis-Gardikas, A.; Kandasamy, K.; Zhao, R.; Gustafsson, T.N.; Engstrand, L.; Hoffner, S.; Engman, L.; Holmgren, A. Inhibition of bacterial thioredoxin reductase: An antibiotic mechanism targeting bacteria lacking glutathione. *FASEB J.* **2013**, *7*, 1394–1403. [CrossRef]

9. Urig, S.; Lieske, J.; Fritz-Wolf, K.; Irmler, A.; Becker, K. Truncated mutants of human thioredoxin reductase 1 do not exhibit glutathione reductase activity. *FEBS Lett.* **2006**, *580*, 3595–3600. [CrossRef]

10. Leone, A.; Roca, M.S.; Ciardiello, C.; Costantini, S.; Budillon, A. Oxidative stress gene expression profile correlates with cancer patient poor prognosis: Identification of crucial pathways might select novel therapeutic approaches. *Oxid. Med. Cell. Longevity* **2017**, *2017*. [CrossRef]

11. Singh, R.P.; Massachi, I.; Manickavel, S.; Singh, S.; Rao, N.P.; Hasan, S.; Mc Curdy, D.K.; Sharma, S.; Wong, D.; Hahn, B.H.; et al. The role of miRNA in inflammation and autoimmunity. *Autoimmun. Rev.* **2013**, *12*, 1160–1165. [CrossRef]

12. Majumder, M.; Landman, E.; Liu, L.; Hess, D.; Lala, P.K. COX-2 elevates oncogenic miR-526b in breast cancer by EP4 activation. *Mol. Cancer Res.* **2015**, *13*, 1022–1033. [CrossRef]

13. Majumder, M.; Dunn, L.; Liu, L.; Hasan, A.; Vincent, K.; Brackstone, M.; Hess, D.; Lala, P.K. COX-2 induces oncogenic microRNA miR655 in human breast cancer. *Sci. Rep.* **2018**, *8*, 327. [CrossRef]

14. Majumder, M.; Nandi, P.; Omar, A.; Ugwuagbo, K.; Lala, P.K. EP4 as a therapeutic target for aggressive human breast cancer. *Int. J. Mol. Sci.* **2018**, *19*, 1019. [CrossRef]

15. Hunter, S.; Nault, B.; Ugwuagbo, K.; Maiti, S.; Majumder, M. Mir526b and Mir655 Promote Tumour Associated Angiogenesis and Lymphangiogenesis in Breast Cancer. *Cancers* **2019**, *11*, 938. [CrossRef]

16. He, J.; Jiang, B.H. Interplay between reactive oxygen species and microRNAs in cancer. *Curr. Pharmacol. Rep.* **2016**, *2*, 82–90. [CrossRef]

17. Longo, A.; Librizzi, M.; Chuckowree, I.; Baltus, C.; Spencer, J.; Luparello, C. Cytotoxicity of

the urokinase-plasminogen activator inhibitor carbamimidothioic acid (4-boronophenyl) methyl ester hydrobromide (BC-11) on triple-negative MDA-MB231 breast cancer cells. *Molecules* **2015**, *20*, 9879–9889. [CrossRef]

18. Librizzi, M.; Longo, A.; Chiarelli, R.; Amin, J.; Spencer, J.; Luparello, C. Cytotoxic effects of Jay Amin hydroxamic acid (JAHA), a ferrocene-based class I histone deacetylase inhibitor, on triple-negative MDA-MB231 breast cancer cells. *Chem. Res. Toxicol.* **2012**, *25*, 2608–2616. [CrossRef]

19. Liang, H.H.; Huang, C.Y.; Chou, C.W.; Makondi, P.T.; Huang, M.T.; Wei, P.L.; Chang, Y.J. Heat shock protein 27 influences the anti-cancer effect of curcumin in colon cancer cells through ROS production and autophagy activation. *Life Sci.* **2018**, *209*, 43–51. [CrossRef]

20. Chang, T.C.; Wei, P.L.; Makondi, P.T.; Chen, W.T.; Huang, C.Y.; Chang, Y.J. Bromelain inhibits the ability of colorectal cancer cells to proliferate via activation of ROS production and autophagy. *PLoS ONE* **2019**, *14*, e0210274. [CrossRef]

21. Wei, P.L.; Huang, C.Y.; Chang, Y.J. Propyl gallate inhibits hepatocellular carcinoma cell growth through the induction of ROS and the activation of autophagy. *PloS ONE* **2019**, *14*, e0210513. [CrossRef]

22. TargetScanHuman 7.1—Predicted miRNA targets of miR-526b-5p. Available online: http://www.targetscan.org/cgi-bin/targetscan/vert_71/targetscan.cgi?mirg=hsa-miR-526b-5p (accessed on 19 July 2019).

23. TargetScanHuman 7.1—Predicted miRNA targets of miR-655-5p. Available online: http://www.targetscan.org/cgi-bin/targetscan/vert_71/targetscan.cgi?mirg=hsa-miR-655-5p (accessed on 19 July 2019).

24. Agarwal, V.; Bell, G.W.; Nam, J.; Bartel, D.P. Predicting effective microRNA target sites in mammalian mRNAs. *eLife* **2015**, *4*, e05005. [CrossRef]

25. miRBase-Release 22.1.—miRNA Entry for MI0003150. Available online: http://www.mirbase.org/cgi-bin/mirna_entry.pl?acc=MI0003150 (accessed on 19 July 2019).

26. miRBase-Release 22.1.—miRNA Entry for MI0003677. Available online: http://www.mirbase.org/cgi-bin/mirna_entry.pl?acc=MI0003677 (accessed on 19 July 2019).

27. Keenan, A.B.; Torre, D.; Lachmann, A.; Leong, A.K.; Wojciechowicz, M.L.; Utti, V.; Jagodnik, K.M.; Kropiwnicki, E.; Wang, Z.; Ma'ayan, A. ChEA3: Transcription factor enrichment analysis by orthogonal omics integration. *Nucleic Acids Res.* **2019**, *2*, 212–224. [CrossRef]

28. Wang, H.; Qu, Y.; Dai, B.; Zhu, Y.; Shi, G.; Zhu, Y.; Shen, Y.; Zhang, H.; Ye, D. PBRM1 regulates proliferation and the cell cycle in renal cell carcinoma through a chemokine/chemokine receptor interaction pathway. *PLoS ONE* **2017**, *12*, e0180862. [CrossRef]

29. Wang, J.; Gao, X.; Wang, M.; Zhang, J. Clinicopathological significance and biological role of TCF21 mRNA in breast cancer. *Tumor Biol.* **2015**, *36*, 8679–8683. [CrossRef]

30. Nourazarian, A.R.; Kangari, P.; Salmaninejad, A. Roles of oxidative stress in the development and progression of breast cancer. *Asian Pac. J. Cancer Prev.* **2014**, *15*, 4745–4751. [CrossRef]

31. Kanchan, R.K.; Tripathi, C.; Baghel, K.S.; Dwivedi, S.K.; Kumar, B.; Sanyal, S.; Sharma, S.; Mitra, K.; Garg, V.; Singh, K.; et al. Estrogen receptor potentiates mTORC2 signaling in breast cancer cells by upregulating superoxide anions. *Free Radical Bio. Med.* **2012**, *53*, 1929–1941. [CrossRef]

32. Yeh, C.C.; Hou, M.F.; Tsai, S.M.; Lin, S.K.; Hsiao, J.K.; Huang, J.C.; Wang, L.H.; Wu, S.H.; Hou, L.A.; Ma, H.; et al. Superoxide anion radical, lipid peroxides and antioxidant status in the blood of patients with breast cancer. *Clin. Chim. Acta.* **2005**, *361*, 104–111. [CrossRef]

33. Chen, H.; Gao, L.; Yang, M.; Zhang, L.; He, F.L.; Shi, Y.K.; Pan, X.H.; Wang, H. MicroRNA-155 affects oxidative damage through regulating autophagy in endothelial cells. *Oncol. Lett.* **2019**, *17*, 2237–2243. [CrossRef]

34. Cheleschi, S.; de Palma, A.; Pascarelli, N.; Giordano, N.; Galeazzi, M.; Tenti, S.; Fioravanti, A. Could oxidative stress regulate the expression of microRNA-146a and microRNA-34a in human osteoarthritic chondrocyte cultures? *Int. J. Mol. Sci.* **2017**, *18*, 2660. [CrossRef]

35. Zhang, X.; Ng, W.L.; Wang, P.; Tian, L.; Werner, E.; Wang, H.; Doetsch, P.; Wang, Y. MicroRNA-21 modulates the levels of reactive oxygen species by targeting SOD3 and TNFα. *Cancer Res.* **2012**, *72*, 4707–4713. [CrossRef]

36. Huang, W.; Luo, S.; Burgess, R.; Yi, Y.H.; Huang, G.; Huang, R.P. New insights into the tumor microenvironment utilizing protein array technology. *Int. J. Mol. Sci.* **2018**, *19*, 559. [CrossRef]

37. Van den Brande, S.; Gijbels, M.; Wynant, N.; Santos, D.; Mingels, L.; Gansemans, Y.; van Nieuwerburgh, F.; Vanden Broeck, J. The presence of extracellular microRNAs in the media of cultured Drosophila cells. *Sci. Rep.* **2018**, *8*, 17312. [CrossRef]

38. Reuter, S.; Gupta, S.C.; Chaturvedi, M.M.; Aggarwal, B.B. Oxidative stress, inflammation, and cancer: How are they linked? *Free Radicals Biol. Med.* **2010**, *49*, 1603–1616. [CrossRef]

39. Zajchowski, D.A.; Bartholdi, M.F.; Gong, Y.; Webster, L.; Liu, H.L.; Munishkin, A.; Beauheim, C.; Harvey, S.; Ethier, S.P.; Johnson, P.H. Identification of gene expression profiles that predict the aggressive behavior of breast cancer cells. *Cancer Res.* **2001**, *61*, 5168–5178.

40. Mo, D.; Li, C.; Liang, J.; Shi, Q.; Su, N.; Luo, S.; Zeng, T.; Li, X. Low PBRM1 identifies tumor progression and poor prognosis in breast cancer. *Int. J. Clin. Exp. Pathol.* **2015**, *8*, 9307–9313.

41. Yang, Z.; Li, D.M.; Xie, Q.; Dai, D.Q. Protein expression and promoter methylation of the candidate biomarker TCF21 in gastric cancer. *J. Cancer Res. Clin. Oncol.* **2015**, *141*, 211–220. [CrossRef]

42. Dai, Y.; Duan, H.; Duan, C.; Zhu, H.; Zhou, R.; Pei, H.; Shen, L. TCF21 functions as a tumor suppressor in colorectal cancer through inactivation of PI3K/AKT signaling. *Onco. Targets Ther.* **2017**, *10*, 1603–1611. [CrossRef]

43. Smith, L.T.; Lin, M.; Brena, R.M.; Lang, J.C.; Schuller, D.E.; Otterson, G.A.; Morrison, C.D.; Smiraglia, D.J.; Plass, C. Epigenetic regulation of the tumor suppressor gene *TCF21* on 6q23-q24 in lung and head and neck cancer. *Proc. Natl. Acad. Sci. USA* **2006**, *103*, 982–987. [CrossRef]

44. Majumder, M.; Xin, X.; Liu, L.; Girish, G.V.; Lala, P.K. Prostaglandin E2 receptor EP4 as the common target on cancer cells and macrophages to abolish angiogenesis, lymphangiogenesis, metastasis and stem-like cell functions. *Cancer Sci.* **2014**, *105*, 1142–1151. [CrossRef]

45. Chua, P.J.; Yip, G.W.C.; Bay, B.H. Cell cycle arrest induced by hydrogen peroxide is associated with modulation of oxidative stress related genes in breast cancer cells. *Exp. Biol. Med.* **2009**, *234*, 1086–1094. [CrossRef]

46. Mahalingaiah, P.K.S.; Ponnusamy, L.; Singh, K.P. Chronic oxidative stress causes estrogen-independent aggressive phenotype, and epigenetic inactivation of estrogen receptor alpha in MCF-7 breast cancer cells. *Breast Cancer Res. Treat.* **2015**, *153*, 41–56. [CrossRef]

47. Majumder, M.; Xin, X.; Liu, L.; Tutunea-Fatan, E.; Rodriguez-Torres, M.; Vincent, K.; Postovit, L.M.; Hess, D.; Lala, P.K. COX-2 induces breast cancer stem cells via EP4/PI3K/AKT/NOTCH/WNT axis. *Cancer Stem Cells* **2016**, *34*, 2290–2305. [CrossRef]

9

MicroRNA Networks Modulate Oxidative Stress in Cancer

Yang-Hsiang Lin

Liver Research Center, Chang Gung Memorial Hospital, Linkou, Taoyuan 333, Taiwan; yhlin0621@cgmh.org.tw

Abstract: Imbalanced regulation of reactive oxygen species (ROS) and antioxidant factors in cells is known as "oxidative stress (OS)". OS regulates key cellular physiological responses through signal transduction, transcription factors and noncoding RNAs (ncRNAs). Increasing evidence indicates that continued OS can cause chronic inflammation, which in turn contributes to cardiovascular and neurological diseases and cancer development. MicroRNAs (miRNAs) are small ncRNAs that produce functional 18-25-nucleotide RNA molecules that play critical roles in the regulation of target gene expression by binding to complementary regions of the mRNA and regulating mRNA degradation or inhibiting translation. Furthermore, miRNAs function as either tumor suppressors or oncogenes in cancer. Dysregulated miRNAs reportedly modulate cancer hallmarks such as metastasis, angiogenesis, apoptosis and tumor growth. Notably, miRNAs are involved in ROS production or ROS-mediated function. Accordingly, investigating the interaction between ROS and miRNAs has become an important endeavor that is expected to aid in the development of effective treatment/prevention strategies for cancer. This review provides a summary of the essential properties and functional roles of known miRNAs associated with OS in cancers.

Keywords: oxidative stress; MicroRNA; signal transduction; therapeutic target

1. Introduction

Imbalanced regulation of reactive oxygen species (ROS) and antioxidant factors in cells is known as "oxidative stress (OS)" (Figure 1). OS drives key cellular physiological regulatory responses through signal transduction, transcription factors (TFs) and noncoding RNAs (ncRNAs) [1]. ROS are oxygen-containing products and are formed during cellular oxidative metabolism. ROS, including superoxide anion (O_2^-), hydroxyl radical (OH^-), hydrogen peroxide (H_2O_2), nitric oxide (NO) and singlet oxygen ($^1O_2$), play important roles in cell differentiation, cell death, cell growth, signal transduction, cell apoptosis and chemoresistance [2,3]. Dual roles have been proposed for ROS in biological phenotypes according to their cellular level [4]. High levels of ROS promote cell apoptosis, while low levels of ROS act as a signal transducer to induce cell survival (Figure 1). Recently, excessive ROS production was identified in several cancers where they were significantly correlated with tumorigenesis. However, the underlying mechanism of ROS regulation in cancer development remains unclear.

MicroRNAs (miRNAs) are small ncRNA comprising 18-25-nucleotide functional RNA molecules that play critical roles in the regulation of target gene expression by binding to complementary regions of mRNA and regulating mRNA degradation or inhibiting translation (Figure 2). Previous studies have demonstrated that miRNAs are significantly associated with tumor growth, metastasis and cancer progression [5,6]. Based on these findings, dysregulated miRNA expression is a hallmark of cancer.

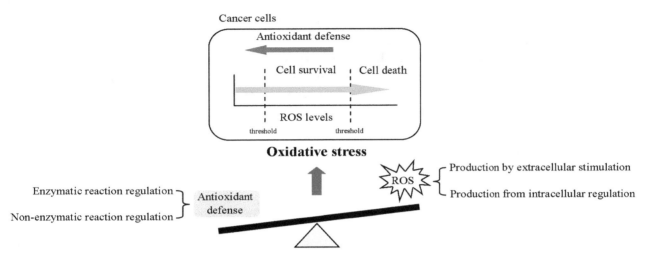

Figure 1. Reactive oxygen species (ROS) production and antioxidant defense in the control of redox homeostasis in cancer cells. Disruption of redox homeostasis by ROS (intra- or extracellular signals) and antioxidant defense (enzymatic or non-enzymatic reactions) induces oxidative stress (OS) and results in various cell functions. The physiological function of ROS is dependent on its concentration. Elevated ROS production and accumulation lead to cell apoptosis. On the other hand, medium levels of ROS promote cell survival and progression.

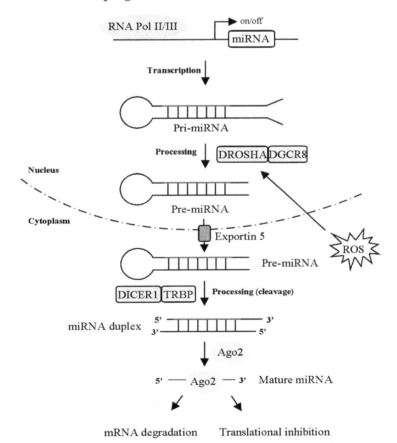

Figure 2. The biogenesis and regulation mechanisms of microRNAs (miRNAs). MiRNAs are transcribed by RNA polymerase II/III and generated the primary miRNA transcript (pri-miRNA). The pri-miRNAs are cleaved into precursor miRNA transcript (pre-miRNA) by the microprocessor complex, a combination of DROSHA and DGCR8. Pre-miRNA is exported to cytoplasm via exportin 5 and further processed by the RNase III enzyme Dicer with the cofactor protein TRBP to generate an approximately 18-25-nt duplex. Either 5p or 3p strand of the mature miRNA (red line) interacts with Argonaute (Ago) protein and forms a miRNA-induced silencing complex (miRISC). There are two models (mRNA degradation and translational repression) of miRNA-mediated gene silencing.

Cross-talk between ROS and miRNAs has been implicated in cancer development, and it is important to identify the nature of this connection. Interestingly, some specific miRNAs, called ROS-miRs or redoximiRs, are regulated by OS and modulate target gene expression in response to ROS [7,8]. Mesenguer et al. [9] demonstrated that the OS/NFκB axis induced miR-9/9* expression and inhibited expression of its target genes, GTPBP3, MTO1 and TRMU, in MELAS cells. On the other hand, a previous study indicated that miR-21 regulated ROS homeostasis and suppressed the antioxidant response in human umbilical vein endothelial cells (HUVECs) [10]. These findings suggest that ROS could be upstream regulators or downstream effectors of miRNAs. In this review, we focus on how ROS affect biological phenotypes through miRNA and how miRNAs regulate ROS-mediated function in cancer.

2. Regulation of ROS Homeostasis in Cells

OS promotes both nuclear and mitochondrial DNA damage and initiates DNA repair pathways [11]. Furthermore, cellular ROS levels can be produced by different mechanisms, such as ionizing radiation, UV radiation, inflammatory cells and chemotherapy. ROS are primarily generated in cells through the byproducts of leaked electrons from the mitochondrial electron transport chain (ETC). Mutations or aberrantly expressed nuclear or mitochondrial genes encoding the ETC components can influence the electron transfer reaction that leads to electron leakage. The electrons are captured by O_2, producing O_2^-, which is usually converted to H_2O_2 by manganese (Mn)-containing mitochondrial superoxide dismutase (MnSOD or SOD2), Cu/Zn-containing cytosolic SOD1 or SOD3 [12]. Subsequently, H_2O_2 can attack chromosomal DNA and subsequently induce DNA damage. On the other hand, O_2^- can be generated through a reaction catalyzed by some enzymes, including the membrane-located NAD(P)H oxidase complex (NOX), which consists of NOX1-4, endoplasmic reticulum-associated xanthine oxidase (XO), cytochrome c oxidase and cyclooxygenase in some cancer cells [13]. In fact, H_2O_2 plays an important role in carcinogenesis because it is capable of diffusing throughout the cell components and producing cellular injury. The injurious effects of ROS in mammalian cells are mediated by the hydroxyl radical ($\cdot$OH). The generation of OH in vivo is produced in the presence of reduced transition metals, including Co, Cu, Fe, or Ni, mainly through the Fenton reaction [14]. Notably, the $\cdot$OH-induced DNA damage includes the generation of 8-hydroxyguanosine (8-OHG), in which the hydrolysis product is 8-hydroxydeoxyguanosine (8-OHdG). 8-OHdG is the most widely used marker of radical attack on DNA. Notably, 8-OHdG is strongly correlated with cancer progression, including that of breast cancer, colorectal cancer, ovarian cancer and hepatocellular carcinoma (HCC) [15–17]. For example, hepatic 8-OHdG levels are useful biomarkers for identifying hepatitis C virus (HCV) infection in patients [18]. Alternatively, cells maintain ROS homeostasis by reducing ROS production and triggering specific antioxidant mechanisms to neutralize ROS or mitigate OS [19]. In fact, antioxidant enzymes include SODs, catalase, peroxiredoxins (PRDXs), thioredoxins, glutathione peroxidase and heme oxygenase. First, SOD converts O_2^- to O_2 or H_2O_2. Then, catalase and glutathione peroxidase subsequently convert H_2O_2 to H_2O and O_2.

3. MiRNAs and Their Roles in Oxidative Stress

Previous studies indicated that ROS can induce or suppress miRNA expression and contribute to downstream biological function through regulation of target genes [20]. Increasing evidence has shown cross-talk between miRNAs and components of redox signaling [21,22]. The transcription, biogenesis, translocation and function of miRNAs are highly correlated with ROS, and miRNAs may regulate the expression of redox sensors and other ROS modulators, such as the key components of cellular antioxidant machinery. Redox sensors have been identified and they include transcription factors (e.g., p53, NFκB, c-Myc and nuclear factor erythroid 2 related factor 2 (NRF2)) and kinases (e.g., Akt and IKK), which trigger cellular redox signaling. Here, we summarize how miRNAs are regulated by ROS at the posttranscriptional and transcriptional levels and how the miRNA/ROS axis controls tumorigenesis.

3.1. MiRNA Processing is Regulated by ROS

Recently, it was reported that miRNAs can be transcribed by RNA polymerase II/III as longer primary transcripts called primary miRNAs (pri-miRNAs). The mature form of miRNA is generated by the two-step processing of pri-miRNA and is subsequently associated with the effector RNA-induced silencing complex (RISC). The biogenesis and function of miRNAs regulated by ROS are described. Two key genes (*Dicer* and *Drosha*) are mediated by the miRNA processing pathway (Figure 2). A report showed that the expression of Dicer was downregulated by aging-related OS in cerebromicrovascular endothelial cells (CMVECs) [23]. Downregulating Dicer dramatically reduced miRNA expression under H_2O_2 treatment compared with the expression of the control. Notably, knocking down Dicer suppressed ROS production in human microvascular endothelial cells (HMECs) [24]. These findings indicated that Dicer expression is part of a feedback loop that modulates ROS production and maintains cellular homeostasis. Upon OS, the expression of pre-miRNA and miRNA in myoblasts is decreased through DGCR8/heme oxygenase-1 (HMOX1) regulation [25]. Heme is required for DGCR8 activity, and the heme-binding domain of DGCR8 plays a crucial role in pri-miRNA recognition for miRNA processing by DROSHA.

3.2. ROS Regulate miRNA Expression through the Modulation of Transcription Factors

Accumulating studies have investigated the miRNAs regulated by ROS/TFs such as c-myc, p53, c-Jun, HIF and NFκB [20,26]. This section summarizes how miRNAs are regulated by ROS/TF at the transcriptional level.

ROS exposure has been shown to be correlated with oncogenic signals such as those transduced by c-Myc and Ras [27,28]. c-Myc, a well-known oncogene, is involved in tumor growth, migration, invasion, metabolism and metastasis through the regulation of gene expression. c-Myc activation induces DNA damage in normal human fibroblasts. This effect has been correlated with ROS generation. Expression levels of miR-15a/16, miR-23a, miR-29 and miR-34 family members were downregulated by c-Myc [29]. Overexpression of miR-15a/16 suppressed cell proliferation, angiogenesis, migration and invasion through inhibition of FGF2 in vitro and in vivo [30]. Furthermore, hypoxia-induced suppression of miR-15/16 expression was directly regulated by c-Myc. By contrast, miR-17-92 and miR-221/222 expression is stimulated by c-Myc [29]. The expression levels of miR-17-92 were remarkably inhibited by triptolide in a c-Myc-dependent manner, which resulted in the induction of target genes, including PTEN, BIM and p21, in HCC cells [31]. Moreover, this suppressive effect contributed to enhanced triptolide-induced cell apoptosis.

P53, a tumor suppressor gene, regulates the cell cycle, apoptosis, growth and metabolism through modulation of target genes. P53 is involved in regulating the drosha-dedicated pri-miRNA processing pathway [32]. In addition, p53 modulates miRNA transcription, such as miR-17-92, miR-34a and miR-200c. Interestingly, stress-regulated miRNAs, namely, miR-34 and miR-200, are upregulated in a p53-dependent manner [33,34]. MiR-34 has been implicated as a tumor suppressor because it suppresses the epithelial-mesenchymal transition (EMT), which promotes cancer cell metastasis. Its expression level is positively associated with p53. Importantly, p53 suppresses Snail expression by interacting with miR-34. A study indicates that miR-200c is upregulated upon H_2O_2 treatment in endothelial cells and that it contributes to cell apoptosis and senescence through inhibition of the target gene ZEB1 [35]. Moreover, knockdown of p53 can reverse H_2O_2-induced miR-200c expression [34].

As mentioned above, exposure to ROS induces chronic inflammation. NFκB acts as a master mediator of the inflammatory response to regulate innate and adaptive immune functions. MiRNA (miR-9, miR-21, miR-30b, miR-146a, miR-155 and miR-17-92 cluster) expression was identified and found to be directly transcriptionally regulated by NFκB [36,37]. Bazzoni et al. [38] indicated that miR-9-1 was induced by lipopolysaccharide (LPS) in a MyD88- and NF-κB-dependent manner. DNA damage activated miR-21 expression through recruitment of NF-κB and signal transducer and activator of transcription 3 (STAT3) to its promoter region and contributed to promoting cell invasion in breast cancer [39]. Another study reported that miR-21 expression and function were mediated by ROS in highly metastatic breast cancer cell lines [40]. In addition, miR-21 induced by ROS via NF-κB activity was involved in arsenic-induced cell transformation [41]. NF-κB bound to the promoter region of the miR-17-92 cluster was identified using chromatin immunoprecipitation (ChIP) assay and was further confirmed by luciferase reporter assay [42]. On the other hand, multiple miRNAs have been identified and have been found to modulate NF-κB activity. MiR-126a was shown to target IκBα, an NFκB inhibitor, and promoted the NFκB signaling pathway [43]. MiR-506 inhibited the expression of the NFκB p65 subunit and led to the production of ROS and p53-dependent apoptosis in lung cancer cells [44]. Notably, miR-506 was regulated by p53. These findings indicated that miR-506 was involved in the p53/NFκB signaling pathway.

NRF2 is a member of the Cap'n'Collar (CNC) family of basic leucine zipper (bZIP) transcription factors [45]. Previously, the actin-binding protein kelch-like ECH-associated protein 1 (KEAP1) was identified as a repressor of NRF2 via proteasomal degradation [46]. NRF2 is involved in antioxidant metabolism, protein degradation, inflammation and radioresistance [47]. Notably, miRNAs can be both indirectly and directly regulated by NRF2 [47,48]. Singh et al. [49] group demonstrated that NRF2 repressed miR-1 and miR-206 expression and led to reprogram glucose metabolism in cancer cells. Furthermore, miR-29 and miR-125b were identified as direct target genes of NRF2 [50,51]. Upregulation of miR-125b by NRF2 resulted in the repression of aryl hydrocarbon receptor repressor and protection of cancer cells from drug-induced toxicity [51]. On the other hand, NRF2 gene was regulated by miRNAs such as miR-28, miR-34a, miR-93 and miR-200a [52–55]. MiR-28 has been shown to interact with NRF2 3′UTR and represses NRF2 expression in breast cancer cells [52]. Overexpression of miR-34a suppressed NRF2 and NRF2 target genes expressions [53]. Functionally, miR-34a was involved in NRF2-dependent antioxidant pathway in liver. These findings suggested that NRF2 and miRNAs formed a regulatory network and regulated cellular functions.

3.3. ROS Regulate miRNA Expression via Epigenetic Regulation

Recently, epigenetic modifications/regulations of the genome have been explored and associated with cancer progression [56]. Changes in the structure or conformation at the nuclear or mitochondrial DNA (nDNA and mtDNA) or RNA level, but not the DNA/RNA sequence, are called epigenetic marks. The main epigenetic alterations in humans are DNA methylation and histone modification, which includes methylation, acetylation and phosphorylation. Aberrant miRNA expression in cancers was discovered and found to be controlled by epigenetic regulation. Promoter regions of miR-125b and miR-199a are hypermethylated through DNMT1 during H_2O_2 treatment, as determined using methylation-specific PCR and bisulfate sequencing [57]. Moreover, these two miRNAs are downregulated by ROS in ovarian cancer cells. The level of histone acetylation has an important role in activating gene expression through chromatin remodeling. In contrast, the gene is silenced by histone deacetylases (HDACs), which promote the deacetylation of lysine residues. MiR-466h-5p acts in a proapoptotic role by directly targeting antiapoptotic genes such as BCL2L2 [58]. ROS induce miR-466h-5p expression through inhibition of HDAC2 and result in increased apoptosis.

4. Interplay between Oxidative Stress, miRNA and Cancer Development

OS has been reported to contribute to neurological disorders, hypertension, diabetes and cancers. This section focuses on the associations of OS and hypoxia, angiogenesis, metastasis, metabolism, cancer stem cell and senescence, which are all involved in cancer progression.

4.1. Association between OS, miRNA and Hypoxia

Hypoxia, known as reduced oxygen availability, mostly occurs in the center of tumors due to the high proliferation ability of cancer cells and abnormal vasculature [59]. Hypoxia-inducible factor 1α (HIF1α) is the master regulator in hypoxia. The activation of HIF1α promotes the expression of several genes, including protein-encoding genes and ncRNAs, and facilitates stem cell renewal, cancer cell survival, metabolism and chemoresistance. The HIF1 transcription factor consists of three hypoxia-induced α subunits (HIF1α/2α/3α) and one β subunit (HIF1β). HIF1α is stabilized and activates a downstream signaling pathway mediated by ROS [60]. Some evidence suggests that telomerase activity is associated with ROS in HCC [61]. Moreover, ROS-mediated telomerase activity is dependent on HIF1α [62]. Expression levels of the human telomerase reverse transcriptase gene (hTERT) are upregulated by HIF1α. Specific binding sites for HIF1α in the hTERT promoter regions were identified by luciferase and ChIP assays. In addition, cancer stem cell (CSC) markers, OCT4 and Notch, are induced by HIF1α and promote stem cell renewal. Expression levels of SOX2 and KLF4 are positively regulated by ROS in glioblastoma cells [63]. HIF1α and ROS activation are responsible for regulating glucose transporter 1 (GLUT1), hexokinase II (HKII) and glutaminase expression and the reprogramming of cancer cell metabolism [64]. Moreover, miR-210 acts in an oncogenic role in cancer development and is induced under hypoxic conditions [65]. HIF1α directly binds to the hypoxia response element (HRE) of the miR-210 promoter. Therefore, miR-210 plays an important role in regulating cellular adaption to hypoxia, suggesting that targeting miR-210 may be a novel approach for the prevention and/or treatment of cancer.

4.2. Association between OS, miRNA and Angiogenesis

Angiogenesis is the process of generating new blood vessels from preexisting vasculature and is required for many functions, such as tissue repair, organ regeneration, cancer development and metastasis [66]. The angiogenesis process is regulated by several cytokines and growth factors, such as vascular endothelial growth factor (VEGF), transforming growth factor β (TGFβ), angiopoietin 1 (Ang-1) and placental growth factor, platelet-derived growth factor (PDGF) β [67–70]. VEGF acts as an effector to control endothelial cell proliferation and new vessel formation. The HIF1α/ROS axis activates tissue-specific angiogenesis through the upregulation of VEGF and its receptors VEGFR1 and VEGFR2. By contrast, VEGF induces ROS production by promoting NADPH oxidase in endothelial cells. ROS can also modulate VEGFR activation, phosphorylation and polymerization. A report indicated that genotoxic stress-induced miR-494 expression suppressed DNA repair and angiogenesis through regulation of MRE11a/RAD50/NBN (MRN) complex in endothelial cells [71]. Moreover, VEGF signaling is regulated by MRN complex in vitro and in vivo. Alternatively, ROS stimulate the MAPK pathway and promote the expression of VEGF. A previous study demonstrated that oxidized phospholipids interact with VEGFR2 and induce angiogenesis through the Src signaling pathway [72]. Other mechanisms of ROS-mediated angiogenesis are the ataxia telangiectasia mutated gene (ATM)/p38α pathway and Sirtuin 1 (SIRT1). Previous studies have indicated that ATM functions in the cell cycle regulation, DNA damage repair and oxidative defense [73]. ATM promotes endothelial cell proliferation and facilitates angiogenesis [74]. Previously, the subtype of histone H2A, called H2AX, can be phosphorylated (γH2AX) and is involved in DNA damage response. Economopoulou et al. [75] group indicated that H2AX is required for endothelial cells to sustain their growth under hypoxia and is important for hypoxia-driven neovascularization. Wilson et al. [76] have shown that miR-103 suppresses developmental and pathological angiogenesis through inhibition of three prime exonucleases 1 in endothelial cells. On the other hand, Yang and co-workers demonstrated that overexpression of miR-328-3p suppressed cell proliferation and promoted radiosensitivity of osteosarcoma cells through suppression of H2AX in vitro and in vivo [77]. Recently, Marampon et al. group demonstrated that NRF2/antioxidant enzymes/H2AX/miRNAs (miR-22, miR-34a, miR-126, miR-146a, miR-210 and miR-375) axis act as potential candidates in radiosensitizing therapeutic strategy for rhabdomyosarcoma clinical treatment [78]. SIRT1, also known as NAD-dependent deacetylase sirtuin-1, has been demonstrated to regulate cellular functions including oxidative stress, apoptosis and aging via deacetylation of a variety of substrates [79]. A report indicates that inhibition of SIRT1 with either an inhibitor or siRNA leads to increased ROS levels, suggesting an association between SIRT1 and ROS. MiR-138, miR-181 and miR-199 have been shown to directly target and inhibit SIRT1 expression in various cell lines [80–82]. MiR-181 is induced by treatment with a high-fat diet and results in repressed SIRT1 expression and insulin sensitivity in the liver [81]. In addition, HIF1α and SIRT1 are upregulated in miR-199a-depleted cells during normoxic conditions [83]. Moreover, SIRT1 is actually a direct target gene of miR-199a and is responsible for suppressing prolyl hydroxylase 2.

4.3. Association between OS, miRNA and Metastasis

Metastasis is a complicated process that includes invasion, intravasation into blood, extravasation to distant organs and growth [84]. Due to these multiple steps, few metastasizing tumor cells can survive and form micrometastases. A typical phenotype that leads to metastasis is EMT, which is a biological event by which epithelial cells undergo alterations that induce the development of a more aggressive mesenchymal phenotype [84]. Increasing evidence suggests that cancer cells during the metastasis process are killed by OS [85,86]. In addition, cancer cells are more sensitive to ROS than normal cells. Reducing ROS levels by treating with antioxidant inhibits tumor promotion of tumor progression in mouse models. ROS-mediated EMT regulation through TGFβ/Smad, E-cadherin, Snail, integrin, β-catenin, matrix metalloproteinases (MMPs) and miRNA has been documented [87–89]. Among these interactions, activation of TGFβ induces ROS production and leads to the promotion of SMAD and ERK1/2 phosphorylation. Moreover, the ROS/TGFβ axis regulates EMT through the interaction of NFκB, HIF1α and cyclooxygenase-2 (COX-2). A previous study indicated that MMP-3, MMP-10 and MMP-13 were directly upregulated by oxidative treatment and promoted cell invasion ability in NMuMG cells [90]. In addition, the activity of MMP-2 and MMP-9 were posttranscriptionally regulated by oxidant treatment [91,92]. These studies suggest that MMP expression or activity is modulated by OS, which is related to chronic inflammation, malignant transformation and the invasive potential of cells. Yoon et al. [93] demonstrated that sustained treatment with H_2O_2 enhances MMP2 activity via the PDGF, VEGF, phosphatidylinositol 3-kinase and NF-κB pathways in HT1080 cell lines. Song and coworkers reported that the expression of miR-509 is significantly more downregulated in breast cancer than it is in normal tissues [94]. Overexpressed miR-509 abrogated cell growth, migration and invasion through inhibition of the target gene SOD2, which is a crucial effector in the production of ROS.

4.4. Association between OS, miRNA and Metabolism

Tumor progression is characterized by the occurrence of metabolic alterations, including those in glycolysis, fatty acid oxidation (FAO) and oxidative phosphorylation [95,96]. The connection and reciprocal regulation between the metabolism and the redox balance of tumor cells have been shown. For this reason, it is important to determine the major metabolic pathways that are the main controllers of the ROS homeostasis of cancer cells. Glucose is converted to glucose-6-phosphate by hexokinase enzyme and triggers a series of downstream enzyme-catalyzed reactions. It is an essential pathway for providing nutrients, metabolites and energy to cells. In 1924, Otto Warburg proposed a theory suggesting that tumor cells tend to exhibit glycolysis regardless of the presence of oxygen [97]. Accumulating evidence has shown that metabolites produced by glucose metabolism are major regulators of the redox homeostasis of tumor cells [98]. Cancer cells demonstrate increased sensitivity to glucose-deprivation-induced cytotoxicity compared with that in normal cells by restricting the burden of ROS. Moreover, inhibition of lactate dehydrogenase-A by a specific inhibitor, FX11, reduced intracellular ATP and promoted OS, which suppressed tumor progression in lymphoma and pancreatic cancer. Sala et al. [10] have shown that miR-21 is upregulated by glucose treatment and inhibits ROS homeostatic genes such as NRF2, SOD2 and KRIT1. Furthermore, other metabolic enzymes, such as TIGAR and ALDH4, decrease ROS production by either inhibiting glycolysis and inducing NAPDH production or enhancing mitochondrial function [99,100]. FAO consists of multiple processes by which fatty acids are broken down by cells to produce ATP and generate biosynthetic pathways. In general, the β-oxidation reaction takes place in mitochondria. FAO causes ROS formation and contributes to the enhanced development of nonalcoholic fatty liver disease (NAFLD) [101]. In hypoxia, HIF-1 suppression of medium-chain acyl-CoA dehydrogenase (MCAD) and light chain acyl-CoA dehydrogenase (LCAD) expression inhibits FAO and ROS production while promoting cell growth of liver cancer cells [102]. The expression levels of LCAD in HCC specimens were analyzed and found to be negatively correlated with survival. These findings indicate the relevance of FAO suppression in the progression of cancer. Previous studies have identified miR-33a/b as an intronic miRNA located with the sterol regulatory element binding factor (SREBP) 1 and 2 genes [103]. These two miRNAs cotranscribe with their host gene and regulate high density lipoprotein (HDL) biosynthesis.

4.5. Association between OS, miRNA and Cancer Stem Cells

Cancer cells are believed to be derived from a small subset of tumor cells that have a high capacity for self-renewal and differentiation—namely, cancer stem cells (CSCs) or tumor-initiating cells [104]. Increasing evidence indicates that miRNAs function as regulators of CSCs and are associated with ROS production during tumor progression and cancer development. Some miRNAs, such as let-7a, miR-21, miR-34a, miR-200 and miR-210, are potentially involved in the modulation of ROS production in CSCs [105–109]. A previous study showed that let-7 acts as a negative regulator of CSC-mediated function by targeting PTEN and LIN28b in prostate and pancreatic cancer. Recently, OS reduced let-7 expression in a p53-dependent manner in various cancer cells. Some experimental studies revealed that the expression of miR-21 is remarkably increased in CSC subpopulations compared to the expression in the hypobromite non-CSC counterparts in vitro and in vivo. Notably, knocking down miR-21 suppressed cell migration, invasion and EMT phenotype in breast cancer CSCs. Moreover, OS induced miR-21 expression and promoted cell migration and self-renewal in prostate and pancreatic CSCs. Another report indicates that miR-21 enhances ROS production via the MAPK pathway and suppresses SOD2, SOD3 and sprouty homolog 2 (SPRY-2) expression [110]. Additionally, a number of studies have revealed that miR-34a suppresses CSC-related genes, such as CD44, and EMT makers and subsequently attenuates cell invasion, metastasis and self-renewal capacity [111]. The interplay between ROS and miR-34 has been documented. The expression of miR-34 is induced by OS in stromal and tumor cells. The first evidence miR-200 was associated with stem cell phenotype, reported in 2009 [112]. Moreover, all five members of the miR-200 family were downregulated in human breast CSCs as well as in normal human and murine mammary stem/progenitor cells [112]. Mechanistically, miR-200 suppresses the expression of B lymphoma Mo-MLV insertion region 1 homolog (Bmil-1), Suz12, and Notch homolog 1 (Notch1), which are known regulators of CSC and EMT phenotypes, and inhibits the CSC self-renewal capacity. MiR-210 expression is enriched in MCF-7 spheroid cells and CD44$^+$/CD24$^-$ MCF7 cells compared with MCF-7 parental cells [113]. Overexpression of miR-210 enhances proliferation, self-renewal capacity, migration and invasion through inhibition of E-cadherin in vitro and in vivo. Thus, these observations indicate that the miRNA/ROS axis plays important roles in multiple events related to CSCs.

4.6. Association between OS, miRNA and Senescence

Cellular senescence is characterized by the expression of senescence-associated β-galactosidase (SA-β-gal), overexpression of the cyclin-dependent kinase (CDK) inhibitor, senescence-associated secretory phenotype (SASP), telomere shortening and persistent DNA damage response (DDR) [114]. ROS cause cell senescence by stimulating the DDR pathway to stabilize p53 and promote CDK inhibitor gene expression. In fact, p53 acts as a master regulator in the cellular response to OS. Mechanically, p53 can decrease ROS levels and repair DNA damage in cells. In contrast, it can also enhance ROS production and promote cell apoptosis or senescence [115]. Several reports indicate that p53 reduces intracellular ROS levels by promoting antioxidant reactions. Several miRNAs, including miR-21, miR-22, miR-29, miR-34a, miR-106b, miR-125b, miR-126, miR-146a, the miR-17-92 cluster, the miR-200 family and miR-210, have been identified to be differentially expressed in senescent cells and to be involved in cellular senescence [116–122]. Notably, miR-34a was found to promote cellular senescence by inhibiting SIRT1 expression in a variety of tissues. Another group indicated that miR-34a and miR-335 promote premature cellular senescence by targeting antioxidative enzymes. Furthermore, miR-217 induces a premature senescence-like phenotype and represses angiogenesis by inhibiting the expression of target gene SIRT1 in endothelial cells [123]. In addition, miR-92a was found to exacerbate endothelial dysfunction under OS exposure by directly targeting SIRT1, Krüppel-like factor 2 (KLF2) and KLF4 genes [124]. Additionally, Liu et al. group demonstrated that knockdown of miR-92a promoted cell growth, decreased caspase 3 activity and ROS through regulation of NRF2-KEAP1/ARE signal pathway [125].

5. ROS-Mediated Therapeutic Strategies in Cancer

OS clearly plays a role in the development of cancer, metastasis and chemotherapeutic resistance. In a strategy to modulate ROS-mediated effects, these biochemical characteristics of tumors are directly impaired. In light of recent studies, the strategy of inhibiting metabolic pathways, targeting NADPH oxidase and ROS scavenging mechanisms represent promising therapeutic options for treatments [126]. The other strategy is to target tumor cells with oxidation-promoting agents that either enhance ROS production or inhibit cellular antioxidants. NADPH oxidase plays an important role in regulating ROS production. Several inhibitors have been demonstrated to reduce NADPH function. In general, diphenylene iodonium (DPI) and apocynin are NADPH inhibitors [127–129]. DPI can inhibit XOD and the proteins of mitochondrial ETC and block flavoprotein. In another strategy, ROS scavenging enzymes are enhanced and used for anticancer therapy [130]. GSH, GST, SOD, GPX, and catalase are able to suppress tumor formation. There are several analogs of GSH drugs, such as N-acetylcysteine (NAC), YM737 and Telcyta, used for cancer treatment [131]. NOV-002, an agent containing oxidized GSH, improved the efficacy of cyclophosphamide to treat colon cancer by controlling the ratio of GSH to GSSG and promoting S-glutathionylation [131]. Yang and coworkers indicated that lithocholic acid treatment and bile duct ligation model promoted c-Myc/miR-27/prohibitin 1 axis, with the consequence of repressing NRF2 expression and ARE binding, resulting in decreased suppressed GSH synthesis and antioxidant ability in chronic cholestatic liver injury [132]. Another study reported that the rate-limiting GSH biosynthetic heterodimeric enzyme γ-glutamyl-cysteine ligase (GCL) was regulated by miR-433 [133]. Ectopic of miR-433 in HUVEC inhibited GCL expression in an NRF2-independent manner. Moreover, inhibition of miR-433 prevented TGFβ-mediated GCL downregulation and fibrogenesis in hepatic cells. Recently, Cheng et al. [134] demonstrated that miR-30e expression was suppressed in an atherosclerosis (AS) model. MiR-30e regulates Snail1/TGF-β/Nox4 expression to modulate ROS. These findings provide novel insights on miRNAs in the anti-ROS pathway, in which miRNA-30e may represent a novel target for AS.

6. Conclusions

Overall, many studies have been conducted to elucidate the molecular mechanisms underlying the ROS/miRNA axis and its role in tumorigenesis (Figure 3). Moreover, miRNAs networks that modulate OS in cancer are comprehensively listed in Table 1. Indeed, ROS and miRNAs exhibit overlapping characteristics in tumorigenesis. ROS, as upstream regulators, modulate miRNA expression through transcriptional, posttranscriptional and epigenetic regulation, respectively. On the other hand, miRNAs disrupt ROS production (downstream mediator) and are involved in ROS-mediated functions. MiRNAs and ROS can act either synergistically or antagonistically to regulate cancer progression. However, many details of their interaction remain unclear and need to be further investigated. MiRNAs/ROS-mediated phenotypes depend on the net result of the downstream molecules and multiple signaling pathways in the specific context. There are still many limitations to treatment because ROS play dual roles in cancer progression. As discussed in this review, the functional roles of miRNA in cellular adaptation to ROS are different in cells based on tissue and cell-type specific effects. These observations raise the possibilities to apply specific miRNAs as therapeutic targets in different contexts. Advantages of using miRNA-target therapy include the conservation of miRNA across multiple species with known sequences and the ability to target multiple genes within defined pathways. Notably, several miRNA-based therapies are being developed. For example, the locked nucleic acid (LNA)-modified anti-miR-122 is the first miRNA-targeted therapy to treat HCV in clinical trials. The association between ROS-mediated function and miRNA regulation provides opportunities for developing novel anticancer strategies.

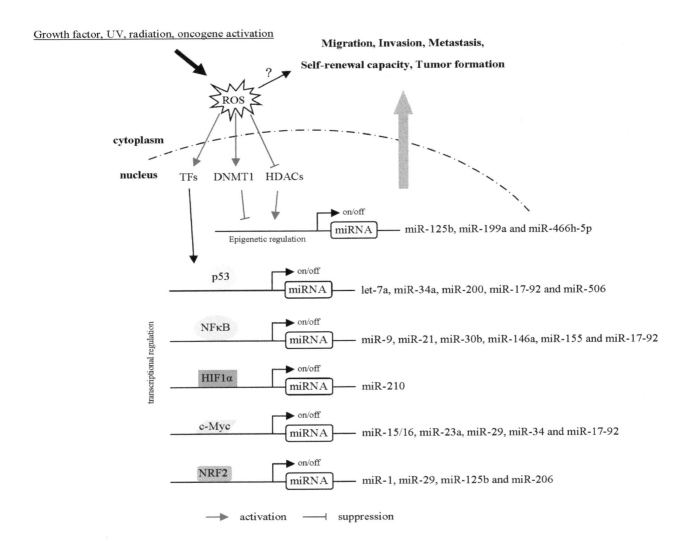

Figure 3. Schematic model showing mechanisms in which ROS regulates the biogenesis and transcription of miRNAs. ROS activate or inhibit epigenetic, transcriptional regulations of miRNA expression. For example, miRNAs are regulated by ROS through modulation of chromatin remodeling factors (DNMT1 and HDACs). In addition, ROS induces or represses transcriptional factor (p53, NFκB, HIF1α, c-Myc and NRF2) to regulate miRNA expressions. Furthermore, ROS/TF/miRNA axis controls cell migration, invasion, metastasis, self-renewal capacity and tumor formation.

Table 1. ROS-related miRNAs and their potential mechanisms in cancers.

miRNA	Regulation Mechanism [a]	ROS Production [b]	Expression in Cancer [c]	Cell/Cancer Types	Molecules, Cellular Processes and Signaling Pathways Involved [d]	References
Let-7a	OS, p53	✓	Down	CSC, prostate cancer, pancreatic cancer	PTEN, LIN28b	[105,135]
miR-1	NRF2, HDAC4	✓	Down	Non-small cell lung cancer	NRF2, KEAP1, glucose metabolism, tumor growth	[49]
miR-15/16	c-Myc	✓	Down	Skin, colon cancer	FGF2, HIF-2α, senescence-like phenotype, angiogenesis, metastasis	[30]
miR-21	Glucose, NFκB, STAT3	✓	Up	CSCs, lung cancer, liver cancer, colorectal cancer	MAPK pathway, cell migration, invasion and EMT phenotype, self-renewal ability	[41,110]
miR-23a	c-Myc	✓	-	Cardiac disease, myeloma	Glutaminase, MnSOD, apoptosis, cell growth	[136–138]
miR-29	c-Myc, H_2O_2, NRF2	✓	Dual role	Ovarian cancer, lung cancer, lymphoma	SIRT1, senescence, proliferation, apoptosis	[50,139–142]
miR-33a/b	-	✓	Down	Liver	HDL biosynthesis, apoptosis, OS resistance	[103]
miR-34	OS, c-Myc, p53	✓	Down	Stromal cells, CSC, bladder cancer, lung cancer	CD44, EMT markers, SIRT1, senescence, metastasis	[33,111,143]
miR-17-92	c-Myc, p53, NFκB	✓	Up	Lung cancer,	Vitamin D, Senescence, apoptosis	[120,144–146]
miR-92a	-	✓	Up	Endothelial cells	SIRT1, KLF2, KLF4	[124,125]
miR-125b	DNMT1, H_2O_2, NRF2	✓	Dual role	Ovarian cancer, liver	Epigenetic regulation	[51,57]
miR-181	-	✓	Up	Macrophagy, HCC	SIRT1, insulin sensitivity, NFκB activity, apoptosis	[80]
miR-199a	DNMT1, H_2O_2	✓	Down, (hypermethylation)	Ovarian cancer	HIF1α, SIRT1, Epigenetic regulation	[57,83]
miR-200	P53, H_2O_2	✓	Down	CSC, breast cancer, liver cancer	Bmi1-1, Suz12, Notch-1, self-renewal capacity, EMT markers, senescence	[34,35]
miR-210	Hypoxia	✓	Up	CSCs	E-cadherin, Hypoxia, proliferation, self-renewal capacity, migration and invasion, senescence	[65,109] [113]

Table 1. *Cont.*

miRNA	Regulation Mechanism [a]	ROS Production [b]	Expression in Cancer [c]	Cell/Cancer Types	Molecules, Cellular Processes and Signaling Pathways involved [d]	References
miR-217	-	-	Dual role	Endothelial cells	SIRT1, Angiogenesis, premature senescence-like phenotype	[123]
miR-466h-5p	ROS, HDAC2	-	-	Mouse ovarian epithelial	BCL2L2, apoptosis	[58]
MiR-506	P53	√	Down	Lung cancer	NFκB signaling pathway	[44]
miR-509	-	√	Down	Breast cancer	SOD2, Cell growth, migration and invasion	[94]

[a]: MiRNAs are regulated by upstream transcriptional factor, ROS or hypoxia, as indicated. -: Information is unavailable. [b]: √: MiRNAs are responsible for producing ROS. -: Information is unavailable. [c]: Expression level of miRNAs in cancer. Up: upregulated in cancer, Down: downregulated in cancer, Dual role: up- or downregulated in cancer. [d]: Downstream molecules, signaling pathways and phenotypes involved in miRNA-mediated functions.

References

1.   Dong, Y.; Xu, W.; Liu, C.; Liu, P.; Li, P.; Wang, K. Reactive oxygen species related noncoding RNAs as regulators of cardiovascular diseases. *Int. J. Biol. Sci.* **2019**, *15*, 680–687. [CrossRef] [PubMed]

2.   Schieber, M.; Chandel, N.S. ROS function in redox signaling and oxidative stress. *Curr. Biol.* **2014**, *24*, R453–R462. [CrossRef] [PubMed]

3.   Engedal, N.; Zerovnik, E.; Rudov, A.; Galli, F.; Olivieri, F.; Procopio, A.D.; Rippo, M.R.; Monsurro, V.; Betti, M.; Albertini, M.C. From oxidative stress damage to pathways, networks, and autophagy via microRNAs. *Oxid. Med. Cell. Longev.* **2018**, *2018*, 4968321. [CrossRef] [PubMed]

4.   Castaneda-Arriaga, R.; Perez-Gonzalez, A.; Reina, M.; Alvarez-Idaboy, J.R.; Galano, A. Comprehensive investigation of the antioxidant and pro-oxidant effects of phenolic compounds: A double-edged sword in the context of oxidative stress? *J. Phys. Chem. B* **2018**, *122*, 6198–6214. [CrossRef] [PubMed]

5.   O'Brien, J.; Hayder, H.; Zayed, Y.; Peng, C. Overview of microRNA biogenesis, mechanisms of actions, and circulation. *Front. Endocrinol. (Lausanne)* **2018**, *9*, 402. [CrossRef]

6.   Tan, W.; Liu, B.; Qu, S.; Liang, G.; Luo, W.; Gong, C. MicroRNAs and cancer: Key paradigms in molecular therapy. *Oncol. Lett.* **2018**, *15*, 2735–2742. [CrossRef]

7.   Liu, Y.; Qiang, W.; Xu, X.; Dong, R.; Karst, A.M.; Liu, Z.; Kong, B.; Drapkin, R.I.; Wei, J.J. Role of miR-182 in response to oxidative stress in the cell fate of human fallopian tube epithelial cells. *Oncotarget* **2015**, *6*, 38983–38998. [CrossRef]

8.   Fierro-Fernandez, M.; Miguel, V.; Lamas, S. Role of redoximiRs in fibrogenesis. *Redox Biol.* **2016**, *7*, 58–67. [CrossRef]

9.   Meseguer, S.; Martinez-Zamora, A.; Garcia-Arumi, E.; Andreu, A.L.; Armengod, M.E. The ROS-sensitive microRNA-9/9* controls the expression of mitochondrial tRNA-modifying enzymes and is involved in the molecular mechanism of MELAS syndrome. *Hum. Mol. Genet.* **2015**, *24*, 167–184. [CrossRef]

10.   La Sala, L.; Mrakic-Sposta, S.; Micheloni, S.; Prattichizzo, F.; Ceriello, A. Glucose-sensing microRNA-21 disrupts ROS homeostasis and impairs antioxidant responses in cellular glucose variability. *Cardiovasc. Diabetol.* **2018**, *17*, 105. [CrossRef]

11.   Saki, M.; Prakash, A. DNA damage related crosstalk between the nucleus and mitochondria. *Free Radic. Biol. Med.* **2017**, *107*, 216–227. [CrossRef] [PubMed]

12.   Case, A.J. On the origin of superoxide dismutase: An evolutionary perspective of superoxide-mediated redox signaling. *Antioxidants (Basel)* **2017**, *6*, 82. [CrossRef] [PubMed]

13.   Meitzler, J.L.; Antony, S.; Wu, Y.; Juhasz, A.; Liu, H.; Jiang, G.; Lu, J.; Roy, K.; Doroshow, J.H. Nadph oxidases: A perspective on reactive oxygen species production in tumor biology. *Antioxid. Redox Signal.* **2014**, *20*, 2873–2889. [CrossRef] [PubMed]

14.   Collin, F. Chemical basis of reactive oxygen species reactivity and involvement in neurodegenerative diseases. *Int. J. Mol. Sci.* **2019**, *20*, 2407. [CrossRef] [PubMed]

15.   Pylvas, M.; Puistola, U.; Laatio, L.; Kauppila, S.; Karihtala, P. Elevated serum 8-OHdG is associated with poor prognosis in epithelial ovarian cancer. *Anticancer Res.* **2011**, *31*, 1411–1415. [PubMed]

16.   Plachetka, A.; Adamek, B.; Strzelczyk, J.K.; Krakowczyk, L.; Migula, P.; Nowak, P.; Wiczkowski, A. 8-hydroxy-2'-deoxyguanosine in colorectal adenocarcinoma–is it a result of oxidative stress? *Med. Sci. Monit.* **2013**, *19*, 690–695. [PubMed]

17.   Ma-On, C.; Sanpavat, A.; Whongsiri, P.; Suwannasin, S.; Hirankarn, N.; Tangkijvanich, P.; Boonla, C. Oxidative stress indicated by elevated expression of NRF2 and 8-OHdG promotes hepatocellular carcinoma progression. *Med. Oncol.* **2017**, *34*, 57. [CrossRef]

18.   Tanaka, H.; Fujita, N.; Sugimoto, R.; Urawa, N.; Horiike, S.; Kobayashi, Y.; Iwasa, M.; Ma, N.; Kawanishi, S.; Watanabe, S.; et al. Hepatic oxidative DNA damage is associated with increased risk for hepatocellular carcinoma in chronic hepatitis c. *Br. J. Cancer* **2008**, *98*, 580–586. [CrossRef]

19.   Akhtar, M.J.; Ahamed, M.; Alhadlaq, H.A.; Alshamsan, A. Mechanism of ROS scavenging and antioxidant signalling by redox metallic and fullerene nanomaterials: Potential implications in ROS associated degenerative disorders. *Biochim. Biophys. Acta Gen. Subj.* **2017**, *1861*, 802–813. [CrossRef]

20.   He, J.; Jiang, B.H. Interplay between reactive oxygen species and microRNAs in cancer. *Curr. Pharmacol. Rep.* **2016**, *2*, 82–90. [CrossRef]

21. Gong, Y.Y.; Luo, J.Y.; Wang, L.; Huang, Y. MicroRNAs regulating reactive oxygen species in cardiovascular diseases. *Antioxid. Redox Signal.* **2018**, *29*, 1092–1107. [CrossRef] [PubMed]

22. Lan, J.; Huang, Z.; Han, J.; Shao, J.; Huang, C. Redox regulation of microRNAs in cancer. *Cancer Lett.* **2018**, *418*, 250–259. [CrossRef] [PubMed]

23. Ungvari, Z.; Tucsek, Z.; Sosnowska, D.; Toth, P.; Gautam, T.; Podlutsky, A.; Csiszar, A.; Losonczy, G.; Valcarcel-Ares, M.N.; Sonntag, W.E.; et al. Aging-induced dysregulation of dicer1-dependent microRNA expression impairs angiogenic capacity of rat cerebromicrovascular endothelial cells. *J. Gerontol. A Biol. Sci. Med. Sci.* **2013**, *68*, 877–891. [CrossRef] [PubMed]

24. Shilo, S.; Roy, S.; Khanna, S.; Sen, C.K. Evidence for the involvement of miRNA in redox regulated angiogenic response of human microvascular endothelial cells. *Arterioscler. Thromb. Vasc. Biol.* **2008**, *28*, 471–477. [CrossRef] [PubMed]

25. Kozakowska, M.; Ciesla, M.; Stefanska, A.; Skrzypek, K.; Was, H.; Jazwa, A.; Grochot-Przeczek, A.; Kotlinowski, J.; Szymula, A.; Bartelik, A.; et al. Heme oxygenase-1 inhibits myoblast differentiation by targeting myomirs. *Antioxid. Redox Signal.* **2012**, *16*, 113–127. [CrossRef] [PubMed]

26. Markopoulos, G.S.; Roupakia, E.; Tokamani, M.; Alabasi, G.; Sandaltzopoulos, R.; Marcu, K.B.; Kolettas, E. Roles of NF-kB signaling in the regulation of miRNAs impacting on inflammation in cancer. *Biomedicines* **2018**, *6*, 40. [CrossRef]

27. Graves, J.A.; Metukuri, M.; Scott, D.; Rothermund, K.; Prochownik, E.V. Regulation of reactive oxygen species homeostasis by peroxiredoxins and c-myc. *J. Biol. Chem.* **2009**, *284*, 6520–6529. [CrossRef]

28. Ferro, E.; Goitre, L.; Retta, S.F.; Trabalzini, L. The interplay between ROS and Ras GTPases: Physiological and pathological implications. *J. Signal. Transduct.* **2012**, *2012*, 365769. [CrossRef]

29. Jackstadt, R.; Hermeking, H. MicroRNAs as regulators and mediators of c-Myc function. *Biochim. Biophys. Acta* **2015**, *1849*, 544–553. [CrossRef]

30. Xue, G.; Yan, H.L.; Zhang, Y.; Hao, L.Q.; Zhu, X.T.; Mei, Q.; Sun, S.H. c-MYC-mediated repression of miR-15-16 in hypoxia is induced by increased HIF-2alpha and promotes tumor angiogenesis and metastasis by upregulating FGF2. *Oncogene* **2015**, *34*, 1393–1406. [CrossRef]

31. Li, S.G.; Shi, Q.W.; Yuan, L.Y.; Qin, L.P.; Wang, Y.; Miao, Y.Q.; Chen, Z.; Ling, C.Q.; Qin, W.X. c-Myc-dependent repression of two oncogenic miRNA clusters contributes to triptolide-induced cell death in hepatocellular carcinoma cells. *J. Exp. Clin. Cancer Res.* **2018**, *37*, 51. [CrossRef] [PubMed]

32. Gurtner, A.; Falcone, E.; Garibaldi, F.; Piaggio, G. Dysregulation of microRNA biogenesis in cancer: The impact of mutant p53 on drosha complex activity. *J. Exp. Clin. Cancer Res.* **2016**, *35*, 45. [CrossRef] [PubMed]

33. Baker, J.R.; Vuppusetty, C.; Colley, T.; Papaioannou, A.I.; Fenwick, P.; Donnelly, L.; Ito, K.; Barnes, P.J. Oxidative stress dependent microRNA-34a activation via pi3kalpha reduces the expression of sirtuin-1 and sirtuin-6 in epithelial cells. *Sci. Rep.* **2016**, *6*, 35871. [CrossRef] [PubMed]

34. Xiao, Y.; Yan, W.; Lu, L.; Wang, Y.; Lu, W.; Cao, Y.; Cai, W. P38/p53/miR-200a-3p feedback loop promotes oxidative stress-mediated liver cell death. *Cell Cycle* **2015**, *14*, 1548–1558. [CrossRef] [PubMed]

35. Magenta, A.; Cencioni, C.; Fasanaro, P.; Zaccagnini, G.; Greco, S.; Sarra-Ferraris, G.; Antonini, A.; Martelli, F.; Capogrossi, M.C. MiR-200c is upregulated by oxidative stress and induces endothelial cell apoptosis and senescence via zeb1 inhibition. *Cell Death Differ.* **2011**, *18*, 1628–1639. [CrossRef] [PubMed]

36. Wu, J.; Ding, J.; Yang, J.; Guo, X.; Zheng, Y. MicroRNA roles in the nuclear factor kappa b signaling pathway in cancer. *Front. Immunol.* **2018**, *9*, 546. [CrossRef] [PubMed]

37. Ma, X.; Becker Buscaglia, L.E.; Barker, J.R.; Li, Y. MicroRNAs in NF-kappaB signaling. *J. Mol. Cell Biol.* **2011**, *3*, 159–166. [CrossRef] [PubMed]

38. Bazzoni, F.; Rossato, M.; Fabbri, M.; Gaudiosi, D.; Mirolo, M.; Mori, L.; Tamassia, N.; Mantovani, A.; Cassatella, M.A.; Locati, M. Induction and regulatory function of miR-9 in human monocytes and neutrophils exposed to proinflammatory signals. *Proc. Natl. Acad. Sci. USA* **2009**, *106*, 5282–5287. [CrossRef] [PubMed]

39. Niu, J.; Shi, Y.; Tan, G.; Yang, C.H.; Fan, M.; Pfeffer, L.M.; Wu, Z.H. DNA damage induces NF-kappaB-dependent microRNA-21 up-regulation and promotes breast cancer cell invasion. *J. Biol. Chem.* **2012**, *287*, 21783–21795. [CrossRef] [PubMed]

40. Chao, J.; Guo, Y.; Li, P.; Chao, L. Role of kallistatin treatment in aging and cancer by modulating miR-34a and miR-21 expression. *Oxid. Med. Cell. Longev.* **2017**, *2017*, 5025610. [CrossRef]

41. Pratheeshkumar, P.; Son, Y.O.; Divya, S.P.; Wang, L.; Zhang, Z.; Shi, X. Oncogenic transformation

of human lung bronchial epithelial cells induced by arsenic involves ROS-dependent activation of STAT3-miR-21-PDCD4 mechanism. *Sci. Rep.* **2016**, *6*, 37227. [CrossRef] [PubMed]

42. Zhou, R.; Hu, G.; Gong, A.Y.; Chen, X.M. Binding of NF-kappaB p65 subunit to the promoter elements is involved in LPS-induced transactivation of miRNA genes in human biliary epithelial cells. *Nucleic Acids Res.* **2010**, *38*, 3222–3232. [CrossRef] [PubMed]

43. Feng, X.; Tan, W.; Cheng, S.; Wang, H.; Ye, S.; Yu, C.; He, Y.; Zeng, J.; Cen, J.; Hu, J.; et al. Upregulation of microRNA-126 in hepatic stellate cells may affect pathogenesis of liver fibrosis through the NF-kappaB pathway. *DNA Cell Biol.* **2015**, *34*, 470–480. [CrossRef] [PubMed]

44. Yin, M.; Ren, X.; Zhang, X.; Luo, Y.; Wang, G.; Huang, K.; Feng, S.; Bao, X.; Huang, K.; He, X.; et al. Selective killing of lung cancer cells by miRNA-506 molecule through inhibiting NF-kappaB p65 to evoke reactive oxygen species generation and p53 activation. *Oncogene* **2015**, *34*, 691–703. [CrossRef] [PubMed]

45. Sykiotis, G.P.; Bohmann, D. Stress-activated cap'n'collar transcription factors in aging and human disease. *Sci. Signal.* **2010**, *3*, re3. [CrossRef] [PubMed]

46. Bellezza, I.; Giambanco, I.; Minelli, A.; Donato, R. Nrf2-Keap1 signaling in oxidative and reductive stress. *Biochim. Biophys. Acta Mol. Cell Res.* **2018**, *1865*, 721–733. [CrossRef]

47. Zhang, C.; Shu, L.; Kong, A.N. MicroRNAs: New players in cancer prevention targeting Nrf2, oxidative stress and inflammatory pathways. *Curr. Pharmacol. Rep.* **2015**, *1*, 21–30. [CrossRef] [PubMed]

48. Cheng, X.H.; Ku, C.H.; Siow, R.C.M. Regulation of the Nrf2 antioxidant pathway by microRNAs: New players in micromanaging redox homeostasis. *Free Radic. Biol. Med.* **2013**, *64*, 4–11. [CrossRef]

49. Singh, A.; Happel, C.; Manna, S.K.; Acquaah-Mensah, G.; Carrerero, J.; Kumar, S.; Nasipuri, P.; Krausz, K.W.; Wakabayashi, N.; Dewi, R.; et al. Transcription factor Nrf2 regulates miR-1 and miR-206 to drive tumorigenesis. *J. Clin. Investig.* **2013**, *123*, 2921–2934. [CrossRef]

50. Kurinna, S.; Schafer, M.; Ostano, P.; Karouzakis, E.; Chiorino, G.; Bloch, W.; Bachmann, A.; Gay, S.; Garrod, D.; Lefort, K.; et al. A novel Nrf2-miR-29-desmocollin-2 axis regulates desmosome function in keratinocytes. *Nat. Commun.* **2014**, *5*, 5099. [CrossRef]

51. Joo, M.S.; Lee, C.G.; Koo, J.H.; Kim, S.G. Mir-125b transcriptionally increased by nrf2 inhibits AHR repressor, which protects kidney from cisplatin-induced injury. *Cell Death Dis.* **2013**, *4*, e899. [CrossRef] [PubMed]

52. Yang, M.; Yao, Y.; Eades, G.; Zhang, Y.; Zhou, Q. Mir-28 regulates Nrf2 expression through a Keap1-independent mechanism. *Breast Cancer Res. Treat.* **2011**, *129*, 983–991. [CrossRef] [PubMed]

53. Huang, X.; Gao, Y.; Qin, J.; Lu, S. The role of mir-34a in the hepatoprotective effect of hydrogen sulfide on ischemia/reperfusion injury in young and old rats. *PLoS ONE* **2014**, *9*, e113305. [CrossRef] [PubMed]

54. Singh, B.; Ronghe, A.M.; Chatterjee, A.; Bhat, N.K.; Bhat, H.K. MicroRNA-93 regulates Nrf2 expression and is associated with breast carcinogenesis. *Carcinogenesis* **2013**, *34*, 1165–1172. [CrossRef] [PubMed]

55. Eades, G.; Yang, M.; Yao, Y.; Zhang, Y.; Zhou, Q. Mir-200a regulates Nrf2 activation by targeting Keap1 mRNA in breast cancer cells. *J. Biol. Chem.* **2011**, *286*, 40725–40733. [CrossRef] [PubMed]

56. Nebbioso, A.; Tambaro, F.P.; Dell'Aversana, C.; Altucci, L. Cancer epigenetics: Moving forward. *PLoS Genet.* **2018**, *14*, e1007362. [CrossRef] [PubMed]

57. He, J.; Xu, Q.; Jing, Y.; Agani, F.; Qian, X.; Carpenter, R.; Li, Q.; Wang, X.R.; Peiper, S.S.; Lu, Z.; et al. Reactive oxygen species regulate ERBB2 and ERBB3 expression via miR-199a/125b and DNA methylation. *EMBO Rep.* **2012**, *13*, 1116–1122. [CrossRef] [PubMed]

58. Druz, A.; Betenbaugh, M.; Shiloach, J. Glucose depletion activates mmu-miR-466h-5p expression through oxidative stress and inhibition of histone deacetylation. *Nucleic Acids Res.* **2012**, *40*, 7291–7302. [CrossRef]

59. Muz, B.; de la Puente, P.; Azab, F.; Azab, A.K. The role of hypoxia in cancer progression, angiogenesis, metastasis, and resistance to therapy. *Hypoxia* **2015**, *3*, 83–92. [CrossRef]

60. Azimi, I.; Petersen, R.M.; Thompson, E.W.; Roberts-Thomson, S.J.; Monteith, G.R. Hypoxia-induced reactive oxygen species mediate n-cadherin and serpine1 expression, EGFR signalling and motility in mda-mb-468 breast cancer cells. *Sci. Rep.* **2017**, *7*, 15140. [CrossRef]

61. Cardin, R.; Piciocchi, M.; Sinigaglia, A.; Lavezzo, E.; Bortolami, M.; Kotsafti, A.; Cillo, U.; Zanus, G.; Mescoli, C.; Rugge, M.; et al. Oxidative DNA damage correlates with cell immortalization and miR-92 expression in hepatocellular carcinoma. *BMC Cancer* **2012**, *12*, 177. [CrossRef]

62. Yatabe, N.; Kyo, S.; Maida, Y.; Nishi, H.; Nakamura, M.; Kanaya, T.; Tanaka, M.; Isaka, K.; Ogawa, S.; Inoue, M. HIF-1-mediated activation of telomerase in cervical cancer cells. *Oncogene* **2004**, *23*, 3708–3715. [CrossRef] [PubMed]

63. Jung, N.; Kwon, H.J.; Jung, H.J. Downregulation of mitochondrial UQCRB inhibits cancer stem cell-like properties in glioblastoma. *Int. J. Oncol.* **2018**, *52*, 241–251. [CrossRef] [PubMed]

64. Corcoran, S.E.; O'Neill, L.A. HIF1alpha and metabolic reprogramming in inflammation. *J. Clin. Investig.* **2016**, *126*, 3699–3707. [CrossRef] [PubMed]

65. Bavelloni, A.; Ramazzotti, G.; Poli, A.; Piazzi, M.; Focaccia, E.; Blalock, W.; Faenza, I. Mirna-210: A current overview. *Anticancer Res.* **2017**, *37*, 6511–6521.

66. Fallah, A.; Sadeghinia, A.; Kahroba, H.; Samadi, A.; Heidari, H.R.; Bradaran, B.; Zeinali, S.; Molavi, O. Therapeutic targeting of angiogenesis molecular pathways in angiogenesis-dependent diseases. *Biomed. Pharmacother.* **2019**, *110*, 775–785. [CrossRef]

67. Hoeben, A.; Landuyt, B.; Highley, M.S.; Wildiers, H.; Van Oosterom, A.T.; De Bruijn, E.A. Vascular endothelial growth factor and angiogenesis. *Pharmacol. Rev.* **2004**, *56*, 549–580. [CrossRef]

68. Ferrari, G.; Cook, B.D.; Terushkin, V.; Pintucci, G.; Mignatti, P. Transforming growth factor-beta 1 (TGF-beta1) induces angiogenesis through vascular endothelial growth factor (VEGF)-mediated apoptosis. *J. Cell. Physiol.* **2009**, *219*, 449–458. [CrossRef]

69. Fagiani, E.; Christofori, G. Angiopoietins in angiogenesis. *Cancer Lett.* **2013**, *328*, 18–26. [CrossRef]

70. Raica, M.; Cimpean, A.M. Platelet-derived growth factor (PDGF)/PDGFreceptors (PDGFR) axis as target for antitumor and antiangiogenic therapy. *Pharmaceuticals* **2010**, *3*, 572–599. [CrossRef]

71. Espinosa-Diez, C.; Wilson, R.; Chatterjee, N.; Hudson, C.; Ruhl, R.; Hipfinger, C.; Helms, E.; Khan, O.F.; Anderson, D.G.; Anand, S. MicroRNA regulation of the MRN complex impacts DNA damage, cellular senescence, and angiogenic signaling. *Cell Death Dis.* **2018**, *9*, 632. [CrossRef] [PubMed]

72. Chu, L.Y.; Ramakrishnan, D.P.; Silverstein, R.L. Thrombospondin-1 modulates VEGF signaling via CD36 by recruiting SHP-1 to VEGFR2 complex in microvascular endothelial cells. *Blood* **2013**, *122*, 1822–1832. [CrossRef] [PubMed]

73. Guo, Z.; Kozlov, S.; Lavin, M.F.; Person, M.D.; Paull, T.T. Atm activation by oxidative stress. *Science* **2010**, *330*, 517–521. [CrossRef] [PubMed]

74. Okuno, Y.; Nakamura-Ishizu, A.; Otsu, K.; Suda, T.; Kubota, Y. Pathological neoangiogenesis depends on oxidative stress regulation by atm. *Nat. Med.* **2012**, *18*, 1208–1216. [CrossRef] [PubMed]

75. Economopoulou, M.; Langer, H.F.; Celeste, A.; Orlova, V.V.; Choi, E.Y.; Ma, M.; Vassilopoulos, A.; Callen, E.; Deng, C.; Bassing, C.H.; et al. Histone h2ax is integral to hypoxia-driven neovascularization. *Nat. Med.* **2009**, *15*, 553–558. [CrossRef] [PubMed]

76. Wilson, R.; Espinosa-Diez, C.; Kanner, N.; Chatterjee, N.; Ruhl, R.; Hipfinger, C.; Advani, S.J.; Li, J.; Khan, O.F.; Franovic, A.; et al. MicroRNA regulation of endothelial TREX1 reprograms the tumour microenvironment. *Nat. Commun.* **2016**, *7*, 13597. [CrossRef]

77. Yang, Z.; Wa, Q.D.; Lu, C.; Pan, W.; Lu, Z.; Ao, J. Mir3283p enhances the radiosensitivity of osteosarcoma and regulates apoptosis and cell viability via H2AX. *Oncol. Rep.* **2018**, *39*, 545–553. [PubMed]

78. Marampon, F.; Codenotti, S.; Megiorni, F.; Del Fattore, A.; Camero, S.; Gravina, G.L.; Festuccia, C.; Musio, D.; De Felice, F.; Nardone, V.; et al. NRF2 orchestrates the redox regulation induced by radiation therapy, sustaining embryonal and alveolar rhabdomyosarcoma cells radioresistance. *J. Cancer Res. Clin. Oncol.* **2019**, *145*, 881–893. [CrossRef] [PubMed]

79. Salminen, A.; Kaarniranta, K.; Kauppinen, A. Crosstalk between oxidative stress and SIRT1: Impact on the aging process. *Int. J. Mol. Sci.* **2013**, *14*, 3834–3859. [CrossRef]

80. Luo, J.; Chen, P.; Xie, W.; Wu, F. MicroRNA-138 inhibits cell proliferation in hepatocellular carcinoma by targeting SIRT1. *Oncol. Rep.* **2017**, *38*, 1067–1074. [CrossRef]

81. Zhou, B.; Li, C.; Qi, W.; Zhang, Y.; Zhang, F.; Wu, J.X.; Hu, Y.N.; Wu, D.M.; Liu, Y.; Yan, T.T.; et al. Downregulation of miR-181a upregulates sirtuin-1 (SIRT1) and improves hepatic insulin sensitivity. *Diabetologia* **2012**, *55*, 2032–2043. [CrossRef] [PubMed]

82. Yamakuchi, M. MicroRNA regulation of SIRT1. *Front. Physiol.* **2012**, *3*, 68. [CrossRef] [PubMed]

83. Rane, S.; He, M.; Sayed, D.; Vashistha, H.; Malhotra, A.; Sadoshima, J.; Vatner, D.E.; Vatner, S.F.; Abdellatif, M. Downregulation of miR-199a derepresses hypoxia-inducible factor-1alpha and Sirtuin 1 and recapitulates hypoxia preconditioning in cardiac myocytes. *Circ. Res.* **2009**, *104*, 879–886. [CrossRef] [PubMed]

84. Lambert, A.W.; Pattabiraman, D.R.; Weinberg, R.A. Emerging biological principles of metastasis. *Cell* **2017**, *168*, 670–691. [CrossRef] [PubMed]

85. Peiris-Pages, M.; Martinez-Outschoorn, U.E.; Sotgia, F.; Lisanti, M.P. Metastasis and oxidative stress: Are antioxidants a metabolic driver of progression? *Cell Metab.* **2015**, *22*, 956–958. [CrossRef] [PubMed]

86. Gill, J.G.; Piskounova, E.; Morrison, S.J. Cancer, oxidative stress, and metastasis. *Cold Spring Harb. Symp. Quant. Biol.* **2016**, *81*, 163–175. [CrossRef] [PubMed]

87. Hsieh, C.L.; Liu, C.M.; Chen, H.A.; Yang, S.T.; Shigemura, K.; Kitagawa, K.; Yamamichi, F.; Fujisawa, M.; Liu, Y.R.; Lee, W.H.; et al. Reactive oxygen species-mediated switching expression of MMP-3 in stromal fibroblasts and cancer cells during prostate cancer progression. *Sci. Rep.* **2017**, *7*, 9065. [CrossRef] [PubMed]

88. Cichon, M.A.; Radisky, D.C. ROS-induced epithelial-mesenchymal transition in mammary epithelial cells is mediated by NF-kB-dependent activation of snail. *Oncotarget* **2014**, *5*, 2827–2838. [CrossRef]

89. Zeller, K.S.; Riaz, A.; Sarve, H.; Li, J.; Tengholm, A.; Johansson, S. The role of mechanical force and ROS in integrin-dependent signals. *PLoS ONE* **2013**, *8*, e64897. [CrossRef]

90. Mori, K.; Shibanuma, M.; Nose, K. Invasive potential induced under long-term oxidative stress in mammary epithelial cells. *Cancer Res.* **2004**, *64*, 7464–7472. [CrossRef]

91. Sawicki, G. Intracellular regulation of matrix metalloproteinase-2 activity: New strategies in treatment and protection of heart subjected to oxidative stress. *Scientifica* **2013**, *2013*, 130451. [CrossRef] [PubMed]

92. Dezerega, A.; Madrid, S.; Mundi, V.; Valenzuela, M.A.; Garrido, M.; Paredes, R.; Garcia-Sesnich, J.; Ortega, A.V.; Gamonal, J.; Hernandez, M. Pro-oxidant status and matrix metalloproteinases in apical lesions and gingival crevicular fluid as potential biomarkers for asymptomatic apical periodontitis and endodontic treatment response. *J. Inflamm.* **2012**, *9*, 8. [CrossRef]

93. Yoon, S.O.; Park, S.J.; Yoon, S.Y.; Yun, C.H.; Chung, A.S. Sustained production of H(2)O(2) activates pro-matrix metalloproteinase-2 through receptor tyrosine kinases/phosphatidylinositol 3-kinase/NF-kappa B pathway. *J. Biol. Chem.* **2002**, *277*, 30271–30282. [CrossRef] [PubMed]

94. Song, Y.H.; Wang, J.; Nie, G.; Chen, Y.J.; Li, X.; Jiang, X.; Cao, W.H. MicroRNA-509-5p functions as an anti-oncogene in breast cancer via targeting sod2. *Eur. Rev. Med. Pharmacol. Sci.* **2017**, *21*, 3617–3625. [PubMed]

95. Yu, L.; Chen, X.; Sun, X.; Wang, L.; Chen, S. The glycolytic switch in tumors: How many players are involved? *J. Cancer* **2017**, *8*, 3430–3440. [CrossRef] [PubMed]

96. Pacella, I.; Procaccini, C.; Focaccetti, C.; Miacci, S.; Timperi, E.; Faicchia, D.; Severa, M.; Rizzo, F.; Coccia, E.M.; Bonacina, F.; et al. Fatty acid metabolism complements glycolysis in the selective regulatory t cell expansion during tumor growth. *Proc. Natl. Acad. Sci. USA* **2018**, *115*, E6546–E6555. [CrossRef]

97. Schwartz, L.; Supuran, C.T.; Alfarouk, K.O. The warburg effect and the hallmarks of cancer. *Anticancer Agents Med. Chem.* **2017**, *17*, 164–170. [CrossRef] [PubMed]

98. Kim, J.; Kim, J.; Bae, J.S. ROS homeostasis and metabolism: A critical liaison for cancer therapy. *Exp. Mol. Med.* **2016**, *48*, e269. [CrossRef]

99. Hu, W.; Zhang, C.; Wu, R.; Sun, Y.; Levine, A.; Feng, Z. Glutaminase 2, a novel p53 target gene regulating energy metabolism and antioxidant function. *Proc. Natl. Acad. Sci. USA* **2010**, *107*, 7455–7460. [CrossRef]

100. Budanov, A.V. The role of tumor suppressor p53 in the antioxidant defense and metabolism. *Subcell. Biochem.* **2014**, *85*, 337–358.

101. Ipsen, D.H.; Lykkesfeldt, J.; Tveden-Nyborg, P. Molecular mechanisms of hepatic lipid accumulation in non-alcoholic fatty liver disease. *Cell. Mol. Life Sci.* **2018**, *75*, 3313–3327. [CrossRef] [PubMed]

102. Mello, T.; Simeone, I.; Galli, A. Mito-nuclear communication in hepatocellular carcinoma metabolic rewiring. *Cells* **2019**, *8*, 417. [CrossRef] [PubMed]

103. Horie, T.; Nishino, T.; Baba, O.; Kuwabara, Y.; Nakao, T.; Nishiga, M.; Usami, S.; Izuhara, M.; Sowa, N.; Yahagi, N.; et al. MicroRNA-33 regulates sterol regulatory element-binding protein 1 expression in mice. *Nat. Commun.* **2013**, *4*, 2883. [CrossRef] [PubMed]

104. Ayob, A.Z.; Ramasamy, T.S. Cancer stem cells as key drivers of tumour progression. *J. Biomed. Sci.* **2018**, *25*, 20. [CrossRef] [PubMed]

105. Saleh, A.D.; Savage, J.E.; Cao, L.; Soule, B.P.; Ly, D.; DeGraff, W.; Harris, C.C.; Mitchell, J.B.; Simone, N.L. Cellular stress induced alterations in microRNA let-7a and let-7b expression are dependent on p53. *PLoS ONE* **2011**, *6*, e24429. [CrossRef] [PubMed]

106. Zhang, X.; Ng, W.L.; Wang, P.; Tian, L.; Werner, E.; Wang, H.; Doetsch, P.; Wang, Y. MicroRNA-21 modulates the levels of reactive oxygen species by targeting sod3 and tnfalpha. *Cancer Res.* **2012**, *72*, 4707–4713. [CrossRef] [PubMed]

107. Li, X.J.; Ren, Z.J.; Tang, J.H. MicroRNA-34a: A potential therapeutic target in human cancer. *Cell Death Dis.* **2014**, *5*, e1327. [CrossRef]

108. Balzano, F.; Cruciani, S.; Basoli, V.; Santaniello, S.; Facchin, F.; Ventura, C.; Maioli, M. Mir200 and miR302: Two big families influencing stem cell behavior. *Molecules* **2018**, *23*, 282. [CrossRef]

109. Kim, J.H.; Park, S.G.; Song, S.Y.; Kim, J.K.; Sung, J.H. Reactive oxygen species-responsive miR-210 regulates proliferation and migration of adipose-derived stem cells via ptpn2. *Cell Death Dis.* **2013**, *4*, e588. [CrossRef]

110. Mei, Y.; Bian, C.; Li, J.; Du, Z.; Zhou, H.; Yang, Z.; Zhao, R.C. Mir-21 modulates the erk-mapk signaling pathway by regulating spry2 expression during human mesenchymal stem cell differentiation. *J. Cell. Biochem.* **2013**, *114*, 1374–1384. [CrossRef]

111. Yu, G.; Yao, W.; Xiao, W.; Li, H.; Xu, H.; Lang, B. MicroRNA-34a functions as an anti-metastatic microRNA and suppresses angiogenesis in bladder cancer by directly targeting cd44. *J. Exp. Clin. Cancer Res.* **2014**, *33*, 779. [CrossRef] [PubMed]

112. Peter, M.E. Let-7 and miR-200 microRNAs: Guardians against pluripotency and cancer progression. *Cell Cycle* **2009**, *8*, 843–852. [CrossRef] [PubMed]

113. Tang, T.; Yang, Z.; Zhu, Q.; Wu, Y.; Sun, K.; Alahdal, M.; Zhang, Y.; Xing, Y.; Shen, Y.; Xia, T.; et al. Up-regulation of miR-210 induced by a hypoxic microenvironment promotes breast cancer stem cells metastasis, proliferation, and self-renewal by targeting e-cadherin. *FASEB J.* **2018**, *32*, fj201801013R. [CrossRef] [PubMed]

114. Maciel-Baron, L.A.; Moreno-Blas, D.; Morales-Rosales, S.L.; Gonzalez-Puertos, V.Y.; Lopez-Diazguerrero, N.E.; Torres, C.; Castro-Obregon, S.; Konigsberg, M. Cellular senescence, neurological function, and redox state. *Antioxid. Redox Signal.* **2018**, *28*, 1704–1723. [CrossRef] [PubMed]

115. Liu, D.; Xu, Y. P53, oxidative stress, and aging. *Antioxid. Redox Signal.* **2011**, *15*, 1669–1678. [CrossRef] [PubMed]

116. Dellago, H.; Preschitz-Kammerhofer, B.; Terlecki-Zaniewicz, L.; Schreiner, C.; Fortschegger, K.; Chang, M.W.; Hackl, M.; Monteforte, R.; Kuhnel, H.; Schosserer, M.; et al. High levels of oncomiR-21 contribute to the senescence-induced growth arrest in normal human cells and its knock-down increases the replicative lifespan. *Aging Cell* **2013**, *12*, 446–458. [CrossRef] [PubMed]

117. Xu, D.; Takeshita, F.; Hino, Y.; Fukunaga, S.; Kudo, Y.; Tamaki, A.; Matsunaga, J.; Takahashi, R.U.; Takata, T.; Shimamoto, A.; et al. Mir-22 represses cancer progression by inducing cellular senescence. *J. Cell Biol.* **2011**, *193*, 409–424. [CrossRef]

118. Hu, Z.; Klein, J.D.; Mitch, W.E.; Zhang, L.; Martinez, I.; Wang, X.H. MicroRNA-29 induces cellular senescence in aging muscle through multiple signaling pathways. *Aging* **2014**, *6*, 160–175. [CrossRef]

119. He, X.; Yang, A.; McDonald, D.G.; Riemer, E.C.; Vanek, K.N.; Schulte, B.A.; Wang, G.Y. Mir-34a modulates ionizing radiation-induced senescence in lung cancer cells. *Oncotarget* **2017**, *8*, 69797–69807. [CrossRef]

120. Hong, L.; Lai, M.; Chen, M.; Xie, C.; Liao, R.; Kang, Y.J.; Xiao, C.; Hu, W.Y.; Han, J.; Sun, P. The miR-17-92 cluster of microRNAs confers tumorigenicity by inhibiting oncogene-induced senescence. *Cancer Res.* **2010**, *70*, 8547–8557. [CrossRef]

121. Nyholm, A.M.; Lerche, C.M.; Manfe, V.; Biskup, E.; Johansen, P.; Morling, N.; Thomsen, B.M.; Glud, M.; Gniadecki, R. Mir-125b induces cellular senescence in malignant melanoma. *BMC Dermatol.* **2014**, *14*, 8. [CrossRef] [PubMed]

122. Olivieri, F.; Lazzarini, R.; Recchioni, R.; Marcheselli, F.; Rippo, M.R.; Di Nuzzo, S.; Albertini, M.C.; Graciotti, L.; Babini, L.; Mariotti, S.; et al. Mir-146a as marker of senescence-associated pro-inflammatory status in cells involved in vascular remodelling. *Age* **2013**, *35*, 1157–1172. [CrossRef] [PubMed]

123. Menghini, R.; Casagrande, V.; Cardellini, M.; Martelli, E.; Terrinoni, A.; Amati, F.; Vasa-Nicotera, M.; Ippoliti, A.; Novelli, G.; Melino, G.; et al. MicroRNA 217 modulates endothelial cell senescence via silent information regulator 1. *Circulation* **2009**, *120*, 1524–1532. [CrossRef] [PubMed]

124. Shang, F.; Wang, S.C.; Hsu, C.Y.; Miao, Y.; Martin, M.; Yin, Y.; Wu, C.C.; Wang, Y.T.; Wu, G.; Chien, S.; et al. MicroRNA-92a mediates endothelial dysfunction in ckd. *J. Am. Soc. Nephrol.* **2017**, *28*, 3251–3261. [CrossRef] [PubMed]

125. Liu, H.; Wu, H.Y.; Wang, W.Y.; Zhao, Z.L.; Liu, X.Y.; Wang, L.Y. Regulation of miR-92a on vascular endothelial aging via mediating nrf2-keap1-are signal pathway. *Eur. Rev. Med. Pharmacol. Sci.* **2017**, *21*, 2734–2742. [PubMed]

126. Sanchez-Sanchez, B.; Gutierrez-Herrero, S.; Lopez-Ruano, G.; Prieto-Bermejo, R.; Romo-Gonzalez, M.; Llanillo, M.; Pandiella, A.; Guerrero, C.; Miguel, J.F.; Sanchez-Guijo, F.; et al. Nadph oxidases as therapeutic targets in chronic myelogenous leukemia. *Clin. Cancer Res.* **2014**, *20*, 4014–4025. [CrossRef]

127. Altenhofer, S.; Radermacher, K.A.; Kleikers, P.W.; Wingler, K.; Schmidt, H.H. Evolution of nadph oxidase inhibitors: Selectivity and mechanisms for target engagement. *Antioxid. Redox Signal.* **2015**, *23*, 406–427. [CrossRef]

128. Suzuki, S.; Pitchakarn, P.; Sato, S.; Shirai, T.; Takahashi, S. Apocynin, an nadph oxidase inhibitor, suppresses progression of prostate cancer via rac1 dephosphorylation. *Exp. Toxicol. Pathol.* **2013**, *65*, 1035–1041. [CrossRef]

129. Fuji, S.; Suzuki, S.; Naiki-Ito, A.; Kato, H.; Hayakawa, M.; Yamashita, Y.; Kuno, T.; Takahashi, S. The nadph oxidase inhibitor apocynin suppresses preneoplastic liver foci of rats. *Toxicol. Pathol.* **2017**, *45*, 544–550. [CrossRef]

130. Kumari, S.; Badana, A.K.; G, M.M.; G, S.; Malla, R. Reactive oxygen species: A key constituent in cancer survival. *Biomark. Insights* **2018**, *13*, 1177271918755391. [CrossRef]

131. Traverso, N.; Ricciarelli, R.; Nitti, M.; Marengo, B.; Furfaro, A.L.; Pronzato, M.A.; Marinari, U.M.; Domenicotti, C. Role of glutathione in cancer progression and chemoresistance. *Oxid. Med. Cell. Longev.* **2013**, *2013*, 972913. [CrossRef] [PubMed]

132. Yang, H.; Li, T.W.; Zhou, Y.; Peng, H.; Liu, T.; Zandi, E.; Martinez-Chantar, M.L.; Mato, J.M.; Lu, S.C. Activation of a novel c-Myc-miR27-prohibitin 1 circuitry in cholestatic liver injury inhibits glutathione synthesis in mice. *Antioxid. Redox Signal.* **2015**, *22*, 259–274. [CrossRef] [PubMed]

133. Espinosa-Diez, C.; Fierro-Fernandez, M.; Sanchez-Gomez, F.; Rodriguez-Pascual, F.; Alique, M.; Ruiz-Ortega, M.; Beraza, N.; Martinez-Chantar, M.L.; Fernandez-Hernando, C.; Lamas, S. Targeting of gamma-glutamyl-cysteine ligase by miR-433 reduces glutathione biosynthesis and promotes tgf-beta-dependent fibrogenesis. *Antioxid. Redox Signal.* **2015**, *23*, 1092–1105. [CrossRef] [PubMed]

134. Cheng, Y.; Zhou, M.; Zhou, W. MicroRNA-30e regulates tgf-beta-mediated nadph oxidase 4-dependent oxidative stress by snai1 in atherosclerosis. *Int. J. Mol. Med.* **2019**, *43*, 1806–1816. [PubMed]

135. Cho, K.J.; Song, J.; Oh, Y.; Lee, J.E. MicroRNA-let-7a regulates the function of microglia in inflammation. *Mol. Cell. Neurosci.* **2015**, *68*, 167–176. [CrossRef] [PubMed]

136. Fulciniti, M.; Amodio, N.; Bandi, R.L.; Cagnetta, A.; Samur, M.K.; Acharya, C.; Prabhala, R.; D'Aquila, P.; Bellizzi, D.; Passarino, G.; et al. Mir-23b/sp1/c-myc forms a feed-forward loop supporting multiple myeloma cell growth. *Blood Cancer J.* **2016**, *6*, e380. [CrossRef] [PubMed]

137. Chen, Q.; Zhang, F.; Wang, Y.; Liu, Z.; Sun, A.; Zen, K.; Zhang, C.Y.; Zhang, Q. The transcription factor c-myc suppresses miR-23b and miR-27b transcription during fetal distress and increases the sensitivity of neurons to hypoxia-induced apoptosis. *PLoS ONE* **2015**, *10*, e0120217. [CrossRef]

138. Gao, P.; Tchernyshyov, I.; Chang, T.C.; Lee, Y.S.; Kita, K.; Ochi, T.; Zeller, K.I.; De Marzo, A.M.; Van Eyk, J.E.; Mendell, J.T.; et al. C-myc suppression of miR-23a/b enhances mitochondrial glutaminase expression and glutamine metabolism. *Nature* **2009**, *458*, 762–765. [CrossRef]

139. Mott, J.L.; Kurita, S.; Cazanave, S.C.; Bronk, S.F.; Werneburg, N.W.; Fernandez-Zapico, M.E. Transcriptional suppression of miR-29b-1/miR-29a promoter by c-myc, hedgehog, and nf-kappab. *J. Cell. Biochem.* **2010**, *110*, 1155–1164. [CrossRef]

140. Hou, M.; Zuo, X.; Li, C.; Zhang, Y.; Teng, Y. Mir-29b regulates oxidative stress by targeting sirt1 in ovarian cancer cells. *Cell. Physiol. Biochem.* **2017**, *43*, 1767–1776. [CrossRef]

141. Mazzoccoli, L.; Robaina, M.C.; Apa, A.G.; Bonamino, M.; Pinto, L.W.; Queiroga, E.; Bacchi, C.E.; Klumb, C.E. Mir-29 silencing modulates the expression of target genes related to proliferation, apoptosis and methylation in burkitt lymphoma cells. *J. Cancer Res. Clin. Oncol.* **2018**, *144*, 483–497. [CrossRef] [PubMed]

142. Wu, D.W.; Hsu, N.Y.; Wang, Y.C.; Lee, M.C.; Cheng, Y.W.; Chen, C.Y.; Lee, H. C-myc suppresses microRNA-29b to promote tumor aggressiveness and poor outcomes in non-small cell lung cancer by targeting fhit. *Oncogene* **2015**, *34*, 2072–2082. [CrossRef] [PubMed]

143. Okada, N.; Lin, C.P.; Ribeiro, M.C.; Biton, A.; Lai, G.; He, X.; Bu, P.; Vogel, H.; Jablons, D.M.; Keller, A.C.; et al. A positive feedback between p53 and miR-34 miRNAs mediates tumor suppression. *Genes Dev.* **2014**, *28*, 438–450. [CrossRef] [PubMed]

144. Mihailovich, M.; Bremang, M.; Spadotto, V.; Musiani, D.; Vitale, E.; Varano, G.; Zambelli, F.; Mancuso, F.M.;

Cairns, D.A.; Pavesi, G.; et al. Mir-17-92 fine-tunes myc expression and function to ensure optimal b cell lymphoma growth. *Nat. Commun.* **2015**, *6*, 8725. [CrossRef] [PubMed]

145. Yan, H.L.; Xue, G.; Mei, Q.; Wang, Y.Z.; Ding, F.X.; Liu, M.F.; Lu, M.H.; Tang, Y.; Yu, H.Y.; Sun, S.H. Repression of the miR-17-92 cluster by p53 has an important function in hypoxia-induced apoptosis. *EMBO J.* **2009**, *28*, 2719–2732. [CrossRef] [PubMed]

146. Borkowski, R.; Du, L.; Zhao, Z.; McMillan, E.; Kosti, A.; Yang, C.R.; Suraokar, M.; Wistuba, I.I.; Gazdar, A.F.; Minna, J.D.; et al. Genetic mutation of p53 and suppression of the miR-17 approximately 92 cluster are synthetic lethal in non-small cell lung cancer due to upregulation of vitamin d signaling. *Cancer Res.* **2015**, *75*, 666–675. [CrossRef] [PubMed]

miR-27a-5p Attenuates Hypoxia-induced Rat Cardiomyocyte Injury by Inhibiting *Atg7*

Jinwei Zhang [1,2,†], **Wanling Qiu** [1,2,†], **Jideng Ma** [1,2,†], **Yujie Wang** [1,2], **Zihui Hu** [1,2], **Keren Long** [1,2], **Xun Wang** [1,2], **Long Jin** [1,2], **Qianzi Tang** [1,2], **Guoqing Tang** [1,2], **Li Zhu** [1,2], **Xuewei Li** [1,2], **Surong Shuai** [1,2,*] and **Mingzhou Li** [1,2,*]

[1] Institute of Animal Genetics and Breeding, College of Animal Science and Technology, Sichuan Agricultural University, Chengdu 611130, Sichuan, China; Jinweizhang50@163.com (J.Z.); qiuwanling2016@163.com (W.Q.); jideng.ma@sicau.edu.cn (J.M.); wangyujie715@163.com (Y.W.); Huzihui2016@163.com (Z.H.); keren.long@sicau.edu.cn (K.L.); xun_wang007@163.com (X.W.); longjin8806@163.com (L.J.); wupie@163.com (Q.T.); tyq003@163.com (G.T.); zhuli7508@163.com (L.Z.); xuewei.li@sicau.edu.cn (X.L.)

[2] Farm Animal Genetic Resource Exploration and Innovation Key Laboratory of Sichuan Province, Sichuan Agricultural University, Chengdu 611130, Sichuan, China

[*] Correspondence: srshuai@sohu.com (S.S.); mingzhou.li@sicau.edu.cn (M.L.)

[†] These authors contributed equally to this work.

Abstract: Acute myocardial infarction (AMI) is an ischemic heart disease with high mortality worldwide. AMI triggers a hypoxic microenvironment and induces extensive myocardial injury, including autophagy and apoptosis. MiRNAs, which are a class of posttranscriptional regulators, have been shown to be involved in the development of ischemic heart diseases. We have previously reported that hypoxia significantly alters the miRNA transcriptome in rat cardiomyoblast cells (H9c2), including miR-27a-5p. In the present study, we further investigated the potential function of miR-27a-5p in the cardiomyocyte response to hypoxia, and showed that miR-27a-5p expression was downregulated in the H9c2 cells at different hypoxia-exposed timepoints and the myocardium of a rat AMI model. Follow-up experiments revealed that miR-27a-5p attenuated hypoxia-induced cardiomyocyte injury by regulating autophagy and apoptosis via *Atg7*, which partly elucidated the anti-hypoxic injury effects of miR-27a-5p. Taken together, this study shows that miR-27a-5p has a cardioprotective effect on hypoxia-induced H9c2 cell injury, suggesting it may be a novel target for the treatment of hypoxia-related heart diseases.

Keywords: miR-27a-5p; acute myocardial infarction; autophagy; apoptosis; hypoxia

1. Introduction

Acute myocardial infarction (AMI) is often the primary pathological cause of death and disability worldwide [1]. During AMI, acute occlusion of the coronary artery deprives the oxygen and nutrients in myocardium and will contributes to cardiac dysfunction, including hypertrophy and remodeling, eventually leads to heart failure [2]. Since cardiomyocytes are terminally differentiated cells that have no or little regenerative potentialities, thus preventing cardiomyocytes loss after AMI injury is clinically a vital therapeutic strategy. Cardiomyocytes death and survival are affected predominantly via three cellular pathways: apoptosis, necrosis and autophagy [3]. Out of these three cellular pathways, apoptosis and necrosis have been extensively researched in AMI, but the effect of autophagy underlying AMI is still controversial to date [4]. Autophagy is an evolutionarily conserved process that maintains homeostasis in a cellular response to stresses by degrading abnormal protein and damaged organelles, which is considered to be closely associated with many heart diseases such as AMI [5]. Recently, autophagy has been considered a double-edged sword in the context of AMI, i.e., autophagy in early stage of AMI is beneficial to cardiomyocytes survival but excessive autophagy after AMI will induce autophagic cell death [6]. Thus, it is indispensable to further elucidate the autophagy regulation mechanism in cardiomyocytes survival after AMI.

MicroRNAs (miRNAs), a class of highly conserved non-coding RNAs, are major posttranscriptional regulators that involving in almost all cellular processes [7]. Currently, accumulating evidence has shown that miRNAs play essential roles in some heart diseases by regulating autophagy-related genes [4]. miRNA-212/132 family induce both cardiac hypertrophy and heart failure by activating pro-hypertrophic calcineurin/NFAT signaling, while inhibiting autophagic response upon starvation by directly targeting the anti-hypertrophic and pro-autophagic FoxO3 transcription factor [8]. miR-188-3p inhibits autophagy and autophagic cell death in the heart by targeting *Atg7* expression, meanwhile this effect can be suppressed by lncRNA APF (autophagy promoting factor) [9]. miR-21 alleviates hypoxia/reoxygenation-induced injury in H9c2 cells through weakening excessive autophagy and apoptosis via the Akt/mTOR pathway [10]. miR-204 has a protective effect against H9c2 cells hypoxia/reoxygenation-induced injury by regulating SIRT1-mediated autophagy [11]. Moreover, a recent study reports that miR-223 alleviates hypoxia-induced excessive autophagy and apoptosis in rat cardiomyocytes via the Akt/mTOR pathway by targeting *PARP-1* [12]. These miRNA may be serve as a potential target for ischemic heart disease treatment. In our previous study, we noted that the expression of miR-27a-5p decreased in acute hypoxia-exposed H9c2 cells using a small RNA-seq [13]. However, whether miR-27a-5p affects hypoxia-induced cardiomyocyte survival through regulating cell autophagy after AMI are still unknown.

In this study, we established a model of hypoxia in H9c2 cells and developed an AMI model in the rat to investigate the miR-27a-5p expression pattern in H9c2 cells and the main visceral tissues of rats (Figure 1). We found that hypoxia induced cell injury in vivo and in vitro and was accompanied by downregulation of miR-27a-5p expression. miR-27a-5p upregulation attenuated hypoxia-induced cardiomyocyte injury by regulating autophagy and apoptosis via *Atg7*, suggesting that miR-27a-5p may be a novel treatment strategy for hypoxia-related heart diseases.

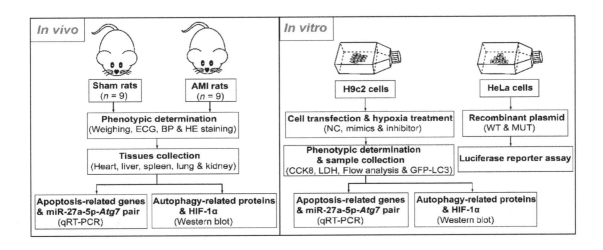

Figure 1. The flow chart of this study. ECG, electrocardiogram; BP, blood pressure; HE staining, hematoxylin & eosin staining; qRT-PCR, quantitative reverse-transcription polymerase chain reaction NC, negative control; CCK8, cell counting kit-8; LDH, lactate dehydrogenase; WT/MUT, wild-type/mutant.

2. Results

2.1. Hypoxia Induces H9c2 Cells Injury and Reduces miR-27a-5p Expression

In this study, we first cultured H9c2 cells in hypoxic condition for 24 h to simulate hypoxia induced by AMI *in vitro*. We found that hypoxia increased HIF-1α protein expression (Figure 2A) and triggered cell injury, including a decrease in cell viability ($p < 0.01$; Figure 2B), increased cell membrane damage ($p < 0.01$; Figure 2C) and apoptosis and necrosis ($p < 0.01$; Figure 2D,E). Meanwhile, hypoxia significantly increased the expression of proapoptotic genes (*Caspase-3*, *BAX*, *Faslg* and *P53*, $p < 0.01$; Figure 2F), but decreased expression of the antiapoptotic gene *Bcl-2* ($p < 0.05$; Figure 2F). Autophagy has previously been observed in ischemic heart disease [14,15] and autophagy levels were assessed in hypoxia-exposed H9c2 cells by western blot and autophagosome formation. These data showed that hypoxia increased autophagosome formation (Figure 2G) and promoted the switch of LC3-I to LC3-II. It also resulted in a reduction in P62 protein expression ($p < 0.01$; Figure 2H). Next, miR-27a-5p expression pattern was assessed in hypoxia-exposed H9c2 cells using qRT-PCR. miR-27a-5p expression decreased in a time-dependent manner (Figure 2I). These results indicate that hypoxia induced cell injury and reduced miR-27a-5p expression levels in H9c2 cells.

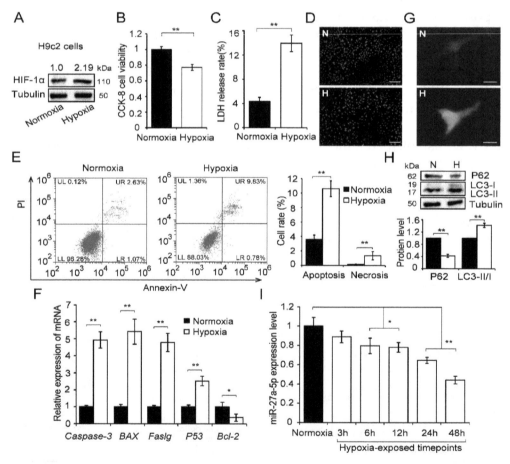

Figure 2. Hypoxia induces H9c2 cell injury and downregulation of miR-27a-5p. H9c2 cells were cultured under hypoxia or normoxia for 24 h. HIF-1α protein increased in H9c2 cells after hypoxia (**A**). Cell viability (**B**), membrane damage (**C**), and cell apoptosis (**D–F**) were evaluated by CCK8 assay, LDH release assays, apoptosis staining (scale bar: 50 μm), flow cytometry, and qRT-PCR analysis, respectively. H9c2 cells were transfected with GFP-LC3 plasmids and exposed to hypoxia for 24 h, fluorescence was observed by confocal fluorescence microscopy (**G**); scale bar: 5 μm. The autophagy-related proteins were detected by western blot (**H**). The expression of miR-27a-5p was tested using qRT-PCR at different hypoxia-exposed timepoints (**I**). Three independent experiments were performed in triplicate. Data are expressed as the mean ± SD. * $p < 0.05$, ** $p < 0.01$. N: normoxia; H: hypoxia.

2.2. AMI Triggers Widespread Injury Accompanied by Downregulation of miR-27a-5p in Rats

To investigate whether the miR-27a-5p expression under hypoxia induced by AMI in vivo was similar to that in hypoxia-exposed cardiomyocytes in vitro, an AMI rat model was established by ligating the coronary artery [16]. We observed S-T segment elevation in the electrocardiogram (ECG) and a reduction in blood pressure (BP) in the AMI group compared with sham, which confirmed successful AMI (Figure 3A,B). A *post hoc* power analysis of Δ BP obtained a power of > 0.90 with $p = 0.05$ in every LAD ligation timepoint (see "Statistical Analysis" for details on power analysis) (Table S1). We also found that the organ index in several main visceral tissues (except lung) was reduced (Table S2), which may be associated with the decreased left ventricular ejection fraction commonly observed after AMI [12]. Meanwhile, HE staining of the left ventricle showed that the cells in sham rat hearts were arranged uniformly with a normal gap, but local necrosis (indicated by arrowhead) and intercellular gaps (indicated by asterisk) were observed in AMI rats (Figure 3C). These data indicate that AMI induced severe damage in the rat myocardium.

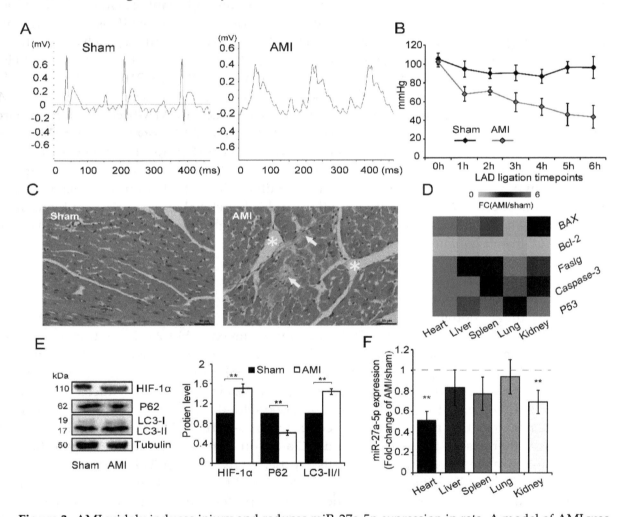

Figure 3. AMI widely induces injury and reduces miR-27a-5p expression in rats. A model of AMI was established in rats by ligating the coronary artery, and confirmed by analyzing ECG (**A**) and BP (**B**). (**C**) HE staining showed morphological differences between sham and AMI rats in coronal sections of the left ventricle; Yellow arrowheads and asterisks highlight local necrosis and intercellular gaps, respectively; scale bar: 50 μm. Expression patterns of apoptosis-related genes (**D**) in main visceral tissues (including heart, liver, spleen, lung and kidney) were determined by qRT-PCR (AMI *vs* Sham). AMI increased HIF-1α expression and promoted the conversion of LC3-I to LC3-II, but decreased P62 expression (**E**). miR-27a-5p expression (**F**) in main visceral tissues by qRT-PCR analysis (AMI *vs* Sham). Data are presented as the means ± SD of three independent experiments. ** $p < 0.01$. LAD: left anterior descending coronary artery.

The expression pattern of apoptosis-related genes showed that AMI triggers widespread apoptosis in the main visceral tissues, especially heart ($p < 0.01$), compared with sham (Figure 3D & Figure S1A). AMI also increased HIF-1α expression, shifted the expression of LC3-I to LC3-II, and decreased the expression of P62 protein (Figure 3E), which indicates that AMI synchronously promotes autophagy and apoptosis in the rat myocardium. In addition, AMI caused a reduction in miR-27a-5p expression in several visceral tissues, in particular the heart and kidney ($p < 0.01$; Figure 3F) when assessed by qRT-PCR analysis. The above results indicate that, similar to the in vitro results, hypoxic injury is widely induced in AMI rats and is accompanied by widespread downregulation of miR-27a-5p. Thus, miR-27a-5p may play a role in AMI-induced hypoxic injury.

2.3. Upregulation of miR-27a-5p Attenuates Hypoxia-Induced Excessive Autophagy and Apoptosis

Several studies have previously reported that autophagy and apoptosis successively appear in the cardiovascular diseases and the crosstalk between them plays an important role in the development of ischemic heart disease [17,18]. Autophagy have bidirectional effects in AMI, as autophagy may have both damaging and protective roles depending on the hypoxic conditions, such as duration or severity [19]. In the present study, assessment of autophagic flux showed that hypoxia-exposed H9c2 cells increased the level of autophagy in a time-dependent manner (Figure S1B). Next, cell viability and membrane damage were assessed after hypoxia in H9c2 cells pretreated with 10 mM 3-MA (a widely-used autophagy inhibitor). Cell viability was decreased in 3-MA-treated cells compared with control at the early stages of hypoxia exposure (within first 12 h); however, cell viability was higher in 3-MA-treated cells than control after hypoxia for 24 h (Figure S1C). Conversely, 3-MA pretreatment increased membrane damage in early stages of hypoxia but then this damage was alleviated after hypoxia for 24 h (Figure S1D). These results indicate that autophagy plays different roles in hypoxia-induced H9c2 cell injury over time and is beneficial in early stage of hypoxia but detrimental after 24 h of hypoxia (excessive autophagy), in keeping with previous reports [12]. Hypoxia for 24 h was used in subsequent experiments.

Based on the miR-27a-5p expression pattern and cell injury in hypoxia-exposed H9c2 cells and AMI rat myocardium, we hypothesized that miR-27a-5p is involved in mediating this biological process. To test this hypothesis, gain and loss of function analyses were performed. Effective overexpression and downregulation of miR-27a-5p was achieved in H9c2 cells by transfecting cells with a miR-27a-5p mimics or inhibitor, respectively, after exposure to hypoxia for 24 h (Figure 4 Overexpression of miR-27a-5p significantly mitigated hypoxic injury, including improved cell viability ($p < 0.01$; Figure 4B), alleviated cell membrane damage ($p < 0.01$; Figure 4C), and reduced cell apoptosis (Figure 4D–F). Meanwhile, miR-27a-5p downregulation yielded the opposite effects (Figure 4D–F). These results demonstrate that miR-27a-5p can reduce hypoxia-induced H9c2 cell injury by inhibiting apoptosis. To further assess the impact of miR-27a-5p on autophagy, the autophagic flux and autophagy-related proteins were assessed in cells exposed to hypoxia for 24 h after transfection. As shown in Figure 4G,H, miR-27a-5p overexpression decreased the level of autophagy, shifted LC3-I expression to LC3-II expression ($p < 0.05$), and increased P62 protein expression compared with control ($p < 0.01$). However, miR-27a-5p downregulation resulted in a more severe autophagy (Figure 4G,H). Altogether, these results indicate that miR-27a-5p has a negative effect on hypoxia-induced autophagy and that miR-27a-5p protects against hypoxia-induced cardiomyocyte injury by reducing apoptosis and excessive autophagy.

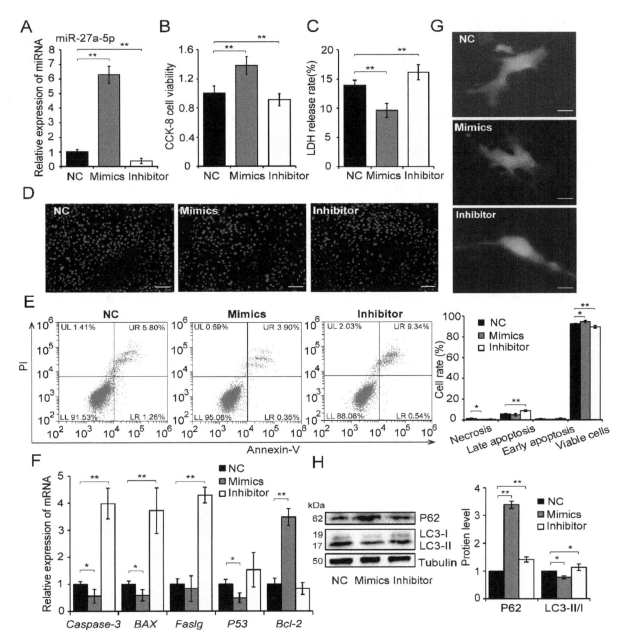

Figure 4. miR-27a-5p attenuates hypoxia-induced excessive autophagy and apoptosis in H9c2 cells. H9c2 cells were exposed to hypoxia for 24 h after transfection of a specific miR-27a -5p mimics or inhibitor. Transfection efficiency was analyzed by qRT-PCR (**A**). Cell viability (**B**), membrane damage (**C**), and cell apoptosis (**D–F**) were assessed by CCK8 assays, LDH release assays, apoptosis staining (scale bar: 50 μm), flow cytometry and qRT-PCR analysis, respectively. The level of autophagy was evaluated by GFP-LC3 fluorescence after hypoxia for 24 h (**G**); scale bar: 5 μm. Autophagy-related proteins were detected by western blot (**H**). Three independent experiments were performed in triplicate. Data are expressed as the mean ± SD. * $p < 0.05$, ** $p < 0.01$. NC: negative control.

2.4. Atg7 is The Target of miR-27a-5p

To explore the mechanism underlying miR-27a-5p regulation of excessive autophagy and inhibition of apoptosis, we analyzed candidate target genes of miR-27a-5p using TargetScan (release 7.2) [20] and RNAhybrid 2.2 prediction [21]. The prediction results showed that the 3'-UTR region of *Atg7* mRNA contained a target site for miR-27a-5p, and *Atg7* has been linked to autophagy [22]. We tested the expression of *Atg7* and miR-27a-5p in hypoxia-exposed H9c2 cells and in the main visceral tissues of AMI rat, and then performed a correlation analysis. We found a strongly negative correlation between the expression of *Atg7* and miR-27a-5p in hypoxia-exposed H9c2 cells at different timepoints ($r = -0.807$; Figure 5A). Meanwhile, a moderate negative correlation was observed in AMI rat visceral tissue ($r = -0.569$; Figure 5B). Additionally, overexpression of miR-27a-5p in hypoxia-exposed H9c2 cells significantly reduced *Atg7* mRNA and protein expression, while miR-27a-5p downregulation showed an opposite trend ($p < 0.01$; Figure 5C,D). The aforementioned results suggest that miR-27a-5p alleviates hypoxia-induced cardiomyocyte injury by targeting *Atg7*.

We subsequently performed a dual-luciferase reporter assay to confirm the potential relationship between *Atg7* and miR-27a-5p. The sequence alignment of miR-27a-5p showed high similarity, and likewise miR-27a-5p-binding site in *Atg7* 3'-UTR among several representative species were also conserved, which suggested the conservative interaction mechanism of miR-27a-5p-*Atg7* pair among species (Figure 5E). *Atg7* 3'-UTR containing the miR-27a-5p binding site (WT or MUT) was inserted into dual luciferase plasmid (pmirGLO-*Atg7*-3'-UTR) (Figure 5E). HeLa cells were co-transfected with the WT or MUT recombinant plasmid and miR-27a-5p mimics. Luciferase activity was detected 48 h after transfection. As shown in Figure 5F, co-transfection of miR-27a-5p and WT pmirGLO reporter significantly inhibited luciferase activity compared with the negative control (0.462 fold-change, $p < 0.01$). This effect was eliminated with the MUT pmirGLO reporter, which indicates that *Atg7* is a direct target for miR-27a-5p. A standard validation reporting for miR-27a-5p-*Atg7* interaction in this study is shown in Table S3 [23,24].

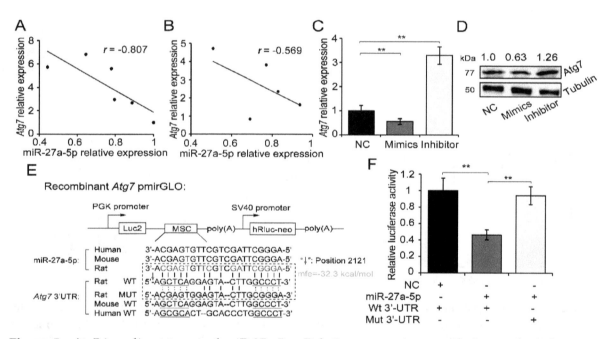

Figure 5. *Atg7* is a direct target of miR-27a-5p. Relative expression correlation analysis between miR-27a-5p and *Atg7* during hypoxia at different timepoints (0, 3, 6, 12, 24 and 48 h after hypoxia) in H9c2 cells (**A**), and in AMI/sham rat visceral tissues (**B**). mRNA (**C**) and protein (**D**) expression of *Atg7* was tested by qRT-PCR and western blotting after miR-27a-5p gain and loss of function in hypoxia-exposed H9c2 cells. (**E**) Schematic diagram showing the structure of dual-luciferase reporter plasmid pmirGLO and the sequence alignment of miR-27a-5p and *Atg7* 3'-UTR among several representative species (human, mouse and rat). *Atg7* 3'-UTR containing the miR-27a-5p binding site (WT or MUT) was inserted into the multiple cloning site (MSC) of pmirGLO plasmid. (**F**) Luciferase activity was analyzed after co-transfection of recombinant plasmid (WT or MUT) with miR-27a-5p mimic or control into HeLa cells. Three independent experiments were performed in triplicate. Data are expressed as the mean ± SD. ** $p < 0.01$. NC: negative control; mfe: minimum free energy.

3. Discussion

In recent years, miRNAs have frequently been reported in cardiovascular disease and play important roles in ischemic heart diseases by regulating the process of autophagy and apoptosis [4]. miR-27a-5p belongs to the miRNA-23a-27a-24 cluster that is reported to be involved in many cardiac diseases [25]. miR-24 has been shown to attenuate mouse AMI and reduces cardiac dysfunction by inhibiting cardiomyocyte apoptosis [26]; miR-23a has been shown to positively regulate cardiac hypertrophy by targeting anti-hypertrophic factor *MuRF1* [27] and *Foxo3a* [28]. Although miRNA-27a has been shown to be involved in the regulation of cardiomyocyte apoptosis, during cardioplegia-induced cardiac arrest through IL10-related pathways [29]; whether it regulates cardiomyocyte survival under hypoxic stress caused by ischemic heart diseases such as AMI, remains to be investigated. Based on previous report that the expression of miR-27a-5p decreased in hypoxia-exposed H9c2 cells, we found in the present study that miR-27a-5p expression likewise decreased in AMI rat myocardium (Figure 3F). More deeply, we revealed the miR-27a-5p-*Atg7* interaction in vivo and in vitro, and functionally, miR-27a-5p attenuated hypoxia-induced cardiomyocyte injury by regulating autophagy and apoptosis via *Atg7*, which further confirmed the crucial roles of miRNA-23a-27a-24 cluster in heart diseases.

Autophagy is an evolutionarily conserved and tightly regulated process that maintains cellular homeostasis in response to stresses, such as hypoxia, by degrading abnormal protein and damaged organelles [30,31]. Nevertheless, autophagy is considered a double-edged sword in the context of AMI, i.e., autophagy may have both damaging and protective roles depending on the hypoxic conditions, such as duration or severity [6,19]. In this study, we found that the degree of autophagy in hypoxia-exposed H9c2 cells increased in a time-dependent manner (Figure S1B). Inhibition of autophagy (hypoxia + 3-MA pretreatment) decreased cell viability and increased hypoxia-induced membrane damage compared with control (hypoxia) at the early stages of hypoxia exposure (within first 12 h), however these effects were alleviated after hypoxia for 24 h (Figure S1C,D). These results indicate that autophagy plays different roles in hypoxia-induced H9c2 cell injury over time and is beneficial in early stage of hypoxia but detrimental after 24 h of hypoxia (excessive autophagy), in keeping with previous reports [12]. Thus, elucidating and manipulating the development of cardiomyocyte autophagy under hypoxia may be beneficial to the clinical treatment of ischemic heart diseases.

Acting as the only E1-like enzyme, *Atg7* is located in the hub of the LC3 and Atg12 ubiquitin-like systems and is essential for the expansion of autophagosomal membranes [22]. Accumulating evidence suggests that *Atg7* is not only a crucial marker of autophagy, but also participates in the regulation of cell death and survival [32,33], including in cardiac progenitor cells [34]. Previously, we noted that the expression of miR-27a-5p decreased in acute hypoxia-exposed H9c2 cells using a small RNA-seq, as a known hypoxamiR, however, its underlying function in the cardiomyocyte hypoxic response is unclear [13]. In this study, we showed for the first time, to our knowledge, the negative correlation of miR-27a-5p-*Atg7* pair in vivo and in vitro, and that miR-27a-5p alleviated hypoxia-induced cardiomyocyte injury through regulation of excessive autophagy and apoptosis by inhibiting *Atg7* in vitro. This further highlights miRNA regulation in hypoxia-related heart diseases and may have potential implications for the treatment of ischemic cardiomyopathy in the future. However, the function of miR-27a-5p in hypoxia-induced cardiomyocyte injury is mainly focused on the cell-based experiments in vitro. Thus, animal studies on miR-27a-5p knock in/out, such as CRISP-Cas9-mediated gene editing, may better demonstrate the function of miR-27a-5p in hypoxia-induced cardiomyocyte injury after AMI and this should be performed in future research. In addition, although the sequence in miR-27a-5p and *Atg7* 3'-UTR has high similarity among several representative species, the function and strength of miR-27a-5p and its clinical application in human remain to be further elucidated.

4. Materials and Methods

4.1. Rat AMI Model

Healthy male Sprague Dawley (SD) rats (308 ± 14 g) were bought from Dashuo Laboratory Animal Center (Chengdu, Sichuan, China) and housed in a standard environment (20 ± 2 °C and 58% ± 2% humidity), with free choice feeding for 1 week before experiment. All animal procedures complied with the Ethics Committee of Sichuan Agricultural University rules (Approval Number DKY-B20171903, 15 February 2018). Coronary artery ligation was performed as previously described, to establish the rat AMI model [35]. Arterial BP and ECG were measured throughout the experiment. A clear elevation of the S-T segment of the ECG indicated successful AMI in the rat ($n = 9$). The same procedure was carried out without coronary artery ligation as sham control ($n = 9$). All rats were anesthetized and euthenized 6 h after coronary artery ligation. Several main visceral tissues were collected and immediately immersed in liquid nitrogen before storing at −80 °C for further experimentation.

4.2. H9c2 Cell Culture and Hypoxia Treatment

H9c2 cells (an embryonic rat heart-derived cell line) were routinely maintained in Dulbecco's Modified Eagle Medium (DMEM) (Hyclone, Logan, UT, USA) with 10% fetal bovine serum (FBS) (GIBCO, Grand Island, NY, USA) at 37 °C in a humidified atmosphere containing 5% CO_2 and 95% air. To establish hypoxia in vitro, cells with 50% confluency received hypoxia treatment for 24 h in a modular incubator chamber with 5% CO_2, 1% O_2 and 95% N_2 (MIC-101, Billups-Rothenberg, Del Mar, CA). Cells in the normoxic group were placed in conventional conditions (5% CO_2 and 95% air) and served as the control.

4.3. H9c2 Cell Transfection

Specific mimics and inhibitor of miR-27a-5p (RIBOBIO, Guangzhou, Guangdong, China) were transfected in cells at 50% confluency to facilitate gain and loss of function. Three groups of cells were designed; a mimic, an inhibitor and a negative control. Transfection solutions were premixed and added to the medium at a final concentration of 50 nM (or 100 nM for the inhibitor) using Lipofectamine 2000 (Invitrogen, Grand Island, NY, USA) in accordance with the manufacturer's protocol. After 6 h in the transfection medium, all groups were replaced with new medium before receiving hypoxia treatment for 24 h for subsequent experimentation.

4.4. Cell Counting Kit-8 (CCK8) and Lactate Dehydrogenase (LDH) Release Assay

To evaluate hypoxia-induced cell injury, cell viability and LDH release were analyzed using a CCK8 and a LDH Cytotoxicity Assay Kit (Beyotime, Shanghai, China), respectively. H9c2 cells were cultured in 96-well plate and received the relevant treatments (such as hypoxia, transfection) at the given time. For CCK8 detection, 10 µL CCK8 reagent was added to the culture medium 4 h before analysis. Optical density (OD)$_{450}$ values were measured using a microplate reader (Thermo Fisher Scientific, Madrid, Spain). For LDH release analysis, the culture medium in each group was premixed with the relevant reagent and incubated in accordance with the manufacturer's protocol. OD$_{490}$ values were measured and LDH release rate presented as the percentage of the maximum enzymatic activity. At least three independent experiments were repeated three times. All values are presented as mean ± standard deviation (SD).

4.5. Cell Apoptosis Analysis

Cell apoptosis was assessed using an Annexin V-FITC and propidium iodide (PI) detection kit (BD Pharmingen, San Diego, CA, USA), in accordance with the manufacturer's protocols. Briefly, cells were digested by trypsin and gently washed with phosphate buffered saline (PBS). Cells were then incubated with Annexin V and PI for 10 min at room temperature and assessed by flow cytometry (Beckman Coulter, Brea, USA). The raw data were analyzed using CytExpert 2.0 software and more than 10,000 cells in each group were used for statistical analysis. All values are presented as mean ± SD.

4.6. HE Staining and Fluorescence Staining of Apoptosis

Tissue sections of the left ventricle were assessed using HE staining. In brief, the rat myocardium was fixed with 4% paraformaldehyde at room temperature, followed by dehydration and embedding in paraffin. The sections were prepared and successively stained using eosin and hematoxylin (Beyotime, Shanghai, China). To observe cell apoptosis, fluorescence staining of H9c2 cells was performed using an apoptosis and necrosis assay kit (Beyotime, Shanghai, China) in accordance with the manufacturer's instructions. Stained tissue sections and cells were imaged using an Olympus IX53 microscope (Olympus, Tokyo, Japan).

4.7. Detection of Autophagosome Formation

H9c2 cells were plated on coverslips. GFP-LC3 plasmids (Beyotime, Shanghai, China) were transfected into H9c2 cells at 50% confluency, before miRNA transfection and exposure to hypoxia. Afterwards, the cells were fixed with 10% formalin and GFP-LC3 fluorescence punctae were imaged using a confocal fluorescence microscope (Olympus, Tokyo, Japan).

4.8. Luciferase Reporter Assay

Luciferase activity assays were performed to validate the potential relationship between miR-27a-5p and *Atg7*. Briefly, HeLa cells were routinely maintained in DMEM with 10% FBS at 37 °C. We synthesized the *Atg7* 3'-UTR sequence containing the miR-27a-5p binding site (WT or MUT) and then cloned into the MCS of pmirGLO plasmid (Figure 5E). The WT or MUT recombinant pmirGLO vector was cotransfected with miR-27a-5p mimic or negative control into HeLa cells using Lipofectamine 3000 (Invitrogen, Grand Island, NY, USA), in accordance with the manufacturer's instructions. Dual luciferase activity was tested by Luciferase Dual Assay Kit (Promega, Madison, WI, USA) 48 h after transfection. Luciferase activity is expressed as an adjusted value (firefly normalized to renilla).

4.9. Total RNA Extraction and qRT-PCR

Total RNA was extracted from tissue or cultured cells using HiPure Total RNA Mini Kit (Magen, Guangzhou, China). The quality of total RNA was assessed by NanoDrop 2000 (Thermo Fisher Scientific, Wilmington, DE) and gel electrophoresis. The reverse transcription of mRNA and miRNA from the qualified total RNA was performed using PrimeScript™ RT Reagent Kit (Takara, Beijing, China) and Mir-X™ miRNA First Strand Synthesis Kit (Clontech, Mountain View, USA), respectively, according to the manufacturers' protocols. qPCR reactions were prepared using an SYBR Premix Ex Taq kit (Takara, Beijing, China) and performed in a Bio-Rad CFX96 PCR System (Bio-Rad, Hercules, USA). The relative expression of mRNA and miRNA was calculated using the $2^{-\Delta\Delta Ct}$ method and expressed as fold-change relative to the corresponding control. *GAPDH* and *U6* served as the reference genes for miRNA and mRNA, respectively. All primers used for qPCR are listed in Table S4.

4.10. Western Blot Analysis

Western blot analysis was performed as previously described [36]. Total protein was extracted from the H9c2 cells and rat myocardium using radioimmunoprecipitation assay lysis buffer containing protease and phosphatase inhibitors (Beyotime, Beijing, China) and quantified using a BCA protein assay. Approximately 30 g of protein was loaded and separated on an 8% SDS-PAGE gel, and then transferred to polyvinylidene difluoride membranes (BIO-RAD, Hercules, USA). The membranes were blocked with nonfat milk for 2 h at room temperature, and then incubated with primary antibodies at 4 °C overnight. Subsequently, the membranes were washed in PBS with Tween-20 before incubating with secondary antibodies for 2 h at room temperature. The antigen–antibody bands were visualized and quantified using ImageJ software (Bethesda, MA, USA). The primary antibodies used in this study and corresponding dilution ratios were as follows: anti-alpha Tubulin (1:1000), anti-Atg7 (1:500), anti-LC3 (1:1000), anti-P62 (1:1000), anti-HIF-1α (1:1000) (Abcam, Cambridge, USA).

4.11. Statistical Analysis

All experiments were performed as at least three independent experiments with three technical repetitions. The data are expressed as mean ± SD. Significance tests were performed using SPSS 22.0 software (SPSS, Chicago, USA). Unpaired Student's *t*-test and one-way ANOVA with Tukey's post-hoc test were used to evaluate the differences between two groups or three or more groups, respectively. The Δ BP were used as a surrogate measure of effect to perform a *post hoc* power analysis. The parameters "($n = 9$, $d = \frac{|\mu_1 - \mu_2|}{\rho}$, sig.level = 0.05, power = , type = "two.sample", alternative = "two.sided")" were performed with R (Version 3.2.0) computed by the *pwr* package [37]. $p < 0.05$ was considered as statistically significant (* $p < 0.05$, ** $p < 0.01$).

5. Conclusions

We have shown that AMI-induced hypoxia causes cell injury and the expression of miR-27a-5p is decreased in hypoxia-exposed H9c2 cells and AMI rat myocardium. miR-27a-5p attenuates hypoxia-induced cardiomyocyte injury by inhibiting excessive autophagy and apoptosis via *Atg7*. Our findings show that miR-27a-5p has a cardioprotective effect on hypoxia-induced H9c2 injury, and may serve as a novel target for the treatment of hypoxia-related heart diseases.

Author Contributions: J.Z., W.Q., S.S. and M.L. conceived and designed the study and drafted the manuscript. J.Z., J.M., Y.W., Z.H., and K.L. performed the experiments. X.W., L.J., Q.T., G.T., and L.Z. analyzed the experiment data. X.L., S.S. and M.L. revised the manuscript. All authors read and approved the final manuscript.

Abbreviations

AMI	acute myocardial infarction
miRNA	microRNA
LAD	left anterior descending
CCK8	cell counting kit-8
LDH	lactate dehydrogenase
ECG	electrocardiogram
BP	blood pressure
NC	negative control
FBS	fetal bovine serum
PBS	phosphate buffered saline
PI	propidium iodide
OD	optical density
HE staning	hematoxylin & eosin staning
MSC	multiple cloning site
kDa	kilodalton
UL	upper left
UR	upper right
LL	lower left
LR	lower right
SD	standard deviation
mfe	minimum free energy
WT/Mut	wild-type/mutant
qRT-PCR	quantitative reverse-transcription polymerase chain reaction
ANOVA	analysis of variance

References

1.   Benjamin, E.J.; Virani, S.S.; Callaway, C.W.; Chamberlain, A.M.; Chang, A.R.; Cheng, S.; de Ferranti, S.D. Heart disease and stroke statistics—2018 update: A report from the American Heart Association. *Circulation* **2018** *137*, e67–e492. [CrossRef]

2. Anderson, J.L.; Morrow, D.A. Acute Myocardial Infarction. *New Engl. J. Med.* **2017**, *376*, 2053. [CrossRef]

3. Nikoletopoulou, V.; Markaki, M.; Palikaras, K.; Tavernarakis, N. Crosstalk between apoptosis, necrosis and autophagy. *BBA-Mol. Cell Res.* **2013**, *1833*, 3448–3459. [CrossRef]

4. Sermersheim, M.A.; Park, K.H.; Gumpper, K.; Adesanya, T.M.; Song, K.; Tan, T.; Ren, X.; Yang, J.M.; Zhu, H. MicroRNA regulation of autophagy in cardiovascular disease. *Front Biosci.* **2017**, *22*, 48.

5. Nishida, K.; Kyoi, S.; Yamaguchi, O.; Sadoshima, J.; Otsu, K. The role of autophagy in the heart. *Cell Death Differ.* **2009**, *16*, 31–38. [CrossRef]

6. Hongxin, Z.; Paul, T.; Johnstone, J.L.; Yongli, K.; Shelton, J.M.; Richardson, J.A.; Vien, L.; Beth, L.; Rothermel, B.A.; Hill, J.A. Cardiac autophagy is a maladaptive response to hemodynamic stress. *J. Clin. Invest.* **2007**, *117*, 1782–1793.

7. Krol, J.; Loedige, I.W. The widespread regulation of microRNA biogenesis, function and decay. *Nat. Rev. Genet.* **2010**, *11*, 597–610. [CrossRef]

8. Ucar, A.; Gupta, S.K.; Fiedler, J.; Erikci, E.; Kardasinski, M.; Batkai, S.; Dangwal, S.; Kumarswamy, R.; Bang, C.; Holzmann, A. The miRNA-212/132 family regulates both cardiac hypertrophy and cardiomyocyte autophagy. *Nat. Commun.* **2012**, *3*, 1078. [CrossRef]

9. Wang, K.; Liu, C.Y.; Zhou, L.Y.; Wang, J.X.; Wang, M.; Zhao, B.; Zhao, W.K.; Xu, S.J.; Fan, L.H.; Zhang, X.J. APF lncRNA regulates autophagy and myocardial infarction by targeting miR-188-3p. *Nat. Commun.* **2015**, *6*, 6779. [CrossRef]

10. Huang, Z.; Wu, S.; Kong, F.; Cai, X.; Ye, B.; Shan, P.; Huang, W. MicroRNA-21 protects against cardiac hypoxia/reoxygenation injury by inhibiting excessive autophagy in H9c2 cells via the Akt/mTOR pathway. *J. Cell Mol. Med.* **2017**, *21*, 467–474. [CrossRef]

11. Qiu, R.; Wen, L.; Liu, Y. MicroRNA-204 protects H9C2 cells against hypoxia/reoxygenation-induced injury through regulating SIRT1-mediated autophagy. *Biomed. Pharmacol.* **2018**, *100*, 15–19. [CrossRef]

12. Liu, X.; Deng, Y.; Xu, Y.; Jin, W.; Li, H. MicroRNA-223 protects neonatal rat cardiomyocytes and H9c2 cells from hypoxia-induced apoptosis and excessive autophagy via the Akt/mTOR pathway by targeting PARP-1. *J. Mol. Cell Cardiol.* **2018**, *118*, 133–146. [CrossRef]

13. Zhang, J.; Ma, J.; Long, K.; Qiu, W.; Wang, Y.; Hu, Z.; Liu, C.; Luo, Y.; Jiang, A.; Jin, L. Overexpression of Exosomal Cardioprotective miRNAs Mitigates Hypoxia-Induced H9c2 Cells Apoptosis. *Int. J. Mol. Sci.* **2017**, *18*, 711. [CrossRef] [PubMed]

14. Gao, Y.H.; Qian, J.Y.; Chen, Z.W.; Fu, M.Q.; Xu, J.F.; Xia, Y.; Ding, X.F.; Yang, X.D.; Cao, Y.Y.; Zou, Y.Z. Suppression of Bim by microRNA-19a may protect cardiomyocytes against hypoxia-induced cell death via autophagy activation. *Toxicol. Lett.* **2016**, *257*, 72–83. [CrossRef]

15. Yutaka, M.; Hiromitsu, T.; Xueping, Q.; Maha, A.; Hideyuki, S.; Tomoichiro, A.; Beth, L.; Junichi, S. Distinct roles of autophagy in the heart during ischemia and reperfusion: Roles of AMP-activated protein kinase and Beclin 1 in mediating autophagy. *Circ. Res.* **2007**, *100*, 914–922.

16. Boon, R.A.; Kazuma, I.; Stefanie, L.; Timon, S.; Ariane, F.; Susanne, H.; David, K.; Karine, T.; Guillaume, C.; Angelika, B. MicroRNA-34a regulates cardiac ageing and function. *Nature* **2013**, *495*, 107–110. [CrossRef] [PubMed]

17. Nishida, K.; Yamaguchi, O.; Otsu, K. Crosstalk between autophagy and apoptosis in heart disease. *Circ. Res.* **2008**, *103*, 343–351. [CrossRef] [PubMed]

18. Li, M.; Gao, P.; Zhang, J. Crosstalk between Autophagy and Apoptosis: Potential and Emerging Therapeutic Targets for Cardiac Diseases. *Int. J. Mol. Sci.* **2016**, *17*, 332. [CrossRef] [PubMed]

19. Sai, M.; Yabin, W.; Yundai, C.; Feng, C. The role of the autophagy in myocardial ischemia/reperfusion injury. *BBA-Mol. Basis Dis.* **2015**, *1852*, 271–276.

20. Maziere, P.; Enright, A.J. Prediction of microRNA targets. *Drug Discov. Today* **2007**, *12*, 452–458. [CrossRef]

21. Jan, K.; Marc, R. RNAhybrid: MicroRNA target prediction easy, fast and flexible. *Nucleic Acids Res.* **2006**, *34*, W451–W454.

22. Xiong, J. Atg7 in development and disease: Panacea or pandora's box? *Protein Cell* **2015**, *6*, 722–734. [CrossRef]

23. Piletič, K.; Kunej, T. Minimal standards for reporting microRNA: Target interactions. *Omics* **2017**, *21*, 197–206. [CrossRef]

24. Desvignes, T.; Batzel, P.; Berezikov, E.; Eilbeck, K.; Eppig, J.T.; McAndrews, M.S.; Singer, A.; Postlethwait, J.

miRNA nomenclature: A view incorporating genetic origins, biosynthetic pathways, and sequence variants. *Trends Genet.* **2015**, *31*, 613–626. [CrossRef]

25. Chhabra, R.; Dubey, R.; Saini, N. Cooperative and individualistic functions of the microRNAs in the miR-23a~27a~24-2 cluster and its implication in human diseases. *Mol. Cancer* **2010**, *9*, 232. [CrossRef]

26. Li, Q.; Van Laake, L.W.; Yu, H.; Siyuan, L.; Wendland, M.F.; Deepak, S. miR-24 inhibits apoptosis and represses Bim in mouse cardiomyocytes. *J. Exp. Med.* **2011**, *208*, 549–560.

27. Zhiqiang, L.; Iram, M.; Kun, W.; Jianqin, J.; Jie, G.; Pei-Feng, L. miR-23a functions downstream of NFATc3 to regulate cardiac hypertrophy. *Proc. Natl. Acad. Sci. USA* **2009**, *106*, 12103–12108.

28. Kun, W.; Zhi-Qiang, L.; Bo, L.; Jian-Hui, L.; Jing, Z.; Pei-Feng, L. Cardiac hypertrophy is positively regulated by MicroRNA miR-23a. *J. Biol. Chem.* **2012**, *287*, 589–599.

29. Yeh, C.H.; Chen, T.P.; Wang, Y.C.; Lin, Y.M.; Fang, S.W. MicroRNA-27a regulates cardiomyocytic apoptosis during cardioplegia-induced cardiac arrest by targeting interleukin 10-related pathways. *Shock* **2012**, *38*, 607–614. [CrossRef] [PubMed]

30. Kroemer, G.; Mariño, G.; Levine, B. Autophagy and the integrated stress response. *Mol. Cell* **2010**, *40*, 280–293. [CrossRef]

31. Chun, Y.; Kim, J. Autophagy: An Essential Degradation Program for Cellular Homeostasis and Life. *Cells* **2018**, *7*, 278. [CrossRef]

32. In Hye, L.; Yoshichika, K.; Fergusson, M.M.; Rovira, I.I.; Bishop, A.J.R.; Noboru, M.; Liu, C.; Toren, F. Atg7 modulates p53 activity to regulate cell cycle and survival during metabolic stress. *Science* **2012**, *336*, 225–228.

33. Li, Y.; Ajjai, A.; Helen, S.; Parmesh, D.; Eric, F.; Sarah, W.; Baehrecke, E.H.; Lenardo, M.J. Regulation of an ATG7-beclin 1 program of autophagic cell death by caspase-8. *Science* **2004**, *304*, 1500–1502.

34. Ma, W.; Ding, F.; Wang, X.; Huang, Q.; Zhang, L.; Bi, C.; Hua, B.; Yuan, Y.; Han, Z.; Jin, M. By Targeting Atg7 MicroRNA-143 Mediates Oxidative Stress-Induced Autophagy of c-Kit + Mouse Cardiac Progenitor Cells. *EBioMedicine* **2018**, *32*, 182–191. [CrossRef] [PubMed]

35. Xin, L.; Baoqiu, W.; Hairong, C.; Yue, D.; Yang, S.; Lei, Y.; Qi, Z.; Fei, S.; Dan, L.; Chaoqian, X. Let-7e replacement yields potent anti-arrhythmic efficacy via targeting beta 1-adrenergic receptor in rat heart. *J. Cell Mol. Med.* **2014**, *18*, 1334–1343.

36. Ma, J.; Zhang, J.; Wang, Y.; Long, K.; Wang, X.; Jin, L.; Tang, Q.; Zhu, L.; Tang, G.; Li, X. MiR-532-5p alleviates hypoxia-induced cardiomyocyte apoptosis by targeting *PDCD4*. *Gene* **2018**, *675*, 36–43. [CrossRef]

37. Basic Functions for Power Analysis. Available online: https://cran.r-project.org/web/packages/pwr/index.html (accessed on 5 May 2019).

11

MicroRNA and Oxidative Stress Interplay in the Context of Breast Cancer Pathogenesis

Giulia Cosentino [1], Ilaria Plantamura [1], Alessandra Cataldo [1,2,*,†] and Marilena V. Iorio [1,2,*,†]

[1]   Molecular Targeting Unit, Research Department, Fondazione IRCCS Istituto Nazionale dei Tumori, 20133 Milan, Italy; giulia.cosentino@istitutotumori.mi.it (G.C.); ilaria.plantamura@istitutotumori.mi.it (I.P.)

[2]   IFOM Istituto FIRC di Oncologia Molecolare, 20139 Milan, Italy

*   Correspondence: alessandra.cataldo@istitutotumori.mi.it (A.C.); marilena.iorio@istitutotumori.mi.it (M.V.I.)

†   These authors contributed equally to this work.

Abstract: Oxidative stress is a pathological condition determined by a disturbance in reactive oxygen species (ROS) homeostasis. Depending on the entity of the perturbation, normal cells can either restore equilibrium or activate pathways of cell death. On the contrary, cancer cells exploit this phenomenon to sustain a proliferative and aggressive phenotype. In fact, ROS overproduction or their reduced disposal influence all hallmarks of cancer, from genome instability to cell metabolism, angiogenesis, invasion and metastasis. A persistent state of oxidative stress can even initiate tumorigenesis. MicroRNAs (miRNAs) are small non coding RNAs with regulatory functions, which expression has been extensively proven to be dysregulated in cancer. Intuitively, miRNA transcription and biogenesis are affected by the oxidative status of the cell and, in some instances, they participate in defining it. Indeed, it is widely reported the role of miRNAs in regulating numerous factors involved in the ROS signaling pathways. Given that miRNA function and modulation relies on cell type or tumor, in order to delineate a clearer and more exhaustive picture, in this review we present a comprehensive overview of the literature concerning how miRNAs and ROS signaling interplay affects breast cancer progression.

Keywords: oxidative stress; miRNAs; breast cancer; ROS

---

1. Oxidative Stress

Reactive oxygen species (ROS) are oxygen-derived small molecules in the form of free radicals (i.e., contains one or more unpaired electrons) or non-radicals [1]. Among the most biologically relevant species there are the superoxide anion radical ($O_2^{-•}$), the hydroxyl radical ($OH·$) and hydrogen peroxide (H_2O_2). At first, it was thought that these molecules were only metabolic waste, deleterious for nucleic acids, lipids and proteins; scientists, however, discovered that ROS are used by the cell as messages to activate different physiological signaling cascades [2,3]. In fact, in a biological system, the balance between the concentration of ROS and the activation of antioxidant mechanisms is finely tuned [4]. When this equilibrium lacks, the phenomenon of oxidative stress occurs, causing the alteration of intracellular molecules, such as DNA and RNA. A shift towards ROS production, thus, triggers a wide range of cellular responses, even apoptosis or phagocytosis, depending on the amplitude of the shift. Several endogenous and exogenous sources can trigger ROS production. In response to stimuli like cytokines and growth factors, NADPH oxidases (NOXs) and mitochondria produce the larger percentage of ROS. NOXs and metabolic complexes I, II and III present on the mitochondrial inner membrane generate, for example, the radical superoxide starting from a molecule of oxygen. Dangerous levels of ROS can be reached also after prolonged exposure to radiations and carcinogens, along with DNA damaging drugs. The major mutagenic product of DNA oxidation is

8-hydroxyl-2'-deoxyguanosine (8-OHdG). Figure 1 summarizes principal sources producing ROS and main regulators and pathways influenced by ROS production (Figure 1). Cancer cells are usually in a chronic state of oxidative stress, which they are able to exploit to sustain a proliferative and aggressive phenotype. Moreover, due to their detrimental action, ROS can also initiate tumorigenesis [5]. It is thus important not to overlook the impact of such phenomenon on every cellular process and, in particular, on those crucial for the development and progression of a neoplastic disease.

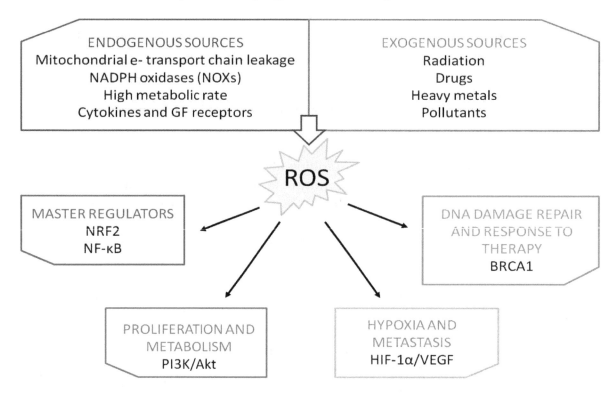

Figure 1. Endogenous and exogenous sources of ROS and pathways influenced by oxidative stress in breast cancer.

2. Breast Cancer

Breast cancer is the second most commonly diagnosed cancer worldwide and the leading cause of cancer death in women [6]. The severity and aggressiveness of breast cancer is evaluated by examining physical and anatomical properties of the disease, in particular by using histological grading and TNM staging, where T (0–4) is used to describe the size and location of the tumor, N (0–3) accounts for the lymph node invasion and M measures the spread of the tumor as distant metastasis [7].

The therapeutic regimen is finally driven by the characterization of the breast cancer subtype according to the immunohistochemical evaluation of three markers: Estrogen receptor (ER), Progesteron Receptor (Pgr) and HER2 (Epidermal Growth Factor Receptor 2) [8]. Tumors lacking the expression of these three markers are called triple negative breast cancers (TNBCs). A major contribution to the increase in survival rate has been provided by the improvement in the therapeutic regimens as well as in early diagnosis. Moreover, it is fundamental to develop always more personalized drugs for different cancers subtypes.

The advent of the genomic era disclosed the complexity of breast cancer. For the first time, in 2000, Perou and colleagues classified the disease in five specific subtypes according to intrinsic gene expression: Luminal-A, Luminal-B, HER2-positive, Basal-like and Normal-like [9]. Further studies later identified a new subtype, the so called claudin-low [10], which accounts for 7–14% of all breast cancers. Moreover, one of the most important applications of the breast cancer molecular classification lies in its ability to identify groups with a different outcome and response to treatments [11–15].

3. Oxidative Stress and Breast Cancer

Breast cancers, in particular estrogen receptor-positive malignancies, are characterized by significant high levels of 8-OHdG, and their detection in blood serum is reported to have prognostic value [16–18].

Estrogen is a major driver of mitochondrial ROS production. It activates redox-sensitive proteins involved in cell proliferation and anti-apoptotic pathways. In order to sustain such signaling without risking cell cycle arrest and apoptosis, estrogen enhances also an antioxidant response by inducing, for example, the transcription factor Nuclear-erythroid-2-related factor 2 (NRF2). This enzyme is the main redox master regulator; under oxidative stress, its inhibitor Kelch-like ECH-associated protein 1 (Keap1) undergoes a conformational change that allows NRF2 dissociation and consequent translocation to the nucleus, where it enhances the transcription of different ROS-counteracting agents [19,20]. Numerous evidence shows that NRF2 is overexpressed in breast cancer, where it promotes cell survival, proliferation, migration and metastasis [21–23].

Additionally, it is important to note the interplay between NRF2 and BRCA1. Gorrini C. et al. demonstrated that BRCA1 enhances and stabilizes NRF2 expression and that estrogen is able to partially mimic this action in BRCA1-null cells [24,25].

Moreover, in 2014, Victorino V. J. et al. analyzed the effect of HER2 overexpression on the oxidative systemic profile in breast cancer patients [26]. The results showed that HER2-overexpressing malignancies are characterized by an enhanced oxidative stress, attenuated by increased SOD and stabilized gluthatione (GSH) levels, which are indicative of an active antioxidant response. In the same year, Kang H. J. et al. reported that also HER2 interacts with NRF2 to promote the transcription of antioxidant and detoxification genes and that this partnership confers drug resistance to human breast cancer cells [27]. Antioxidants can, thus, favor breast neoplastic transformation: by reducing ROS concentrations they can prevent ROS-dependent cell death [28,29]. Therefore, the role of antioxidants in breast cancer is often controversial; for example antioxidant superoxide dismutase 2 (SOD2), which converts the highly toxic radical superoxide into more stable hydrogen peroxide in the mitochondria, can act both as an oncogene and as a tumor suppressor. In fact, it is found downmodulated in early-stage breast cancer while upregulated in advanced tumors [30,31]. Despite these results, SOD mimics have been proposed for therapeutic purposes [32,33]. Catalase, glutathione peroxidases, and peroxiredoxins are among the other antioxidant enzymes which balance ROS production. In 2017, Bao B. et al. demonstrated that the addition of a re-engineered protein form of the catalase enzyme to EGFR-inhibitor erlotinib treatment helps overcoming resistance by specifically targeting the stem-like portion of TNBC cells [34]. Conversely, peroxiredoxin-1 (PRDX1) downmodulation was shown to be beneficial for breast cancer therapy, especially in concomitance with prooxidant agents [35]. Moreover, specific acquaporins allow H_2O_2 to cross cell membranes more rapidly than by sole diffusion [36]. In breast cancer, Aquaporin-3 has been proposed as target for therapy due to its role in CXCL12/CXCR4-dependent cancer cell migration [37].

Finally, damages to RNA molecules have to be considered equally harmful [38]. For example, it is of particular relevance for this review the impact on miRNA biology and the consequent influence on different regulation networks [39].

4. MicroRNAs

MicroRNAs (miRNAs) are small single strand molecules (~18–25 nucleotides), they are non-coding RNAs that are able to control gene expression at post-transcriptional level [40]. MiRNA biogenesis starts when RNA polymerase II/III transcribes for a long primary transcript with a hairpin structure, called pri-miRNAs [41]. The pri-miRNA is the substrate of Drosha and Dicer, two members of the

RNase III family enzymes. First, Drosha cleaves the pri-miRNA in a ~70-nucleotide pre-miRNA into the nucleus, which is then exported into the cytoplasm by the Exportin-5 Ran-GTPase, where Dicer catalyzes its conversion to a short miRNA/miRNA* duplex (~20 bp). To complete the miRNA biogenesis, the transactivation-responsive RNA-binding protein (TRBP) leads to the assembling of the miRNA-induced silencing complex (miRISC), mediating the interaction between DICER and Argonaute protein (AGO1, AGO2, AGO3 or AGO4). Finally, the miRISC complex selects one single strand of the duplex (mature miRNA), which recognizes the "seed" region on the target mRNA, usually placed at the 5' UTR, inducing translational repression or deadenylation and degradation. The small RNA lin-4 was the first no-coding RNA discovered in *Caenorhabditis elegans*, involved in the larval development [42]. Afterwards, several studies have pointed out the importance of these small molecules; currently it is well known that miRNAs are involved in almost every biological process in mammals, including oxidative stress and cancer [43]. Indeed, miRNAs can act as oncosuppressors or oncogenes, which are generally found respectively downregulated and upregulated in tumor cells (e.g., miR-205 and miR-21, respectively). In 2005, Iorio M.V. et al. discovered a panel of dysregulated microRNAs in breast cancer: miR-10b, miR-125b, and miR-145 were down-regulated, and miR-21 and miR-155 were up-regulated, suggesting that they could have a role in breast cancer disease [44].

Recently, we reported that miRNAs have a relevant role in DNA damage response, occurring following an exogenous oxidative stress, such as chemotherapy [45,46]. In fact, miRNAs have the capability to target several genes involved in the DNA repair machinery, regulating therapy responsiveness. Here, we review the literature concerning the role of miRNAs in the regulation of the major actors and principal pathways altered by oxidative stress in breast cancer.

5. MiRNAs Modulate Oxidative Stress Master Regulators: NRF2 and NF-κB

NRF2 is an important transcription factor which induction, or derepression, depends on the redox status of the cell. Normally, NRF2 is found inactive in the cytoplasm bound to its homodimeric repressor Keap1, which anchors the protein Cullin-3 (CUL3) to form an E3 ubiquitin ligase complex; the complex is responsible for NRF2 ubiquitination and consequent proteasomal degradation [47]. When cellular ROS concentrations increase, specific Keap1 cystenyl residues are modified and NRF2 is released and free to translocate into the nucleus, where it recognizes the so called "Antioxidant Responsive Elements (ARE)" sequences on target gene promoters and enhances the transcription process [48]. NRF2 promotes the expression of antioxidants and detoxifying enzymes and, initially, it was thought to act as a defensive agent against tumorigenesis. However, as previously explained for SOD2, an excessive reduction of ROS levels can prove counterproductive. Therefore, it is not unusual to find contradictory literature concerning the prospective of using NRF2 inhibitors for therapeutic purposes [49,50]. NRF2 pathogenic activation and accumulation can be triggered by different events; one of the most frequent alterations concerns Keap1 expression or its ability to stably bind and degrade NRF2 [51]. MiRNAs were found to exert this oncogenic activity, NRF2 induction, in different malignancies [52–54]. In 2011, Eades G. et al. demonstrated for the first time a miRNA-dependent Keap1 regulation in breast cancer: miR-200a targets Keap1 mRNA and induces its degradation [55]. Interestingly, the same group published the same year an additional paper describing NRF2 inhibition by miR-28 in MCF7 breast cancer cell line [56]. Two other miRNAs, miR-93 and miR-153, have been reported to target NRF2 and their overexpression is associated with breast carcinogenesis [57,58]. This evidence validates once more the context-specific value of NRF2 modulation (Figure 2A).

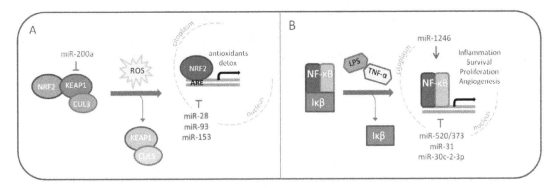

Figure 2. MiRNAs modulating oxidative stress master regulators NRF2 (**A**) and NF-κB (**B**) in breast cancer (The red arrow indicates upmodulation, the red "T" stands for inhibition).

The same concept can be translated to the other redox master regulator, the nuclear factor-kB (NF-κB). NF-κB can be found as both homo- and heterodimer of five distinct proteins, RelA, RelB, c-Rel, p50 and p52. It is inhibited in the cytoplasm by the IκB families, which interfere with the target activity by interacting with its important Rel homology domain (RHD), implicated in the formation of dimers and DNA binding [59].

IκB proteins are generally degraded in response to inflammatory cues like TNFα and lipopolysaccharide (LPS). The consequent NF-κB signaling, modulated by ROS, is cell type and context specific. This is probably due to the transcription factor wide range of action: cell growth, proliferation, migration and apoptosis are among the pathways it influences [60,61]. NF-κB signaling is frequently found dysregulated in human cancers [62]. In breast cancer, the protein is reported as constitutively activated and associated to aggressive and chemoresistant malignances [63–65]. MiRNAs play an important part also in this scenario. First of all, NF-κB favors breast cancer cell invasion by inducing the expression of the oncomiR miR-21 in response to DNA damage [66]. Wiemann S. and his group, instead, published different papers on NF-κB-regulating miRNAs over the years [67–70]. In 2012, they demonstrated the tumor suppressive role of miR-520/373 family in ER-negative breast cancer, through the targeting of NF-κB and TGF-β signaling pathways. In the same context, in 2013, miR-31 was seen to sensitize cancer cells to apoptosis by impairing NF-κB pathway. In 2015, miR-30c-2-3p was shown to reduce proliferation and invasion of MDA-MB-231 cells through the downmodulation of TNFR/NF-κB signaling and cell cycle proteins. Conversely, in 2017, a role as an oncomiR was attributed to miR-1246, which was reported to induce the NF-κB pro-inflammatory signaling in breast cancer cells. The same year, another group discovered that miR-221/222 promote stem-like properties and tumor growth of breast cancer via targeting PTEN and sustained Akt/NF-κB/COX-2 activation (Figure 2B) [71].

Despite having an oscillatory expression in a physiological context, NF-κB thus emerges from the literature presented as a proper oncogene in breast cancer. Moreover, due to its broad spectrum of interactions, numerous are the miRNAs involved in the regulation of the signaling cascade and, consequently, many are the hints for therapeutic interventions.

6. MiRNAs Modulate Pathways Altered by Oxidative Stress

6.1. Metabolism

The main goal of cancer cells is proliferation and survival. Such activities require a great amount of energy in a short period of time. Therefore, cancer cells tend to modify their metabolism in order to respond to this demand. According to the known Warburg effect, cancers prefer a rapid glycolysis to the more efficient mitochondrial oxidative phosphorylation. This switch also allows avoiding an excessive mitochondria-related production of ROS. Interestingly, it has been suggested that the latter could be the primary reason for the metabolic reprogramming [72]. In breast cancers, the metabolic status seems to be linked to the molecular subtype. In fact, the more aggressive TNBCs are characterized by a glycolytic phenotype, while luminal malignancies retain oxidative phosphorylation as the major source of energy [73]. It is important to note that it is not unusual to find heterogeneity also among cells of the same tumor mass, a scenario that can be as deleterious as a predominant Warburg setting. In 2018, our group indeed proposed that, starting from a mixed population of TNBC cells, pushing all the cells towards a glycolytic phenotype could become counterproductive for the tumor. Through the downmodulation of the lactate transporter MCT1, miR-342-3p is able to disrupt the energetic fluxes between neighboring glycolytic and oxidative cells, promoting the shift and ultimately triggering a competition for glucose [74]. It has been demonstrated that glucose deprivation induces oxidative stress in cancer cells [75]. One of the most cited mechanisms of breast cancer cell metabolic reprogramming that involves miRNAs is miR-155 promotion of hexokinase II (HKII) expression, necessary to start glycolysis. This miRNA modulates multiple pathways that control HKII: first, miR-155, through the direct downmodulation of C/EBPβ, reduces miR-143, a HKII inhibitor; second, the miRNA frees STAT3 from its suppressor SOCS1 to enhance HKII; third, miR-155 positively regulates HKII by interfering with the PIK3R1-FOXO3a-cMYC axis [76–78]. Another recent example is the work by Eastlack S. C. et al. that demonstrated miR-27b promotes breast cancer progression by targeting Pyruvate Dehydrogenase Protein X (PDHX), thus altering cell's metabolic configuration [79]. PI3K/Akt pathway, which players are frequently mutated in breast cancers, deeply impacts on metabolism and ROS production by directly regulating mitochondrial bioenergetics and NOX enzymes. Vice versa, oxidative stress activates PI3K and suppresses the activity of PTEN, inhibitor of PI3K/Akt signaling [80–82]. Due to the relevance of the pathway, numerous are the miRNAs found implicated in its regulation in breast cancer. Among the latest reported, there are the tumor suppressor miR-204-5p, which targets PIK3CB, and the PTEN-inhibiting oncomiRs miR-1297 and miR-498 (Figure 3A) [83–85].

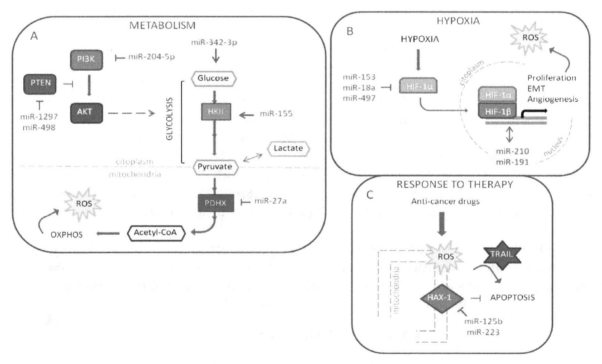

Figure 3. MiRNAs involved in the regulation of hallmarks of cancer influenced by oxidative stress in breast cancer: metabolism (**A**), hypoxia (**B**) and response to therapy (**C**) (The red arrow indicates upmodulation, the red "T" stands for inhibition).

6.2. Hypoxia

Hypoxia refers to a pathological level of oxygen tension, caused by the high proliferative rates of cancer cells and insufficient vasculature. The lack of oxygen supply, thus, induces cancer cells to undergo epithelial-to-mesenchymal transition (EMT), which corresponds to the acquisition of migratory and invasive properties, stem-like features and resistance to apoptosis. The main player in this context is the transcription factor HIF-1α, which stimulates angiogenesis and triggers a positive feedback loop on proliferation pathways. As a consequence, oxidative stress increases upon re-oxygenation and mitochondrial electron leaks. Numerous studies showed a link between the miRNA's role and the hypoxia in the breast cancer initiation and progression. The first miRNA to be pointed out is miR-210. In 2007, this miRNA emerged as part of the miRNA signature of hypoxia and the year after it was elected as an independent prognostic factor in breast cancer [86,87]. Moreover, Liang H. and colleagues investigated miR-153 mechanism of action in breast cancer; showing that this miRNA acts as a tumor suppressor by targeting HIF-1α [88]. In fact, miR-153 inhibits migration, proliferation and tube formation in HUVEC cells and angiogenesis in MDA-MB-231 in vivo model through the inhibition of the HIF-1α/VEGFA axis. In another paper, the high expression of miR-191 in breast cancer cell lines induces a more aggressive tumor under hypoxia [89]. Consequently, the authors suggest that miR-191 inhibition may be exploited as a new therapeutic option for hypoxic breast cancer. In addition, miR-18a targets HIF-1α, which high expression is associated with shorter DMFS (distant metastasis-free survival) in patients with basal-like breast tumors [90]. In metastatic MDA-MB-231 cells, ectopic miR-18a expression reduces both primary tumor and lung metastasis. Another miRNA reported targeting HIF-1α is miR-497, thus, it represses the hypoxic conditions and for this reason it is usually downregulated in breast cancer cells [91]. MiR-497 also targets a pro-angiogenic molecule, VEGF (vascular endothelial growth factor) and its ectopic expression reduces tumor growth and angiogenesis in breast cancer tumor model (Figure 3B). In conclusion, we could support the strategic role of miRNAs in the tumor progression and in particular in hypoxia and metastasis and we could speculate the possibility to use miRNAs as therapeutic tools to reduce tumor aggressiveness and dissemination.

6.3. Response To Therapy

It is well known that miRNAs can influence response to therapy in breast cancer. Moreover, they are under investigation as potential therapeutic tools, alone or in combination with standard therapy to impair cancer progression. Chemotherapy and radiotherapy still represent the standard therapy for breast cancer; miRNAs are able to target different genes reducing drug resistance and promoting therapeutic response. Indeed, in 2016, we reported that miR-302b, by targeting E2F1 and DNA repair, enhances cisplatin response in breast cancer cells [45]. Chemotherapy drugs, such as platinum compounds and anthracyclines, and also ionizing radiation induce oxidative stress generating high levels of ROS [92]. The induction of oxidative stress can lead to the preferential killing of cancer cells. Currently, the main problem of chemotherapy and radiotherapy is the development of resistance mechanisms; recent works report the role of miRNAs in the response to these therapies by targeting oxidative stress molecules. Recently, it was demonstrated that miR-125b is involved in chemotherapy resistance by affecting oxidative stress pathways in breast cancer [93]. MiR-125b, by targeting HAX-1, an anti-apoptotic gene, impacts on doxorubicin resistance. The mechanism behind this phenomenon is a decrease in the levels of MMP following HAX-1 downregulation and the release of ROS from the mitochondria into the cytoplasm. Thus, miR-125b is able to re-sensitize breast cancer cells to doxorubicin treatment using ROS pathway (Figure 3C). Concerning chemoresistance, Roscigno G. et al. have reported that miR-24, up-regulated in breast cancer stem cells, induces resistance to cisplatin by targeting the pro-apoptotic factor BimL [94]. Furthermore, miR-24 targets FIH1 that induces the repression of HIF-1α. Thus, the authors have shown that miR-24 is induced in hypoxic conditions, leading to cancer stem cell growth and consequently inducing chemotherapy resistance. Breast cancer patients often poorly respond to radiotherapy, and the mechanisms of radioresistance have not been elucidated yet. MiR-668 was found increased in breast cancer cells resistant to radiotherapy; this phenomenon occurs because IκBα is a direct target of miR-668, leading to the activation of NF-κB [95]. Generally, drug resistance is an important challenge in the treatment of breast cancer, especially for

TNBC, which still don't have target therapy. To date, novel therapeutic strategies have been tested mainly in the treatment of TNBC. MiR-223 is related to resistance to TRAIL-induced apoptosis in cancer stem cells of TNBC [96]. Indeed, reintroduction of miR-223 and treatment with TRAIL in MDA-MB-231 cell line induces a strong generation of ROS, through the targeting of HAX-1 into the mitochondria, and TNBC stem cells are more sensitive to TRAIL treatment. Moreover, the miR-223/HAX-1 axis enhances the sensitivity to doxorubicin and cisplatin in TNBC stem cells (Figure 3C).

7. Conclusive Remarks

In this review, we have illustrated what emerges from the literature about the important role of oxidative stress in the pathogenesis of breast cancer, influencing most of the pathways usually altered in tumors, affecting also response to therapy. Moreover, many of the proteins involved in this process, such as SOD2 and NRF2, can exert opposite roles depending on the context, complicating the scenario. Thus, it is important to explore in more detail the mechanisms behind the regulation of the redox status in relation to a specific scenario in order to better define which pathways can be proposed as therapeutic targets. MiRNAs act as regulative elements in almost every biological process, including oxidative stress and cancer. Here, we have mainly reviewed the literature concerning miRNAs involved in the regulation of oxidative stress players in breast cancer disease. MiRNA role in the regulation of redox status makes them as hypothetical and crucial targets or tools for therapy since they could provide the treatment context specificity. MiRNA general use as therapy option has yet to show relevant results, but an increasing body of evidence has been provided through the years in favor of such a solution, especially in oncology. Additionally, breast cancer is one of the most studied neoplasia and many are the miRNAs which mechanism of action is consolidated in this framework. Hopefully, therefore, it will be soon possible to have major improvements in this research field.

Author Contributions: G.C., I.P. and A.C. wrote the manuscript. M.V.I. and A.C. edited and revised the manuscript.

References

1. Li, R.; Jia, Z.; Trush, M.A. Defining ROS in Biology and Medicine. *React. Oxyg. Species (Apex, NC)* **2016**, *1*, 9–21. [CrossRef] [PubMed]
2. Finkel, T. Signal transduction by reactive oxygen species. *J. Cell Biol.* **2011**, *194*, 7–15. [CrossRef] [PubMed]
3. Zhang, J.; Wang, X.; Vikash, V.; Ye, Q.; Wu, D.; Liu, Y.; Dong, W. ROS and ROS-Mediated Cellular Signaling. *Oxid. Med. Cell. Longev.* **2016**, *2016*, 4350965. [CrossRef] [PubMed]
4. Sies, H.; Berndt, C.; Jones, D.P. Oxidative Stress. *Annu. Rev. Biochem.* **2017**, *86*, 715–748. [CrossRef] [PubMed]
5. Sosa, V.; Moliné, T.; Somoza, R.; Paciucci, R.; Kondoh, H.; LLeonart, M.E. Oxidative stress and cancer: An overview. *Ageing Res. Rev.* **2013**, *12*, 376–390. [CrossRef]
6. Bray, F.; Ferlay, J.; Soerjomataram, I.; Siegel, R.L.; Torre, L.A.; Jemal, A. Global cancer statistics 2018: GLOBOCAN estimates of incidence and mortality worldwide for 36 cancers in 185 countries. *CA Cancer J. Clin.* **2018**, *68*, 394–424. [CrossRef]
7. Giuliano, A.E.; Edge, S.B.; Hortobagyi, G.N. Eighth Edition of the AJCC Cancer Staging Manual: Breast Cancer. *Ann. Surg. Oncol.* **2018**, *25*, 1783–1785. [CrossRef]
8. Waks, A.G.; Winer, E.P. Breast Cancer Treatment: A Review. *JAMA* **2019**, *321*, 288–300. [CrossRef]
9. Perou, C.M.; Sørlie, T.; Eisen, M.B.; van de Rijn, M.; Jeffrey, S.S.; Rees, C.A.; Pollack, J.R.; Ross, D.T.; Johnsen, H.; Akslen, L.A.; et al. Molecular portraits of human breast tumours. *Nature* **2000**, *406*, 747–752. [CrossRef]
10. Herschkowitz, J.I.; Simin, K.; Weigman, V.J.; Mikaelian, I.; Usary, J.; Hu, Z.; Rasmussen, K.E.; Jones, L.P.; Assefnia, S.; Chandrasekharan, S.; et al. Identification of conserved gene expression features between murine mammary carcinoma models and human breast tumors. *Genome Biol.* **2007**, *8*, 76. [CrossRef]
11. Sørlie, T.; Perou, C.M.; Tibshirani, R.; Aas, T.; Geisler, S.; Johnsen, H. Gene expression patterns of breast carcinomas distinguish tumor subclasses with clinical implications. *Proc. Natl. Acad. Sci. USA* **2001**, *98*, 10869–10874. [CrossRef] [PubMed]

12. Rouzier, R.; Perou, C.M.; Symmans, W.F.; Ibrahim, N.; Cristofanilli, M.; Anderson, K.; Hess, K.R.; Stec, J.; Ayers, M.; Wagner, P.; et al. Breast cancer molecular subtypes respond differently to preoperative chemotherapy. *Clin. Cancer Res.* **2005**, *11*, 5678–5685. [CrossRef] [PubMed]

13. Hu, Z.; Fan, C.; Oh, D.S.; Marron, J.S.; He, X.; Qaqish, B.F.; Livasy, C.; Carey, L.A.; Reynolds, E.; Dressler, L.; et al. The molecular portraits of breast tumors are conserved across microarray platforms. *BMC Genom.* **2006**, *7*, 96.

14. Carey, L.A.; Dees, E.C.; Sawyer, L.; Gatti, L.; Moore, D.T.; Collichio, F.; Ollila, D.W.; Sartor, C.I.; Graham, M.L.; Perou, C.M. The triple negative paradox: Primary tumor chemosensitivity of breast cancer subtypes. *Clin. Cancer Res.* **2007**, *13*, 2329–2334. [CrossRef]

15. Shehata, M.; Teschendorff, A.; Sharp, G.; Novcic, N.; Russell, I.A.; Avril, S.; Prater, M.; Eirew, P.; Caldas, C.; Watson, C.J.; et al. Phenotypic and functional characterisation of the luminal cell hierarchy of the mammary gland. *Breast Cancer Res.* **2012**, *14*, R134. [CrossRef]

16. Musarrat, J.; Arezina-Wilson, J.; Wani, A.A. Prognostic and aetiological relevance of 8-hydroxyguanosine in human breast carcinogenesis. *Eur. J. Cancer* **1996**, *32*, 1209–1214. [CrossRef]

17. Jakovcevic, D.; Dedic-Plavetic, N.; Vrbanec, D.; Jakovcevic, A.; Jakic-Razumovic, J. Breast Cancer Molecular Subtypes and Oxidative DNA Damage. *Appl. Immunohistochem. Mol. Morphol.* **2015**, *23*, 696–703. [CrossRef]

18. Nour Eldin, E.E.M.; El-Readi, M.Z.; Nour Eldein, M.M.; Alfalki, A.A.; Althubiti, M.A.; Mohamed Kamel, H.F.; Eid, S.Y.; Al-Amodi, H.S.; Mirza, A.A. 8-Hydroxy-2′-deoxyguanosine as a Discriminatory Biomarker for Early Detection of Breast Cancer. *Clin. Breast Cancer* **2019**, *19*, 385–393. [CrossRef]

19. Tian, H.; Gao, Z.; Wang, G.; Li, H.; Zheng, J. Estrogen potentiates reactive oxygen species (ROS) tolerance to initiate carcinogenesis and promote cancer malignant transformation. *Tumour Biol.* **2016**, *37*, 141–150. [CrossRef]

20. Rojo de la Vega, M.; Chapman, E.; Zhang, D.D. NRF2 and the Hallmarks of Cancer. *Cancer Cell* **2018**, *34*, 21–43. [CrossRef]

21. Zhang, C.; Wang, H.J.; Bao, Q.C.; Wang, L.; Guo, T.K.; Chen, W.L.; Xu, L.L.; Zhou, H.S.; Bian, J.L.; Yang, Y.R.; et al. NRF2 promotes breast cancer cell proliferation and metastasis by increasing RhoA/ROCK pathway signal transduction. *Oncotarget* **2016**, *7*, 73593–73606. [CrossRef] [PubMed]

22. Lu, K.; Alcivar, A.L.; Ma, J.; Foo, T.K.; Zywea, S.; Mahdi, A.; Huo, Y.; Kensler, T.W.; Gatza, M.L.; Xia, B. NRF2 Induction Supporting Breast Cancer Cell Survival Is Enabled by Oxidative Stress-Induced DPP3-KEAP1 Interaction. *Cancer Res.* **2017**, *77*, 2881–2892. [CrossRef] [PubMed]

23. Zhang, H.S.; Zhang, Z.G.; Du, G.Y.; Sun, H.L.; Liu, H.Y.; Zhou, Z.; Gou, X.M.; Wu, X.H.; Yu, X.Y.; Huang, Y.H. Nrf2 promotes breast cancer cell migration via up-regulation of G6PD/HIF-1α/Notch1 axis. *J. Cell. Mol. Med.* **2019**, *23*, 3451–3463. [CrossRef] [PubMed]

24. Gorrini, C.; Baniasadi, P.S.; Harris, I.S.; Silvester, J.; Inoue, S.; Snow, B.; Joshi, P.A.; Wakeham, A.; Molyneux, S.D.; Martin, B.; et al. BRCA1 interacts with Nrf2 to regulate antioxidant signaling and cell survival. *J. Exp. Med.* **2013**, *210*, 1529–1544. [CrossRef]

25. Gorrini, C.; Gang, B.P.; Bassi, C.; Wakeham, A.; Baniasadi, S.P.; Hao, Z.; Li, W.Y.; Cescon, D.W.; Li, Y.T.; Molyneux, S.; et al. Estrogen controls the survival of BRCA1-deficient cells via a PI3K-NRF2-regulated pathway. *Proc. Natl. Acad. Sci. USA* **2014**, *111*, 4472–4477. [CrossRef]

26. Victorino, V.J.; Campos, F.C.; Herrera, A.C.; Colado Simão, A.N.; Cecchini, A.L.; Panis, C.; Cecchini, R. Overexpression of HER-2/neu protein attenuates the oxidative systemic profile in women diagnosed with breast cancer. *Tumour Biol.* **2014**, *35*, 3025–3034. [CrossRef]

27. Kang, H.J.; Yi, Y.W.; Hong, Y.B.; Kim, H.J.; Jang, Y.J.; Seong, Y.S.; Bae, I. HER2 confers drug resistance of human breast cancer cells through activation of NRF2 by direct interaction. *Sci. Rep.* **2014**, *4*, 7201. [CrossRef]

28. Harris, I.S.; Treloar, A.E.; Inoue, S.; Sasaki, M.; Gorrini, C.; Lee, K.C.; Yung, K.Y.; Brenner, D.; Knobbe-Thomsen, C.B.; Cox, M.A.; et al. Glutathione and thioredoxin antioxidant pathways synergize to drive cancer initiation and progression. *Cancer Cell* **2015**, *27*, 211–222. [CrossRef]

29. Monteiro, H.P.; Ogata, F.T.; Stern, A. Thioredoxin promotes survival signaling events under nitrosative/oxidative stress associated with cancer development. *Biomed. J.* **2017**, *40*, 189–199. [CrossRef]

30. Becuwe, P.; Ennen, M.; Klotz, R.; Barbieux, C.; Grandemange, S. Manganese superoxide dismutase in breast cancer: From molecular mechanisms of gene regulation to biological and clinical significance. *Free Radic. Biol. Med.* **2014**, *77*, 139–151. [CrossRef]

31. Wang, Y.; Branicky, R.; Noë, A.; Hekimi, S. Superoxide dismutases: Dual roles in controlling ROS damage and regulating ROS signaling. *J. Cell Biol.* **2018**, *217*, 1915–1928. [CrossRef] [PubMed]

32. Batinić-Haberle, I.; Rebouças, J.S.; Spasojević, I. Superoxide dismutase mimics: Chemistry, pharmacology, and therapeutic potential. *Antioxid. Redox Signal.* **2010**, *13*, 877–918. [CrossRef] [PubMed]

33. Fernandes, A.S.; Saraiva, N.; Oliveira, N.G. Redox Therapeutics in Breast Cancer: Role of SOD Mimics. In *Redox-Active Therapeutics: Oxidative Stress in Applied Basic Research and Clinical Practice*; Batinić-Haberle, I., Rebouças, J., Spasojević, I., Eds.; Springer: Cham, Switzerland, 2016.

34. Bao, B.; Mitrea, C.; Wijesinghe, P.; Marchetti, L.; Girsch, E.; Farr, R.L.; Boerner, J.L.; Mohammad, R.; Dyson, G.; Terlecky, S.R.; et al. Treating triple negative breast cancer cells with erlotinib plus a select antioxidant overcomes drug resistance by targeting cancer cell heterogeneity. *Sci. Rep.* **2017**, *7*, 44125. [CrossRef]

35. Bajor, M.; Zych, A.O.; Graczyk-Jarzynka, A.; Muchowicz, A.; Firczuk, M.; Trzeciak, L.; Gaj, P.; Domagala, A.; Siernicka, M.; Zagozdzon, A.; et al. Targeting peroxiredoxin 1 impairs growth of breast cancer cells and potently sensitises these cells to prooxidant agents. *Br. J. Cancer* **2018**, *119*, 873–884. [CrossRef]

36. Bienert, G.P.; Chaumont, F. Aquaporin-facilitated transmembrane diffusion of hydrogen peroxide. *Biochim. Biophys. Acta* **2014**, *1840*, 1596–1604. [CrossRef] [PubMed]

37. Satooka, H.; Hara-Chikuma, M. Aquaporin-3 Controls Breast Cancer Cell Migration by Regulating Hydrogen Peroxide Transport and Its Downstream Cell Signaling. *Mol. Cell. Biol.* **2016**, *36*, 1206–1218. [CrossRef]

38. Kong, Q.; Lin, C.G. Oxidative damage to RNA: Mechanisms, consequences, and diseases. *Cell. Mol. Life Sci.* **2010**, *67*, 1817–1829. [CrossRef]

39. He, J.; Jiang, B.H. Interplay between Reactive oxygen Species and MicroRNAs in Cancer. *Curr. Pharmacol. Rep.* **2016**, *2*, 82–90. [CrossRef]

40. Lee, Y.; Kim, M.; Han, J.; Yeom, K.H.; Lee, S.; Baek, S.H.; Kim, V.N. MicroRNA genes are transcribed by RNA polymerase II. *EMBO J.* **2004**, *23*, 4051–4060. [CrossRef]

41. Krol, J.; Loedige, I.; Filipowicz, W. The widespread regulation of microRNA biogenesis, function and decay. *Nat. Rev. Genet.* **2010**, *11*, 597–610. [CrossRef]

42. Lee, R.C.; Feinbaum, R.L.; Ambros, V. Elegans heterochronic gene lin-4 encodes small RNAs with antisense complementarity to lin-14. *Cell* **1993**, *75*, 843–854. [CrossRef]

43. Iorio, M.V.; Casalini, P.; Piovan, C.; Braccioli, L.; Tagliabue, E. Breast cancer and microRNAs: Therapeutic impact. *Breast* **2011**, *20*, 63–70. [CrossRef]

44. Iorio, M.V.; Ferracin, M.; Liu, C.G.; Veronese, A.; Spizzo, R.; Sabbioni, S.; Magri, E.; Pedriali, M.; Fabbri, M.; Campiglio, M.; et al. MicroRNA gene expression deregulation in human breast cancer. *Cancer Res.* **2005**, *65*, 7065–7070. [CrossRef] [PubMed]

45. Cataldo, A.; Cheung, D.G.; Balsari, A.; Tagliabue, E.; Coppola, V.; Iorio, M.V.; Palmieri, D.; Croce, C.M. miR-302b enhances breast cancer cell sensitivity to cisplatin by regulating E2F1 and the cellular DNA damage response. *Oncotarget* **2016**, *7*, 786–797. [CrossRef] [PubMed]

46. Plantamura, I.; Cosentino, G.; Cataldo, A. MicroRNAs and DNA-Damaging Drugs in Breast Cancer: Strength in Numbers. *Front. Oncol.* **2018**, *8*, 352. [CrossRef]

47. McMahon, M.; Itoh, K.; Yamamoto, M.; Hayes, J.D. Keap1-dependent proteasomal degradation of transcription factor Nrf2 contributes to the negative regulation of antioxidant response element-driven gene expression. *J. Biol. Chem.* **2003**, *278*, 21592–21600. [CrossRef]

48. Suzuki, T.; Yamamoto, M. Molecular basis of the Keap1-Nrf2 system. *Free Radic. Biol. Med.* **2015**, *88*, 93–100. [CrossRef]

49. Sporn, M.B.; Liby, K.T. NRF2 and cancer: The good, the bad and the importance of context. *Nat. Rev. Cancer* **2012**, *12*, 564–571. [CrossRef]

50. Panieri, E.; Saso, L. Potential Applications of NRF2 Inhibitors in Cancer Therapy. *Oxid. Med. Cell. Longev.* **2019**, *2019*, 8592348. [CrossRef]

51. Taguchi, K.; Yamamoto, M. The KEAP1-NRF2 System in Cancer. *Front. Oncol.* **2017**, *7*, 85. [CrossRef]

52. Kabaria, S.; Choi, D.C.; Chaudhuri, A.D.; Jain, M.R.; Li, H.; Junn, E. MicroRNA-7 activates Nrf2 pathway by targeting Keap1 expression. *Free Radic. Biol. Med.* **2015**, *89*, 548–556. [CrossRef] [PubMed]

53. Shi, L.; Wu, L.; Chen, Z.; Yang, J.; Chen, X.; Yu, F.; Zheng, F.; Lin, X. MiR-141 Activates Nrf2-Dependent Antioxidant Pathway via Down-Regulating the Expression of Keap1 Conferring the Resistance of Hepatocellular Carcinoma Cells to 5-Fluorouracil. *Cell. Physiol. Biochem.* **2015**, *35*, 2333–2348. [CrossRef] [PubMed]

54. Akdemir, B.; Nakajima, Y.; Inazawa, J.; Inoue, J. miR-432 induces NRF2 Stabilization by Directly Targeting KEAP1. *Mol. Cancer Res.* **2017**, *15*, 1570–1578. [CrossRef] [PubMed]

55. Eades, G.; Yang, M.; Yao, Y.; Zhang, Y.; Zhou, Q. miR-200a regulates Nrf2 activation by targeting Keap1 mRNA in breast cancer cells. *J. Biol. Chem.* **2011**, *286*, 40725–40733. [CrossRef] [PubMed]

56. Yang, M.; Yao, Y.; Eades, G.; Zhang, Y.; Zhou, Q. MiR-28 regulates Nrf2 expression through a Keap1-independent mechanism. *Breast Cancer Res. Treat.* **2011**, *129*, 983–991. [CrossRef]

57. Singh, B.; Ronghe, A.M.; Chatterjee, A.; Bhat, N.K.; Bhat, H.K. MicroRNA-93 regulates NRF2 expression and is associated with breast carcinogenesis. *Carcinogenesis* **2013**, *34*, 1165–1172. [CrossRef]

58. Wang, B.; Teng, Y.; Liu, Q. MicroRNA-153 Regulates NRF2 Expression and is Associated with Breast Carcinogenesis. *Clin. Lab.* **2016**, *62*, 39–47. [CrossRef]

59. Hayden, M.S.; Ghosh, S. Shared principles in NF-kappaB signaling. *Cell* **2008**, *132*, 344–362. [CrossRef]

60. Morgan, M.J.; Liu, Z.G. Crosstalk of reactive oxygen species and NF-κB signaling. *Cell Res.* **2011**, *21*, 103–115. [CrossRef]

61. Lingappan, K. NF-κB in Oxidative Stress. *Curr. Opin. Toxicol.* **2018**, *7*, 81–86. [CrossRef]

62. Xia, Y.; Shen, S.; Verma, I.M. NF-κB, an active player in human cancers. *Cancer Immunol. Res.* **2014**, *2*, 823–830. [CrossRef] [PubMed]

63. Smith, S.M.; Lyu, Y.L.; Cai, L. NF-κB affects proliferation and invasiveness of breast cancer cells by regulating CD44 expression. *PLoS ONE* **2014**, *9*, e106966. [CrossRef] [PubMed]

64. Pires, B.R.; Mencalha, A.L.; Ferreira, G.M.; de Souza, W.F.; Morgado-Díaz, J.A.; Maia, A.M.; Corrêa, S.; Abdelhay, E.S. NF-kappaB Is Involved in the Regulation of EMT Genes in Breast Cancer Cells. *PLoS ONE* **2017**, *12*, e0169622. [CrossRef] [PubMed]

65. Kim, J.Y.; Jung, H.H.; Ahn, S.; Bae, S.; Lee, S.K.; Kim, S.W.; Lee, J.E.; Nam, S.J.; Ahn, J.S.; Im, Y.H.; et al. The relationship between nuclear factor (NF)-κB family gene expression and prognosis in triple-negative breast cancer (TNBC) patients receiving adjuvant doxorubicin treatment. *Sci. Rep.* **2016**, *6*, 31804. [CrossRef]

66. Niu, J.; Shi, Y.; Tan, G.; Yang, C.H.; Fan, M.; Pfeffer, L.M.; Wu, Z.H. DNA damage induces NF-κB-dependent microRNA-21 up-regulation and promotes breast cancer cell invasion. *J. Biol. Chem.* **2012**, *287*, 21783–21795. [CrossRef]

67. Keklikoglou, I.; Koerner, C.; Schmidt, C.; Zhang, J.D.; Heckmann, D.; Shavinskaya, A.; Allgayer, H.; Gückel, B.; Fehm, T.; Schneeweiss, A.; et al. MicroRNA-520/373 family functions as a tumor suppressor in estrogen receptor negative breast cancer by targeting NF-κB and TGF-β signaling pathways. *Oncogene* **2012**, *31*, 4150–4163. [CrossRef]

68. Körner, C.; Keklikoglou, I.; Bender, C.; Wörner, A.; Münstermann, E.; Wiemann, S. MicroRNA-31 sensitizes human breast cells to apoptosis by direct targeting of protein kinase C epsilon (PKCepsilon). *J. Biol. Chem.* **2013**, *288*, 8750–8761. [CrossRef]

69. Shukla, K.; Sharma, A.K.; Ward, A.; Will, R.; Hielscher, T.; Balwierz, A.; Breunig, C.; Münstermann, E.; König, R.; Keklikoglou, I.; et al. MicroRNA-30c-2-3p negatively regulates NF-κB signaling and cell cycle progression through downregulation of TRADD and CCNE1 in breast cancer. *Mol. Oncol.* **2015**, *9*, 1106–1119. [CrossRef]

70. Bott, A.; Erdem, N.; Lerrer, S.; Hotz-Wagenblatt, A.; Breunig, C.; Abnaof, K.; Wörner, A.; Wilhelm, H.; Münstermann, E.; Ben-Baruch, A.; et al. miRNA-1246 induces pro-inflammatory responses in mesenchymal stem/stromal cells by regulating PKA and PP2A. *Oncotarget* **2017**, *8*, 43897–43914. [CrossRef]

71. Li, B.; Lu, Y.; Yu, L.; Han, X.; Wang, H.; Mao, J.; Shen, J.; Wang, B.; Tang, J.; Li, C.; et al. miR-221/222 promote cancer stem-like cell properties and tumor growth of breast cancer via targeting PTEN and sustained Akt/NF-κB/COX-2 activation. *Chem. Biol. Interact.* **2017**, *277*, 33–42. [CrossRef]

72. Rodic, S.; Vincent, M.D. Reactive oxygen species (ROS) are a key determinant of cancer's metabolic phenotype. *Int. J. Cancer* **2018**, *142*, 440–448. [CrossRef] [PubMed]

73. Choi, J.; Kim, D.H.; Jung, W.H.; Koo, J.S. Metabolic interaction between cancer cells and stromal cells according to breast cancer molecular subtype. *Breast Cancer Res.* **2013**, *15*, 78. [CrossRef] [PubMed]

74. Romero-Cordoba, S.L.; Rodriguez-Cuevas, S.; Bautista-Pina, V.; Maffuz-Aziz, A.; D'Ippolito, E.; Cosentino, G.; Baroni, S.; Iorio, M.V.; Hidalgo-Miranda, A. Loss of function of miR-342-3p results in MCT1 over-expression and contributes to oncogenic metabolic reprogramming in triple negative breast cancer. *Sci. Rep.* **2018**, *8*, 12252. [CrossRef] [PubMed]

75. Spitz, D.R.; Sim, J.E.; Ridnour, L.A.; Galoforo, S.S.; Lee, Y.J. Glucose deprivation-induced oxidative stress in human tumor cells. A fundamental defect in metabolism? *Ann. N. Y. Acad. Sci.* **2000**, *899*, 349–362. [CrossRef]

76. Jiang, S.; Zhang, L.F.; Zhang, H.W.; Hu, S.; Lu, M.H.; Liang, S.; Li, B.; Li, Y.; Li, D.; Wang, E.D.; et al. A novel miR-155/miR-143 cascade controls glycolysis by regulating hexokinase 2 in breast cancer cells. *EMBO J.* **2012**, *31*, 1985–1998. [CrossRef]

77. Lei, K.; Du, W.; Lin, S.; Yang, L.; Xu, Y.; Gao, Y.; Xu, B.; Tan, S.; Xu, Y.; Qian, X.; et al. 3B, a novel photosensitizer, inhibits glycolysis and inflammation via miR-155-5p and breaks the JAK/STAT3/SOCS1 feedback loop in human breast cancer cells. *Biomed. Pharmacother.* **2016**, *82*, 141–150. [CrossRef]

78. Kim, S.; Lee, E.; Jung, J.; Lee, J.W.; Kim, H.J.; Kim, J.; Yoo, H.J.; Lee, H.J.; Chae, S.Y.; Jeon, S.M.; et al. microRNA-155 positively regulates glucose metabolism via PIK3R1-FOXO3a-cMYC axis in breast cancer. *Oncogene* **2018**, *37*, 2982–2991. [CrossRef]

79. Eastlack, S.C.; Dong, S.; Ivan, C.; Alahari, S.K. Suppression of PDHX by microRNA-27b deregulates cell metabolism and promotes growth in breast cancer. *Mol. Cancer* **2018**, *17*, 100. [CrossRef]

80. Yuan, T.L.; Cantley, L.C. PI3K pathway alterations in cancer: Variations on a theme. *Oncogene* **2008**, *27*, 5497–5510. [CrossRef]

81. Leslie, N.R.; Bennett, D.; Lindsay, Y.E.; Stewart, H.; Gray, A.; Downes, C.P. Redox regulation of PI 3-kinase signalling via inactivation of PTEN. *EMBO J.* **2003**, *22*, 5501–5510. [CrossRef]

82. Koundouros, N.; Poulogiannis, G. Phosphoinositide 3-Kinase/Akt Signaling and Redox Metabolism in Cancer. *Front. Oncol.* **2018**, *8*, 160. [CrossRef] [PubMed]

83. Hong, B.S.; Ryu, H.S.; Kim, N.; Kim, J.; Lee, E.; Moon, H.; Kim, K.H.; Jin, M.S.; Kwon, N.H.; Kim, S.; et al. Tumor Suppressor miRNA-204-5p Regulates Growth, Metastasis, and Immune Microenvironment Remodeling in Breast Cancer. *Cancer Res.* **2019**, *79*, 1520–1534. [PubMed]

84. Liu, C.; Liu, Z.; Li, X.; Tang, X.; He, J.; Lu, S. MicroRNA-1297 contributes to tumor growth of human breast cancer by targeting PTEN/PI3K/AKT signaling. *Oncol. Rep.* **2017**, *38*, 2435–2443. [CrossRef] [PubMed]

85. Chai, C.; Wu, H.; Wang, B.; Eisenstat, D.D.; Leng, R.P. MicroRNA-498 promotes proliferation and migration by targeting the tumor suppressor PTEN in breast cancer cells. *Carcinogenesis* **2018**, *39*, 1185–1196. [CrossRef] [PubMed]

86. Kulshreshtha, R.; Ferracin, M.; Wojcik, S.E.; Garzon, R.; Alder, H.; Agosto-Perez, F.J.; Davuluri, R.; Liu, C.G.; Croce, C.M.; Negrini, M.; et al. A microRNA signature of hypoxia. *Mol. Cell. Biol.* **2007**, *27*, 1859–1867. [CrossRef]

87. Camps, C.; Buffa, F.M.; Colella, S.; Moore, J.; Sotiriou, C.; Sheldon, H.; Harris, A.L.; Gleadle, J.M.; Ragoussis, J. hsa-miR-210 Is induced by hypoxia and is an independent prognostic factor in breast cancer. *Clin. Cancer Res.* **2008**, *14*, 1340–1348. [CrossRef]

88. Liang, H.; Xiao, J.; Zhou, Z.; Wu, J.; Ge, F.; Li, Z.; Zhang, H.; Sun, J.; Li, F.; Liu, R.; et al. Hypoxia induces miR-153 through the IRE1α-XBP1 pathway to fine tune the HIF1α/VEGFA axis in breast cancer angiogenesis. *Oncogene* **2018**, *37*, 1961–1975. [CrossRef]

89. Nagpal, N.; Ahmad, H.M.; Chameettachal, S.; Sundar, D.; Ghosh, S.; Kulshreshtha, R. HIF-inducible miR-191 promotes migration in breast cancer through complex regulation of TGFβ-signaling in hypoxic microenvironment. *Sci. Rep.* **2015**, *5*, 9650. [CrossRef]

90. Krutilina, R.; Sun, W.; Sethuraman, A.; Brown, M.; Seagroves, T.N.; Pfeffer, L.M.; Ignatova, T.; Fan, M. MicroRNA-18a inhibits hypoxia-inducible factor 1α activity and lung metastasis in basal breast cancers. *Breast Cancer Res.* **2014**, *16*, 78. [CrossRef]

91. Wu, Z.; Cai, X.; Huang, C.; Xu, J.; Liu, A. miR-497 suppresses angiogenesis in breast carcinoma by targeting HIF-1α. *Oncol. Rep.* **2016**, *35*, 1696–1702. [CrossRef]

92. Gorrini, C.; Harris, I.S.; Mak, T.W. Modulation of oxidative stress as an anticancer strategy. *Nat. Rev. Drug Discov.* **2013**, *12*, 931–947. [CrossRef] [PubMed]

93. Hu, G.; Zhao, X.; Wang, J.; Lv, L.; Wang, C.; Feng, L.; Shen, L.; Ren, W. miR-125b regulates the drug-resistance of breast cancer cells to doxorubicin by targeting HAX-1. *Oncol. Lett.* **2018**, *15*, 1621–1629. [CrossRef] [PubMed]

94. Roscigno, G.; Puoti, I.; Giordano, I.; Donnarumma, E.; Russo, V.; Affinito, A.; Adamo, A.; Quintavalle, C.; Todaro, M.; Vivanco, M.D.; et al. MiR-24 induces chemotherapy resistance and hypoxic advantage in breast cancer. *Oncotarget* **2017**, *8*, 19507–19521. [CrossRef] [PubMed]

95. Luo, M.; Ding, L.; Li, Q.; Yao, H. miR-668 enhances the radioresistance of human breast cancer cell by targeting IκBα. *Breast Cancer* **2017**, *24*, 673–682. [CrossRef] [PubMed]

12

MicroRNA Mediate Visfatin and Resistin Induction of Oxidative Stress in Human Osteoarthritic Synovial Fibroblasts Via NF-κB Pathway

Sara Cheleschi [1,*,†], Ines Gallo [1], Marcella Barbarino [2], Stefano Giannotti [3], Nicola Mondanelli [3], Antonio Giordano [2], Sara Tenti [1,†] and Antonella Fioravanti [1,†]

[1] Department of Medicine, Surgery and Neuroscience, Rheumatology Unit, Azienda Ospedaliera Universitaria senese, Policlinico Le Scotte, 53100 Siena, Italy; ins.gll3@gmail.com (I.G.); sara_tenti@hotmail.it (S.T.); fioravanti7@virgilio.it (A.F.)
[2] Sbarro Institute for Cancer Research and Molecular Medicine, Department of Biology, College of Science and Technology, Temple University, Philadelphia, PA 19122, USA; marcella.barbarino@unisi.it (M.B.); giordano@temple.edu (A.G.)
[3] Department of Medicine, Surgery and Neurosciences, Section of Orthopedics and Traumatology, University of Siena, Policlinico Le Scotte, 53100 Siena, Italy; stefano.giannotti@unisi.it (S.G.); nicola@nicolamondanelli.it (N.M.)
*   Correspondence: saracheleschi@hotmail.com or sara.cheleschi@unisi.it
†   These authors contributed equally to this work.

Abstract: Synovial membrane inflammation actively participate to structural damage during osteoarthritis (OA). Adipokines, miRNA, and oxidative stress contribute to synovitis and cartilage destruction in OA. We investigated the relationship between visfatin, resistin and miRNA in oxidative stress regulation, in human OA synovial fibroblasts. Cultured cells were treated with visfatin and resistin. After 24 h, we evaluated various pro-inflammatory cytokines, metalloproteinases (*MMPs*), type II collagen (*Col2a1*), *miR-34a, miR-146a, miR-181a,* antioxidant enzymes, and B-cell lymphoma (*BCL)2* by qRT-PCR, apoptosis and mitochondrial superoxide production by cytometry, p50 nuclear factor (NF)-κB by immunofluorescence. Synoviocytes were transfected with miRNA inhibitors and oxidative stress evaluation after adipokines stimulus was performed. The implication of NF-κB pathway was assessed by the use of a NF-κB inhibitor (BAY-11-7082). Visfatin and resistin significantly up-regulated gene expression of interleukin *(IL)-1β, IL-6, IL-17,* tumor necrosis factor *(TNF)-α, MMP-1, MMP-13* and reduced *Col2a1.* Furthermore, adipokines induced apoptosis and superoxide production, the transcriptional levels of *BCL2,* superoxide dismutase (*SOD*)-2, catalase (*CAT*), nuclear factor erythroid 2 like 2 (*NRF2*), *miR-34a, miR-146a,* and *miR-181a.* MiRNA inhibitors counteracted adipokines modulation of oxidative stress. Visfatin and resistin effects were suppressed by BAY-11-7082. Our data suggest that miRNA may represent possible mediators of oxidative stress induced by visfatin and resistin via NF-κB pathway in human OA synoviocytes.

Keywords: microRNA; visfatin; resistin; osteoarthritis; oxidative stress; apoptosis; synovial fibroblasts; synovitis; NF-κB

1. Introduction

Osteoarthritis (OA) is the most prevalent musculoskeletal disease characterized by a progressive degradation of articular cartilage, osteophyte formation, subchondral sclerosis and synovitis [1,2]. Increasing evidence suggests that synovial membrane inflammation is implicated in the pathophysiology of the disease; prostaglandins, leukotrienes, reactive oxygen species (ROS), cytokines, chemokines and adipokines, produced by inflamed synovium, induced cartilage degradation and further bolster inflammation [3–5].

Adipokines, including adiponectin, chemerin, leptin, resistin, and visfatin, are secreted by white adipose tissue and are known to be involved in multiple biological processes, as immunity, inflammation, cartilage and bone metabolism. Much attention has been paid regarding their implication in the pathogenesis of many rheumatic diseases, even OA [6–10].

Visfatin has originally identified as an insulin-mimetic factor, with pro-inflammatory and immunomodulating functions [11], while resistin is implicated in obesity-associated insulin resistance and involved in inflammatory response [12].

Visfatin and resistin serum levels and synovial fluid were found to be increased in patients with knee and hand OA [9,13–15]; moreover, it has been highlighted the pro-inflammatory effect of these adipokines on the expression of different cytokines and chemokines, as well as their role in mediating the production of matrix degrades enzymes in human OA chondrocytes and synovial fibroblasts [16–19].

Recent studies demonstrated a complex interaction between adipokines and microRNAs (miRNA) [17,18,20,21]. miRNA are an abundant class of conserved double stranded non-coding RNA molecules of 22–25 nucleotides that are classified as important post-transcriptional regulators of gene expression of target gene messenger RNA [22]. They are implicated in important physiological cellular processes as well as in the pathophysiology of different disorders, including OA [23–26]. Some miRNA, also known as oxidative stress-responsive factors, can be induced or suppressed by ROS, and their biological function, through regulation of target genes, should be influenced [27]; besides, a specific modulation of oxidative stress balance by specific miRNA has been postulated [28].

In the present study, we investigated the complex cross-talk between visfatin, resistin and some miRNA (*miR-34a, miR-146a,* and *miR-181a*) in the regulation of oxidative stress, in human OA synovial fibroblasts.

In particular, we analyzed the effect of visfatin and resistin in gene expression of interleukin (*IL*)-1β, *IL-6, IL-17A,* tumor necrosis factor (*TNF*)-α, metalloproteinases (*MMP*)-1, *MMP-13,* collagen type II (*Col2a1*). Furthermore, the apoptotic cells and the transcriptional levels of the anti-apoptotic marker B-cell lymphoma (*BCL*) 2, as well as the production of mitochondrial superoxide anion and the gene levels of antioxidant enzymes [superoxide dismutase (*SOD*)-2, catalase (*CAT*)] and nuclear factor erythroid 2 like 2 (*NRF2*) were also investigated.

To examine the potential role of *miR-34a, miR-146a,* and *miR-181a* as mediators of the visfatin and resistin effects on oxidative stress, we transfected synovial fibroblasts with miRNA specific inhibitors.

Finally, the possible implication of nuclear factor (NF)-κB pathway in adipokines-mediated effects was assessed.

2. Results

2.1. Cell viability Evaluation in Visfatin and Resistin Treated Cells

Cell viability assay was analyzed by 3-(4,4-dimethylthiazol-2-yl)-2,5-diphenyl-tetrazoliumbromide (MTT) test and the results are represented in Figure S1. A significant reduction of the percentage of survival cells was observed in human OA synovial fibroblasts incubated with visfatin 5 μg/mL and 10 μg/mL ($p < 0.05$) and resistin 50 ng/mL and 100 ng/mL ($p < 0.05$), in comparison to basal condition.

2.2. Visfatin and Resistin Promote Inflammation and Regulate Cartilage Turnover

The effect of adipokines on gene expression of the main pro-inflammatory mediators IL-1β, IL-6, Il-17A and TNF-α in human OA synovial fibroblasts is reported in Figure 1.

Visfatin, tested at both concentrations, 5 μg/mL and 10 μg/mL, significantly increased the mRNA expression of *IL-1β, IL-6, IL-17A,* and *TNF-α* ($p < 0.01$, $p < 0.001$) (Figure 1A), in a dose dependent manner. Similarly, resistin 50 and 100 ng/mL induced a significant up-regulation ($p < 0.001$) of gene levels of the studied cytokines compared with the un-stimulated cells (Figure 1B).

In Figure 1C,D we summarized the regulation of the main extracellular matrix (ECM) degrading enzyme, MMP-1, MMP-13, and of the main component of articular ECM, Col2a1.

In human OA synovial fibroblasts stimulated with visfatin 5 and 10 µg/mL (Figure 1C) and resistin 50 ng/mL and 100 ng/mL (Figure 1D) we showed a significant increase of *MMP-1*, *MMP-13* ($p < 0.01$, $p < 0.001$) and a reduction of *Col2a1* ($p < 0.01$, $p < 0.001$) expression levels, in comparison to basal time.

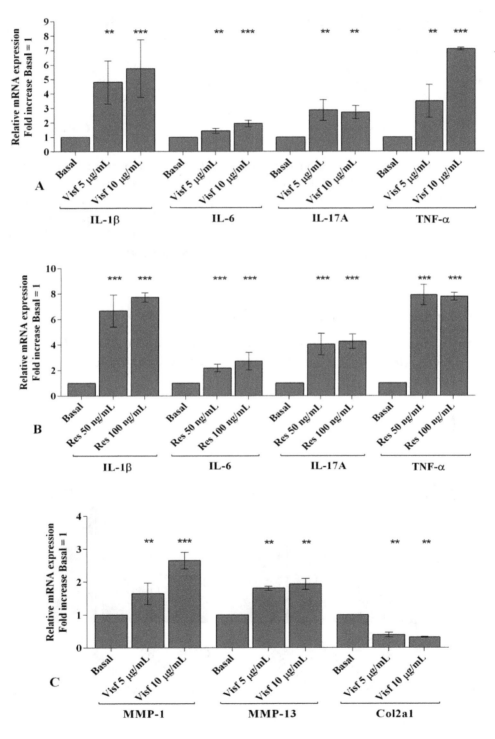

Figure 1. *Cont.*

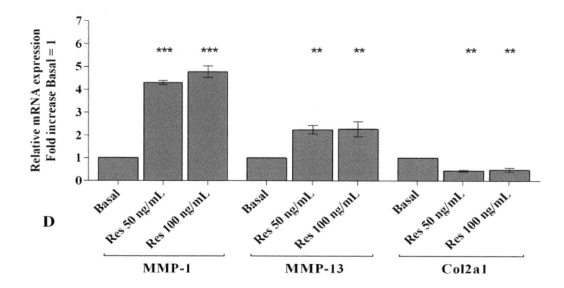

Figure 1. (**A–D**) Expression levels of interleukin (*IL*)-*1β*, *IL-6*, *IL-17A*, tumor necrosis factor (*TNF*)-α, metalloproteinases (*MMP*)-*1*, *MMP-13*, and collagen type II (*Col2a1*) by real-time PCR. Human osteoarthritic (OA) synovial fibroblasts were evaluated at basal condition and after incubation with visfatin (5 and 10 μg/mL) and resistin (50 and 100 ng/mL) for 24 h. The gene expression was referenced to the ratio of the value of interest and the value of basal condition (basal, cells without treatment), reported equal to 1. Data were expressed as mean ± SD of triplicate values. ** $p < 0.01$, *** $p < 0.001$ versus basal condition. Visf = visfatin, Res = resistin.

2.3. Adipokines Induce Apoptosis and Regulate BCL2 Expression

Visfatin (5 and 10 μg/mL) and resistin (50 and 100 ng/mL) stimulation induced a significant and dose-dependent increase ($p < 0.01$, $p < 0.001$) of apoptotic OA synovial fibroblasts in comparison to baseline (Figure S2 and Figure 2A).

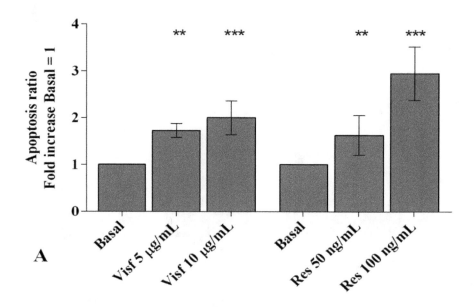

Figure 2. *Cont.*

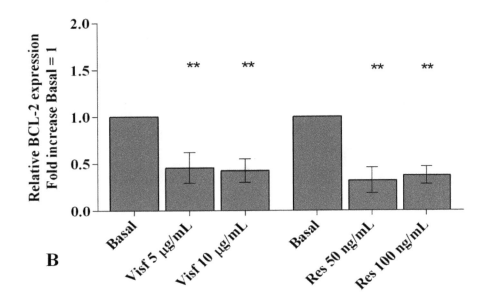

Figure 2. (A) Apoptosis detection performed by the analysis at flow cytometry and measured with Annexin Alexa fluor 488 assay. Data were expressed as the percentage of positive cells for Annexin-V and propidium iodide (PI) staining. **(B)** Expression levels of gene B-cell lymphoma (*BCL*)2 by real-time PCR. Human osteoarthritic (OA) synovial fibroblasts were evaluated at basal condition and after incubation with visfatin (5 and 10 µg/mL) and resistin (50 and 100 ng/mL) for 24 h. The apoptosis ratio and the gene expression were referenced to the ratio of the value of interest and the value of basal condition (basal, cells without treatment), reported equal to 1. Data were expressed as mean ± SD of triplicate values. ** $p < 0.01$, *** $p < 0.001$ versus basal condition. Visf = visfatin, Res = resistin.

Real-time PCR analysis underlines a significant reduction of the expression levels of the anti-apoptotic marker *BCL2* ($p < 0.01$) in cells incubated with visfatin and resistin, at both tested concentrations, when compared to un-treated cells (Figure 2B).

2.4. Visfatin and Resistin Regulate Oxidant/Antioxidant Balance

To investigate the potential role of the studied adipokines in the regulation of oxidant/antioxidant balance, we assessed the production of superoxide anion and the analysis of the gene expression of the main antioxidant enzymes implicated in ROS scavenge (Figure S3 and Figure 3).

The stimulus of the cells with the higher concentration of visfatin (10 µg/mL) caused a significant increase of mitochondrial superoxide anion production ($p < 0.05$, Figure 3A); resistin 50 and 100 ng/mL significantly induced a dose-dependent activation of oxidative stress condition ($p < 0.05$, $p < 0.01$, respectively) in comparison to basal time (Figure 3A).

Both concentrations of the tested adipokines significantly up-regulated the expression levels of the antioxidant enzymes *SOD-2* ($p < 0.01$, $p < 0.001$), *CAT* ($p < 0.01$, $p < 0.001$), and *NRF2* ($p < 0.001$) (Figure 3B,C).

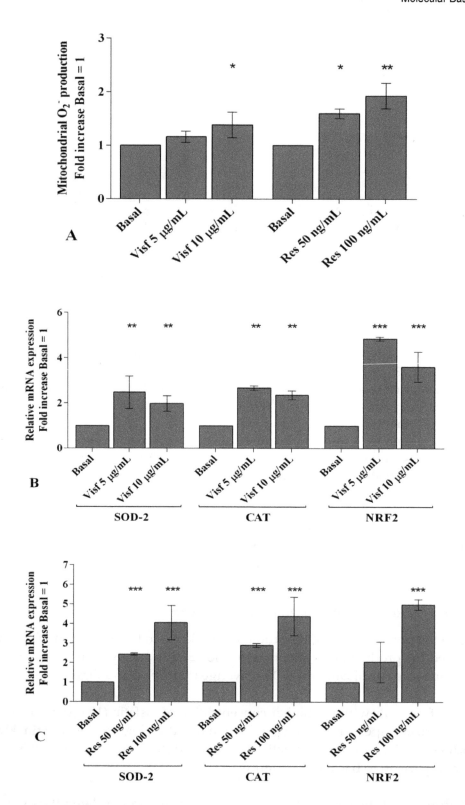

Figure 3. (**A**) Mitochondrial superoxide anion production was assessed by the analysis at flow cytometry using MitoSox Red staining. (**B,C**) Expression levels of superoxide dismutase (*SOD-2*), catalase (*CAT*), nuclear factor erythroid 2 like 2 (*NRF2*) by real-time PCR. Human osteoarthritic (OA) synovial fibroblasts were evaluated at basal condition and after incubation with visfatin (5 and 10 μg/mL) and resistin (50 and 100 ng/mL) for 24 h. The superoxide anion production and the gene expression were referenced to the ratio of the value of interest and the value of basal condition (basal, cells without treatment), reported equal to 1. Data were expressed as mean ± SD of triplicate values. * $p < 0.05$, ** $p < 0.01$, *** $p < 0.001$ versus basal condition. Visf = visfatin, Res = resistin.

2.5. Visfatin and Resistin Modulate miRNA Gene Expression

A real-time PCR analysis has been performed in order to evaluate the modulation of *miR-34a*, *miR-146a*, and *miR-181a* gene expression induced by adipokines. Visfatin at a concentration of 5 and 10 µg/mL ($p < 0.01$, $p < 0.001$) up-regulated *miR-34a* and *miR-146a* transcriptional levels in comparison to basal condition, while it did not influence *miR-181a* levels (Figure 4A). Resistin 50 and 100 ng/mL significantly increased the gene expression of *miR-34a* ($p < 0.001$), *miR-146a* ($p < 0.01$), and *miR-181a* ($p < 0.05$, $p < 0.01$) (Figure 4B).

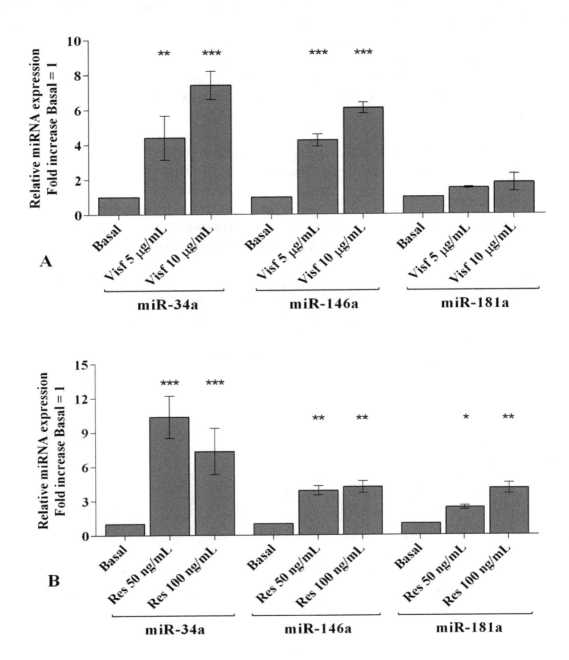

Figure 4. (**A,B**) Expression levels of *miR-34a*, *miR-146a*, and *miR-181a* by real-time PCR. Human osteoarthritic (OA) synovial fibroblasts were evaluated at basal condition and after incubation with visfatin (5 and 10 µg/mL) and resistin (50 and 100 ng/mL) for 24 h. The gene expression was referenced to the ratio of the value of interest and the value of basal condition (basal, cells without treatment), reported equal to 1. Data were expressed as mean ± SD of triplicate values. * $p < 0.05$, ** $p < 0.01$, *** $p < 0.001$ versus basal condition. Visf = visfatin, Res = resistin.

2.6. MiRNA Regulate Oxidative Stress Induced by Visfatin and Resistin

To confirm the involvement of miRNA in modulating oxidative stress induced by visfatin and resistin, we transfected OA synoviocytes with *miR-34a, miR-146a,* and *miR-181a* specific inhibitors (Figure 5).

Real-time PCR showed a significant reduction of gene expression levels of the studied miRNA ($p < 0.01$) in transfected OA cells with respect to basal condition and NC (Figure 5A).

Visfatin (5 and 10 μg/mL) and resistin (50 and 100 ng/mL) significantly up-regulated transcriptional levels of *miR-34a, miR-146a,* and *miR-181a* ($p < 0.01$, Figure 5B–G) in OA synoviocytes incubated with NC. After the transfection with miRNA inhibitors, the treatment with visfatin or resistin did not show any significant modification in *miR-34a, miR-146a,* and *miR-181a* expression in comparison to what is observed in synoviocytes transfected with the inhibitors alone (Figure 5B–G). In addition, the inhibition of *miR-34a, miR-146a,* and *miR-181a* significantly reduced the increase of miRNA transcriptional levels induced by visfatin and resistin incubation ($p < 0.01$, Figure 5B–G).

In Figures 6–8 we reported the modulation of redox balance induced by visfatin and resistin after the transfection of OA synoviocytes with *miR-34a, miR-146a,* and *miR-181a* inhibitors.

MiRNA silencing determined a significant reduction of mitochondrial superoxide anion production ($p < 0.05$, $p < 0.01$, Figures 6A, 7A and 8A) as well as a down-regulation of *SOD-2, CAT,* and *NRF2* expression levels ($p < 0.05$, $p < 0.01$, Figures 6B, 7B and 8B) in comparison to basal condition and NC.

The production of superoxide anion and the expression of *SOD-2, CAT,* and *NRF2* were increased, in a significant manner, in OA cells transfected with NC after stimulus with visfatin ($p < 0.01$, $p < 0.001$, Figure 6C,E, Figure 7C,E and Figure 8C,E) and resistin ($p < 0.01$, $p < 0.001$, Figure 6D,F, Figure 7D,F and Figure 8D,F), while their effect was significantly inhibited by *miR-34a, miR-146a,* and *miR-181a* specific inhibitors ($p < 0.01$, Figure 6C–F, Figure 7C–F and Figure 8C–F).

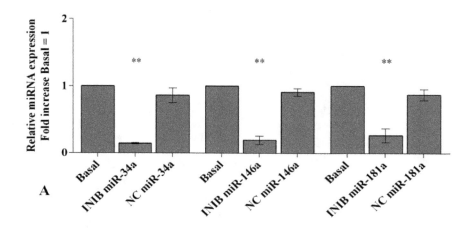

Figure 5. *Cont.*

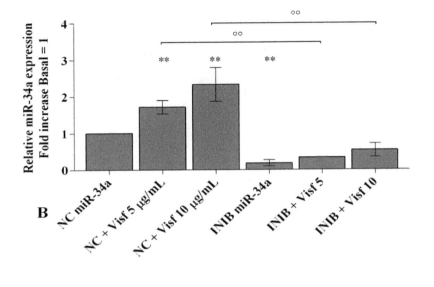

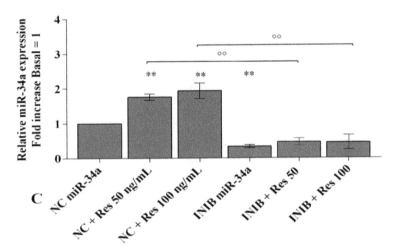

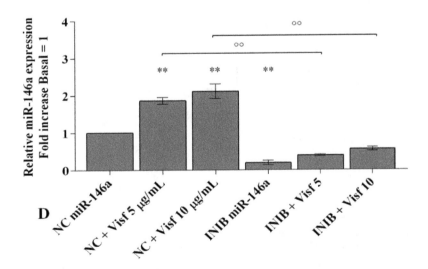

Figure 5. *Cont.*

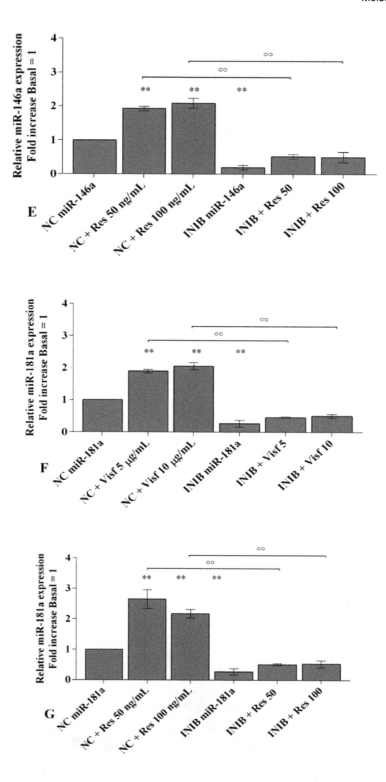

Figure 5. (A–G) Expression levels of *miR-34a*, *miR-146a*, and *miR-181a* by real-time PCR. Human osteoarthritic (OA) synovial fibroblasts were evaluated at basal condition, after 24 h of transfection with *miR-34a*, *miR-146a*, and *miR-181a* inhibitors or NC, and after incubation with visfatin (5 and 10 µg/mL) and resistin (50 and 100 ng/mL). The gene expression was referenced to the ratio of the value of interest and the value of basal condition (basal, cells without treatment) or NC, reported equal to 1. Data were expressed as mean ± SD of triplicate values. ** $p < 0.01$ versus basal condition or NC. °° $p < 0.01$ versus inhibitor. INIB= inhibitor, NC= negative control siRNA, Visf= visfatin, Res = resistin.

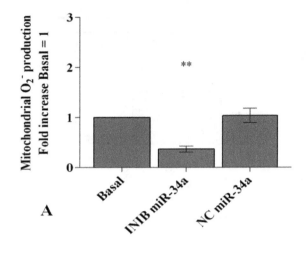

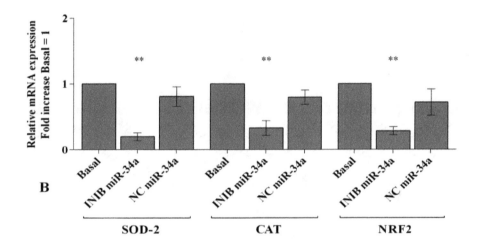

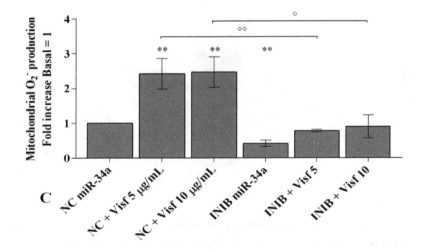

Figure 6. *Cont.*

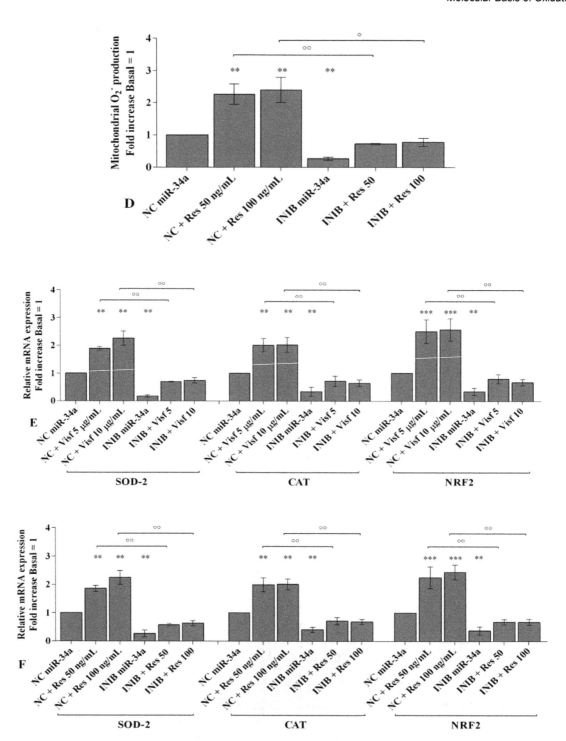

Figure 6. (**A,C,D**) Mitochondrial superoxide anion production was assessed by the analysis at flow cytometry using MitoSox Red staining. (**B,E,F**) Expression levels of superoxide dismutase (*SOD-2*), catalase (*CAT*), nuclear factor erythroid 2 like 2 (*NRF2*) by real-time PCR. Human osteoarthritic (OA) synovial fibroblasts were evaluated at basal condition, after 24 h of transfection with *miR-34a* inhibitor or NC, and after incubation with visfatin (5 and 10 μg/mL) and resistin (50 and 100 ng/mL). The superoxide anion production and the gene expression were referenced to the ratio of the value of interest and the value of basal condition (basal, cells without treatment) or NC, reported equal to 1. Data were expressed as mean ± SD of triplicate values. ** $p < 0.01$, *** $p < 0.001$ versus basal condition or NC. ° $p < 0.05$, °° $p < 0.01$ versus inhibitor. INIB= inhibitor, NC= negative control siRNA, Visf= visfatin, Res = resistin.

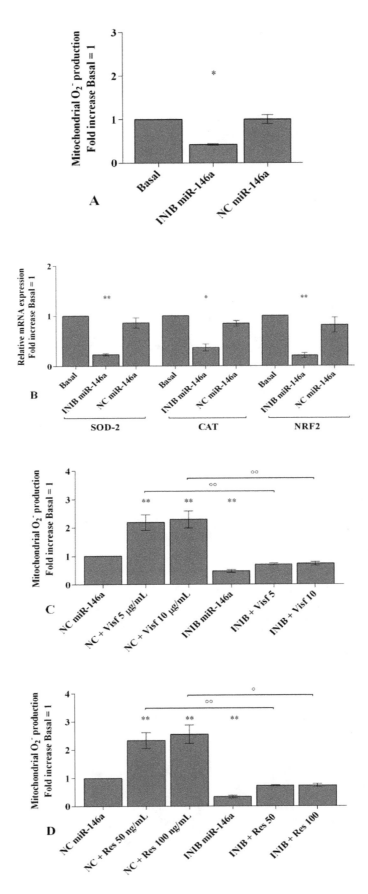

Figure 7. *Cont.*

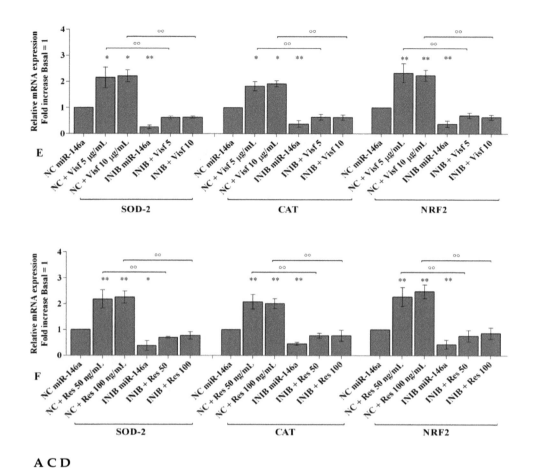

A C D

Figure 7. (, ,) Mitochondrial superoxide anion production was assessed by flow cytometry using MitoSox Red staining. (**B,E,F**) Expression levels of superoxide dismutase (*SOD-2*), catalase (*CAT*), nuclear factor erythroid 2 like 2 (*NRF2*) by real-time PCR. Human osteoarthritic (OA) synovial fibroblasts were evaluated at basal condition, after 24 h of transfection with *miR-146a* inhibitor or NC, and after incubation with visfatin (5 and 10 µg/mL) and resistin (50 and 100 ng/mL). The superoxide anion production and the gene expression were referenced to the ratio of the value of interest and the value of basal condition (basal, cells without treatment) or NC, reported equal to 1. Data were expressed as mean ± SD of triplicate values. * $p < 0.05$, ** $p < 0.01$ versus basal condition or NC. ° $p < 0.05$, °° $p < 0.01$ versus inhibitor. INIB= inhibitor, NC= negative control siRNA, Visf= visfatin, Res = resistin.

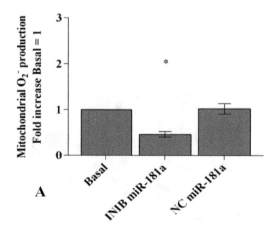

Figure 8. *Cont.*

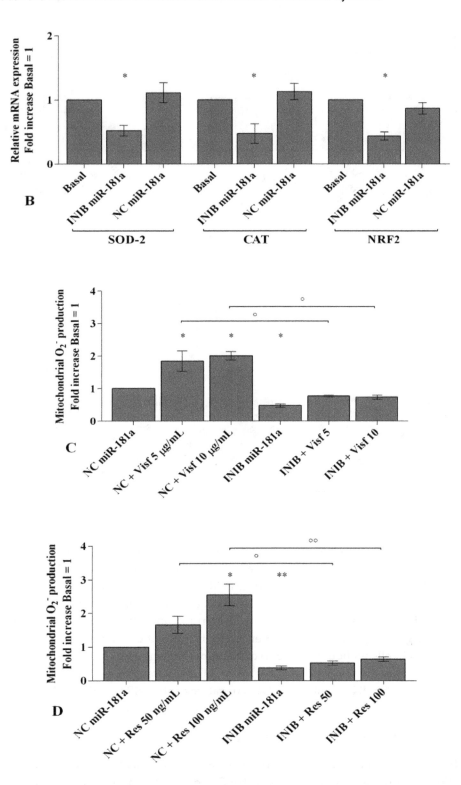

Figure 8. *Cont.*

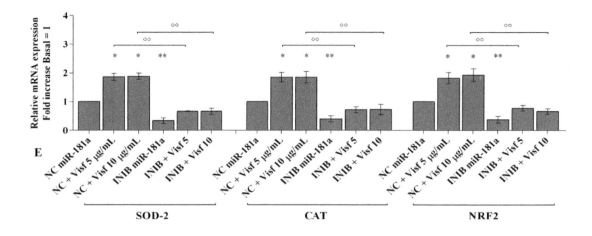

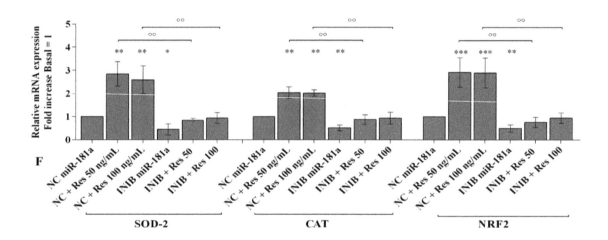

Figure 8. (**A,C,D**) Mitochondrial superoxide anion production was assessed by flow cytometry using MitoSox Red staining. (**B,E,F**) Expression levels of superoxide dismutase (*SOD-2*), catalase (*CAT*), nuclear factor erythroid 2 like 2 (*NRF2*) by real-time PCR. Human osteoarthritic (OA) synovial fibroblasts were evaluated at basal condition, after 24 h of transfection with *miR-181a* inhibitor or NC, and after incubation with visfatin (5 and 10 µg/mL) and resistin (50 and 100 ng/mL). The superoxide anion production and the gene expression were referenced to the ratio of the value of interest and the value of basal condition (basal, cells without treatment) or NC reported equal to 1. Data were expressed as mean ± SD of triplicate values. * $p < 0.05$, ** $p < 0.01$, *** $p < 0.001$ versus basal condition or NC. ° $p < 0.05$, °° $p < 0.01$ versus inhibitor. INIB= inhibitor, NC= negative control siRNA, Visf= visfatin, Res = resistin.

2.7. Visfatin and Resistin Activate NF-κB Signaling Pathway

Figure 9A,B shows the cytoplasmic and nuclear signal intensity of p50 NF-κB subunit in synovial fibroblasts stimulated with visfatin and resistin for 30 min and 4 h. The signal of p50 NF-κB was low mainly detected in the cytoplasm of the cells, with a minimum translocation into the nucleus, at basal condition. After 30 min of incubation with visfatin and resistin we observed a significant increase of p50 subunit cytoplasmic synthesis and nuclear translocation ($p < 0.05$, $p < 0.01$, respectively), in comparison to baseline, while no significant modifications of p50 subunit signal were found after 4 h of adipokines incubation.

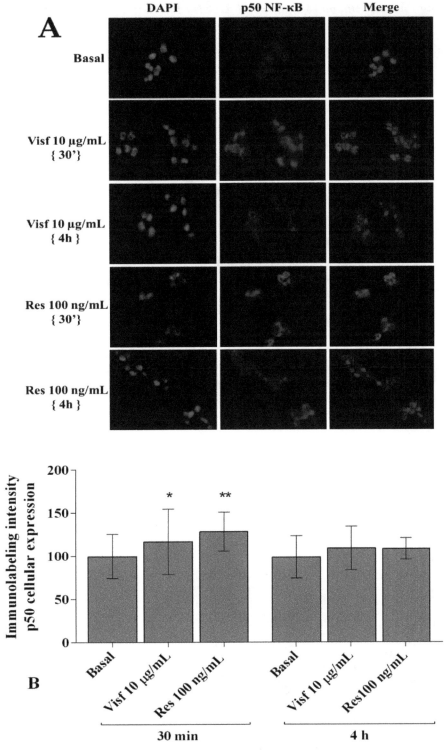

Figure 9. Immunofluorescence labeling of p50 NF-κB subunit localization. Human osteoarthritic (OA) synovial fibroblasts were evaluated at basal condition and after 30 min or 4 h of incubation with visfatin (10 µg/mL) and resistin (100 ng/mL). (**A**) Representative immunocytochemical images of the cells showing localization of p50 NF-κB (red); nuclei were stained with DAPI (blue). Original Magnification 400×. Scale bar: 20 µm. (**B**) The histogram of immunolabeling intensity was plotted for the nuclear and cytoplasmic expression for p50 subunit. Data were expressed as mean ± SD of triplicate values. * $p < 0.05$, ** $p < 0.01$ versus basal condition. Visf = visfatin, Res = resistin.

2.8. NF-κB Signaling Pathway Inhibits Visfatin and Resistin Effects

The involvement of NF-κB pathway in mediating the adipokines-induced effects on inflammatory, apoptotic and oxidative stress mediators is summarized in Figure 10.

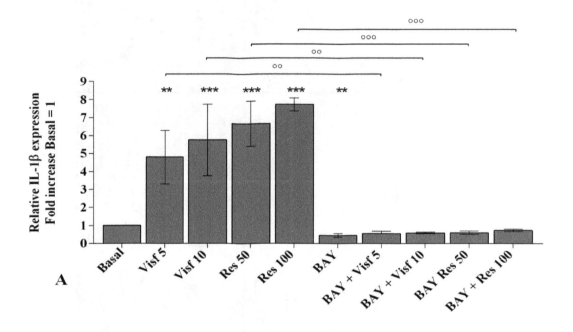

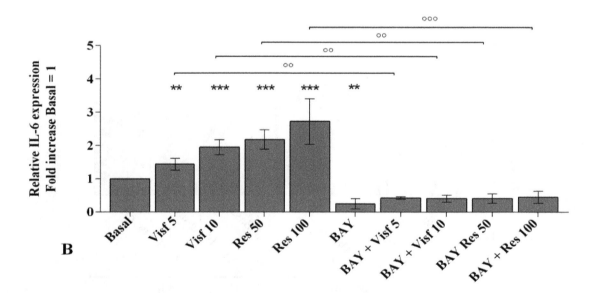

Figure 10. *Cont.*

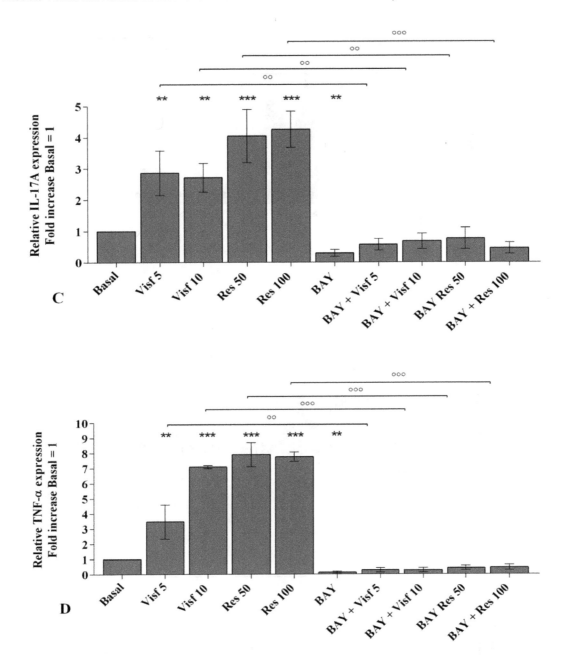

Figure 10. Expression levels of interleukin (*IL*)-*1β* (**A**), *IL-6* (**B**), *IL-17A* (**C**), tumor necrosis factor (*TNF*)-*α* (**D**) by real-time PCR. Human osteoarthritic (OA) synovial fibroblasts were evaluated at basal condition, after 2 h pre-incubation with a specific nuclear factor (NF)-κB inhibitor (BAY 11-7082, IKKα/β, 1 μM) and after 24 h of stimulus with visfatin (5 and 10 μg/mL) and resistin (50 and 100 ng/mL). The gene expression was referenced to the ratio of the value of interest and the value of basal condition (basal, cells without treatment) reported equal to 1. Data were expressed as mean ± SD of triplicate values, ** $p < 0.01$, *** $p < 0.001$ versus basal condition. °° $p < 0.01$, °°° $p < 0.001$ versus BAY. BAY = BAY 11-7082, Visf = visfatin, Res= resistin.

A specific NF-κB inhibitor (IKKα/β, BAY 11-7082) was used to analyze the modulation of the signaling pathway in the gene expression of selected target genes (Figures 10–12) and the studied miRNA (Figure 13).

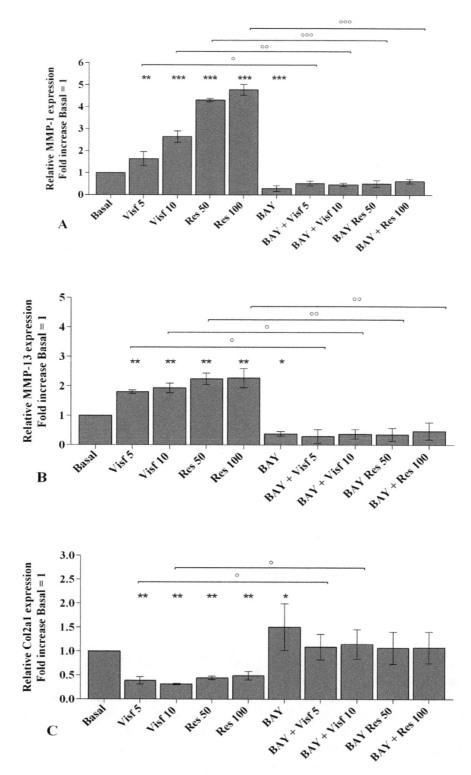

Figure 11. Expression levels metalloproteinases (*MMP*)-*1* (**A**), *MMP-13* (**B**), collagen type II (*Col2a1*) (**C**) by real-time PCR. Human osteoarthritic (OA) synovial fibroblasts were evaluated at basal condition, after 2 h pre-incubation with a specific nuclear factor (NF)-κB inhibitor (BAY 11-7082, IKKα/β, 1 μM) and after 24 h of stimulus with visfatin (5 and 10 μg/mL) and resistin (50 and 100 ng/mL). The gene expression was referenced to the ratio of the value of interest and the value of basal condition (basal, cells without treatment) reported equal to 1. Data were expressed as mean ± SD of triplicate values. * $p < 0.05$, ** $p < 0.01$, *** $p < 0.001$ versus basal condition. ° $p < 0.05$, °° $p < 0.01$, °°° $p < 0.001$ versus BAY. BAY = BAY 11-7082, Visf = visfatin, Res= resistin.

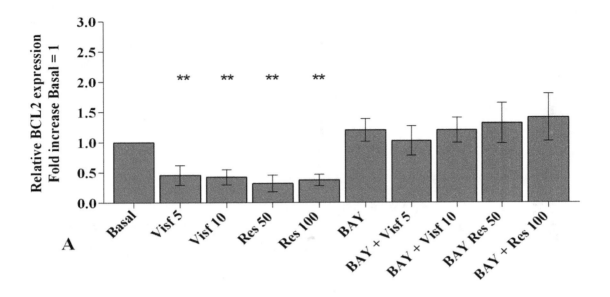

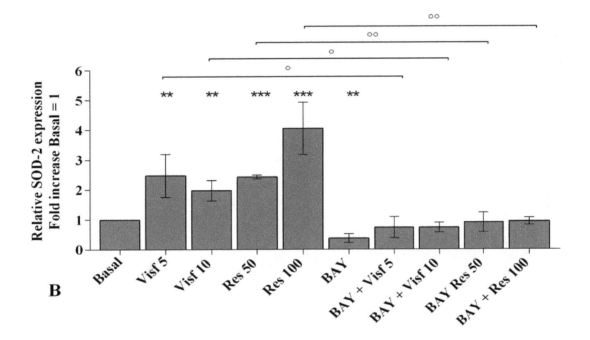

Figure 12. *Cont.*

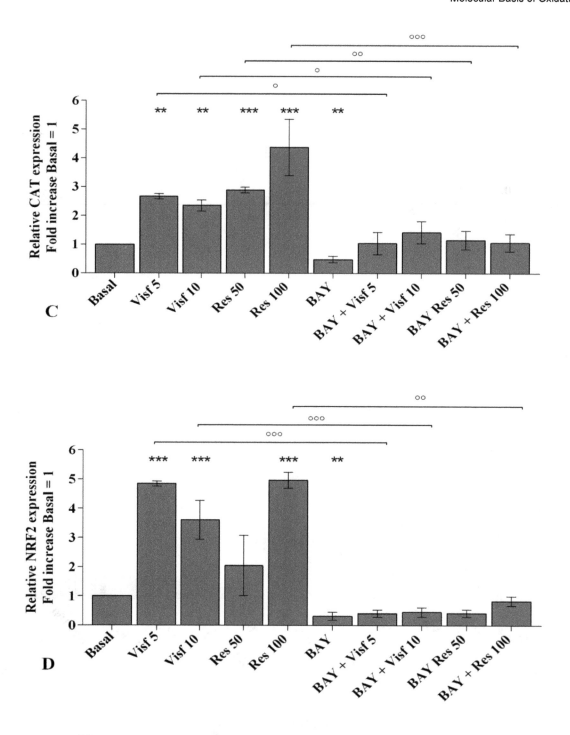

Figure 12. Expression levels of B-cell lymphoma (*BCL*)2 (**A**), superoxide dismutase (*SOD-2*) (**B**), catalase (*CAT*) (**C**), nuclear factor erythroid 2 like 2 (*NRF2*) (**D**) by real-time PCR. Human osteoarthritic (OA) synovial fibroblasts were evaluated at basal condition, after 2 h pre-incubation with a specific nuclear factor (NF)-κB inhibitor (BAY 11-7082, IKKα/β, 1 μM) and after 24 h of stimulus with visfatin (5 and 10 μg/mL) and resistin (50 and 100 ng/mL). The gene expression was referenced to the ratio of the value of interest and the value of basal condition (basal, cells without treatment) reported equal to 1. Data were expressed as mean ± SD of triplicate values. ** $p < 0.01$, *** $p < 0.001$ versus basal condition. ° $p < 0.05$, °° $p < 0.01$, °°° $p < 0.001$ versus BAY. BAY = BAY 11-7082, Visf = visfatin, Res= resistin.

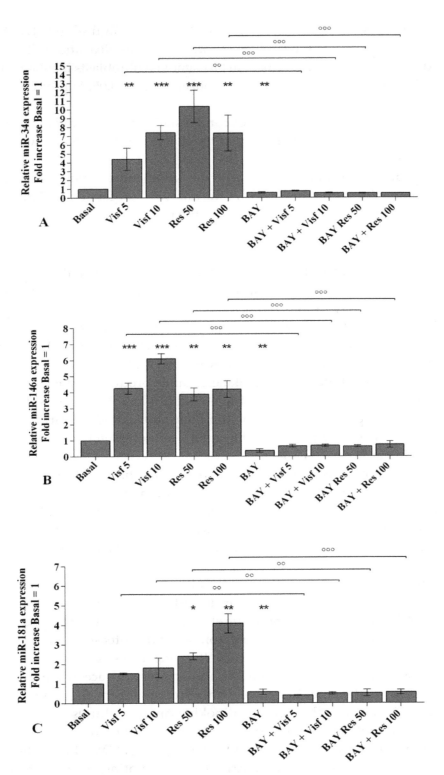

Figure 13. Expression levels of *miR-34a* (**A**), *miR-146a* (**B**), and *miR-181a* (**C**) by real-time PCR. Human osteoarthritic (OA) synovial fibroblasts were evaluated at basal condition, after 2 h pre-incubation with a specific nuclear factor (NF)-κB inhibitor (BAY 11-7082, IKKα/β, 1 μM) and after 24 h of stimulus with visfatin (5 and 10 μg/mL) and resistin (50 and 100 ng/mL). The gene expression was referenced to the ratio of the value of interest and the value of basal condition (basal, cells without treatment) reported equal to 1. Data were expressed as mean ± SD of triplicate values. * $p < 0.05$, ** $p < 0.01$, *** $p < 0.001$ versus basal condition. °° $p < 0.01$, °°° $p < 0.001$ versus BAY. BAY = BAY 11-7082, Visf = visfatin, Res= resistin.

The transcriptional levels of *IL-1β, IL-6, IL-17A, TNF-α* (Figure 10A–D), *MMP-1, MMP-13* (Figure 11A,B), *SOD-2, CAT, NRF2* (Figure 12B–D), *miR-34a, miR-146a,* and *miR-181a* (Figure 13A–C) were significantly decreased ($p < 0.01$, $p < 0.001$) in OA synovial fibroblasts incubated with BAY 11-7082, while an up-regulation of *Col2a1* mRNA levels was observed ($p < 0.05$, Figure 11C), in comparison to basal condition.

The co-treatment of the cells with BAY 11-7082 and visfatin or resistin did not exhibit any difference in miRNA and target genes expression with respect to what is observed in OA synoviocytes incubated with BAY 11-7082 alone (Figures 10–13).

Furthermore, the pre-treatment of the cells with the NF-κB inhibitor significantly limited the effect of visfatin and resistin on the expression levels of the analyzed target genes (Figures 10–13).

No modifications in mRNA levels of *BCL2*, after the treatment, were observed (Figure 12A).

3. Discussion

OA is a musculoskeletal condition mainly characterized by articular cartilage degeneration, however, in recent years, the role of synovial inflammation in the development and in the progression of the disease has been gradually recognized [2,4].

Fibroblast-like synoviocytes actively participate in the synovitis-structural damage cycle of OA through the production of inflammatory cytokines, including IL-6, IL-1β, and TNF-α, and cartilage-degrading enzymes and proteases, such as MMPs [2,29].

Growing evidence demonstrated that adipokines, mainly produced by adipose tissue and by other adipose tissue depots as infrapatellar fat pad, are potentially involved in OA pathophysiology [30]. Indeed, the adipokines may participate in synovium-bone and synovium-cartilage interactions [7,31], however, their exact effect in OA synovial cells have not been completely elucidated [16,32,33] and the results on *in vitro* studies are sparse [19,34].

In the present study, performed in human OA synovial fibroblast cultures, we confirmed previous evidence about the role of visfatin and resistin in inflammation. Furthermore, we demonstrated their impact on apoptosis and oxidative stress processes, as well as in the modulation of some miRNA and target genes, implicated in OA pathogenesis, through the activation of NF-κB pathway. Finally, we hypothesized the direct cross-talk between miRNA and adipokines in mediating oxidative stress induction, via NF-κB signaling.

It is well established that IL-1β, IL-6, IL-17, and TNF-α are the main important cytokines involved in the pathogenesis of OA [35]; they have been found elevated in serum and synovial fluid of patients with knee OA [36,37] and play synergistic effects in OA chondrocytes and synovial fibroblasts stimulating the synthesis and secretion of other cytokines and proteases [16,29].

Our data showed a significant increase of *IL-1β, IL-6,* and *TNF-α* gene expression levels in human OA synovial fibroblast cultures stimulated with visfatin and resistin, according to what is observed by other authors [16,18,38]. On the other hand, we demonstrated, for the first time, the up-regulation of *IL-17* expression levels induced by the studied adipokines in our cultures.

MMPs are the main proteases implicated in cartilage turnover, playing a significant role in the degradation of cartilage ECM that occur during OA damage [39]. MMP-1 and MMP-13 are expressed in chondrocytes and in synoviocytes and contribute to promoting cartilage breakdown inducing the destruction of proteoglycans and Col2a1, the major structural protein of articular ECM [40]. The exposure of OA chondrocytes and fibroblast-like synoviocytes to pro-inflammatory cytokines, such as IL-1β, and adipokines, as visfatin and resistin, determined a markedly increase of matrix-degrading enzymes and a down-regulation of *Col2a1* gene levels [16,17,29,41,42]. In agreement with the current literature we reported the up-regulation of *MMP-1, MMP-13* and a reduction of *Col2a1* expression levels in visfatin and resistin-stimulated OA synovial cells.

These results highlight the role of the studied adipokines in mediating the pro-inflammatory cascade in synovial cells and their consequent implication in articular cartilage destruction that occur in course of OA. Previous evidence reporting that chondrocytes and synovial cells express membrane toll-like receptors (TLRs) which are identified as putative receptors for visfatin and resistin mechanism of action. Adipokines bind to TLRs and stimulate phosphorylation of ERK/p38/mitogen-activated protein kinase (MAPK) signaling, inducing the expression of cytokines, chemokines and degrading proteases [10,14,38,43,44].

The regulation of chondrocytes and fibroblast-like synoviocytes survival is important for the maintenance of a proper cartilage and synovium structure and function [17,45]. Indeed, apoptosis is a complex multi-step process playing a critical role in maintaining the homeostasis of various tissues and cells, and an increasing number of genes have been identified as controller and inductors of this mechanism. Among them, BCL-2 family, anti-apoptotic proteins, are responsible for many biochemical processes driving apoptosis [45].

Dysregulation of apoptosis, thus, is related to a variety of diseases including autoimmune and degenerative disorders as rheumatoid arthritis (RA) and OA [45,46]. The over-expression of BCL-2 family proteins protects OA chondrocytes and human synovial fibroblasts from the programmed cell death [47,48].

The results of our research revealed an increased percentage of apoptosis and a down-regulation of *BCL-2* gene expression in human OA synovial fibroblasts stimulated with visfatin and resistin. Similar data were previously obtained by other authors in endothelial cell lines and in human OA chondrocyte cultures [17,49]. However, we first observed the effect of resistin in the regulation of BCL-2 protein in this cell type.

Oxidative stress and inflammation have been increasingly recognized as being closely integrated with OA pathology. Under physiological conditions, the production of endogenous ROS is balanced by the antioxidant defense system, mainly controlled by NRF2 [50]. The latter is translocated to the nucleus, when released from its repressive cytosolic protein Kelch-like ECH associated protein 1 (KEAP1), and activates the expression of cytoprotective genes, including enzymes involved in the biosynthesis, activity, and detoxification of different ROS species, such as SOD-2 and CAT [50,51]. Various inflammatory mediators, such as cytokines, chemokines, prostaglandins, and growth factors participate to increase oxidative stress in the joint with accumulation of ROS, and nitric oxide (NO), and concomitant failure in the expression of antioxidant scavenging systems [50]. At the cellular level, oxidative stress causes mitochondrial and nuclear DNA damage, lipid peroxidation, alterations in cell signaling and transcription, and epigenetic changes in gene expression contributing to exacerbate synovitis, destruction of matrix components and cell apoptosis [50,52].

In this paper, the analysis of endogenous production of ROS reported an increase of mitochondrial superoxide anion content in OA synoviocytes cultures after visfatin and resistin stimulation, with a concomitant up-regulation of *SOD-2*, *CAT*, and *NRF2* gene expression. There is no evidence about the effects of the studied adipokines on oxidative stress induction in synovial fibroblasts; however, a number of studies, performed in different cell lines incubated with visfatin, resistin and leptin, are in agreement with our data [53–55].

The observed rapid increase of the studied detoxificant factors and NRF2 in adipokines-stimulated human synoviocytes confirm what is observed in a previous study on OA chondrocyte cultures [56]. In our opinion, this result could be explained as an acute adaptive response to protect mitochondria from the deleterious effects of the raised oxidant agents after adipokines stimulus [27,52,56].

Taken together, these findings underline the involvement of visfatin and resistin in the regulation of apoptosis and oxidative stress balance. This conclusion could be supported by the effects of adipokines in stimulating p38 phosphorylation to further activate PI3K/Art signaling and NADPH oxidase (NOX), a major source of ROS generation. Indeed, NOX activation cause the ROS-forming cascade signaling, induces NF-κB translocation into the nucleus, leading to likewise inflammation, cell proliferation, survival and apoptosis [53,55].

MiRNA has been widely investigated for their role in gene regulation; by binding to mRNA 3'-UTRs, miRNA can affect many protein-encoding genes at the post-transcriptional levels [22,24,57].

It is proved that some miRNA are differentially expressed in OA cartilage samples with respect to normal ones, demonstrating their role in the development and progression of OA [23,24,26].

MiR-34a is largely known to be an anti-proliferative factor regulating cell cycle arrest or senescence [58]. Some authors reported the involvement of *miR-34a* in activating apoptosis signaling and limiting cell proliferation in human OA chondrocytes and RA synovial fibroblasts [59,60], as well as its role in modulation of oxidative stress balance in HUVEC lines [61].

MiR-181a was found highly expressed in circulating PBMC of OA patients and in human OA chondrocytes [62,63], and its results implicated the regulation of apoptosis and oxidative stress signaling by targeting multiple anti-apoptotic BCL2 members and modulating mitochondria metabolism in different cell types [63–65].

Data from the current literature concerning the involvement of miR-146a in OA pathogenesis are controversial [27,66,67]. Yamasaki et al. [66] demonstrated that this miRNA is up-regulated in OA cartilage with a low grade on the Mankin scale, or after the stimulus of OA chondrocytes with IL-1β [67]. On the contrary, its reduced expression in hydrogen peroxide-stimulated OA cells was observed [27]. Additionally, this miRNA resulted implicated in oxidative stress regulation by its direct effect on NRF2 transcriptional factor [68].

In this study we showed a significant increase of *miR-34a*, *miR-146a*, and *miR-181a* gene expression after the incubation of OA synoviocytes with visfatin and resistin, consistently with the results of other in vitro studies [17,18,69–71]. On the basis of the results obtained by Wu et al. [18] we can hypothesize the modulation of miRNA gene expression through the phosphorylation of ERK/p38/MAPK signaling induced by visfatin and resistin.

Accumulating evidence has shown a cross-talk between miRNA and components of redox signaling [27,28,57,72]. The transcription, biogenesis, translocation, and function of miRNA are highly correlated with ROS, and, meanwhile, miRNA can regulate the expression of redox factors and other ROS modulators, such as the key components of cellular antioxidant machinery [27,28,57].

Recently, some miRNA were identified as oxidative stress-responsive factors after the treatment of OA chondrocytes with H_2O_2 [27,73], on the other hand, cellular mechanisms regulating oxidative stress were fine-tuned by particular miRNA [28,56].

A number of studies demonstrated the regulation of *miR-34a*, *miR-146a*, and *miR-181a* expression by oxidative stress in PC12, cardiac and carcinoma cell lines and in OA chondrocytes [27,74–76]; furthermore, the inhibition of these miRNA decreased the expression of the main antioxidant enzymes and reduced the mitochondrial intracellular ROS levels [56,61,64,74,76]. According to this evidence, in the present study, the transient transfection of OA synovial fibroblasts with *miR-34a*, *miR-146a*, and *miR-181a* specific inhibitors significantly reduced the production of mitochondrial superoxide anion as well as the expression of *SOD-2*, *CAT*, and *NRF2*, limiting the negative effects of visfatin and resistin. In a similar manner, other authors revealed the involvement of miRNA in mediating visfatin and resistin effects in HepG2 cells and in human synovial fibroblasts [18,71]. The ability of these miRNA in regulating oxidative stress has been reported in different in vitro studies and seems to be related to the regulation of NRF2 activity [57]. Huang et al. [77] showed the implication of miR-34a in modulating *NRF2* expression and NRF2-dependent antioxidant pathway through the direct targeting of *miR-34a* with the 3'UTR of *NRF2* mRNA. Furthermore, *miR-146a* resultingly involved in the regulation of *NRF2* activation by targeting the 3'-UTR of IL-1R-associated kinase (*IRAK)1* and TNFR-associated factor (*TRAF)6* mRNA, the downstream adaptors of TLRs [68]. These data suggest the presence of a regulatory network between miRNA and NRF2 in regulating oxidative stress.

However, in the present study we observed a reduction in the gene expression of antioxidant enzymes when the miRNA were inhibited. This finding could be due to the fact that *miR-34a* and *miR-181a* also directly bind the 3'UTR of silent mating type information regulation 2 homolog (*SIRT)1* mRNA, inducing a decrease in the protein and/or mRNA expression of this gene.

SIRT1 and SIRT6 are putative anti-ageing molecules that regulate the expression of several antioxidant genes and are classified as regulator of oxidative stress balance. Elevated oxidative stress decreased both the protein and mRNA levels of *SIRT1*, whilst up-regulating the expression of miR-34a and miR-181a.

In view of these reports, we can postulate that the obtained results concerning the gene expression of antioxidant enzymes could be related to the up-regulation of *SIRT1* after of *miR-34* and *miR-181a* inhibition [77].

We finally supposed that the complex crosstalk found between adipokines and miRNA, in OA synovial fibroblasts, could be regulated by NF-κB signaling pathway.

NF-κB proteins constitute a family of ubiquitously expressed transcription factors playing essential roles in phlogistic events, immune and stress responses, and in cartilage degradation [78,79]. Accumulation data indicate NF-κB signaling as the most prominent mechanism in the pathogenesis of OA [78,79]. Furthermore, the importance of NF-κB signaling pathway for visfatin and resistin-induced inflammation, as well as for miRNA-related post-transcriptional regulation has been reported [16–19, 56,70].

Our results showed an increase of NF-κB activation and of p50 subunit nuclear translocation in OA synoviocytes stimulated with visfatin and resistin, in agreement with other researches performed in various cell cultures [16,17,33,49,55,80,81]. Besides, these studies also affirmed that NF-κB is involved in regulation of visfatin and resistin-mediated effects in human OA chondrocytes and endothelial progenitor cells incubated with a specific NF-κB inhibitor [16,33,49,55]. Our data support these findings demonstrating that the inhibition of NF-κB signaling limits inflammation and oxidative stress induced by visfatin and resistin, in human OA synovial fibroblasts. The current literature establishes the activation of NF-κB signaling after phosphorylation of ERK/p38/ MAPK pathway induced by visfatin and resistin, triggering the downstream up-regulation of pro-inflammatory and pro-catabolic-related genes, which contribute to inflammatory and degrading processes of OA. Hence, the inhibition of NF-κB transcriptional factor could represent one of the molecular mechanisms to limit adipokines effects on joint injury.

In addition, we also observed that the modulation of *miR-34a*, *miR-146a*, and *miR-181a* expression induced by the studied adipokines was strongly limited by NF-κB inhibition. Similar results were found by other authors, showing an increased gene expression of *miR-34a* and *miR-146a* after IL-1β stimulus through activation of NF-kB; in turn, *miR-34a* and *miR-146a* were found to be able to inhibit the activation of NF-kB via suppressing their target genes expression such as *NRF2*, *IRAK1* and *TRAF6* [68,77,82].

These data suggest that the cross-talk between visfatin, resistin and miRNA could be mediated by NF-κB signaling pathway, highlighting the mutual interaction between miRNA and NF-kB.

However, the present study presents some limitations that need to take into consideration.

First of all, additional experiments on healthy primary cells are recommended; further transfection experiments with specific miRNA mimic could be useful to confirm the regulation induced by the studied miRNA. In addition, the protein levels of the antioxidant enzymes and of the transcriptional factor NRF2 should be detected as well to elucidate if transcription modifications reflect a translational regulation.

Finally, a simultaneous miRNA and NF-κB inhibition could help to deeper investigate their direct interaction in mediating adipokines effects.

4. Materials and Methods

4.1. Sample Collection and Cell Culture

Synovial tissue samples were obtained from three non-obese (BMI from 20 to 25 Kg/m^2) and non-diabetic patients (two men and three women, age from 67 to 75) with primary knee OA defined by the clinical and radiological ACR criteria [83], during their total knee arthroplasty. The tissues were supplied by the Orthopaedic Surgery, University of Siena, Italy. The human articular samples protocols used in this work were evaluated and approved by the Ethic Committee of Azienda Ospedaliera

Universitaria Senese/Siena University Hospital (Prot n 13931_2018, 15 October 2018), and all patients signed a free and informed consent form.

Synovial tissue was separated from adjacent cartilaginous and adipose structures, and isolated immediately after surgery. Briefly, samples were aseptically dissected from each donor, cut into small thick pieces and processed by an enzymatic digestion by using trypsin-EDTA Solution 10× (Sigma–Aldrich, Milan, Italy) for 15 min at 37 °C and then, washed and incubated with type IV collagenase (Sigma–Aldrich, Milan, Italy) in Dulbecco's Modified Eagle Medium (DMEM) (Euroclone, Milan, Italy) medium with shaking for 12–16 h at 37 °C.

The obtained cell suspension was filtered using 70-µm nylon meshes, washed, and centrifuged for 5 min at 700× g. The viability was assessed by Trypan Blue (Sigma–Aldrich, Milan, Italy) test and a percentage of 90% to 95% of cell survival was assessed. Cells were collected, seeded into 10-cm diameter tissue culture plates, and expanded for a minimum of two weeks in a monolayer in incubator with 5% CO_2 and 90% humidified atmosphere at 37 °C, until a confluence of 80% to 85% was reached.

Human OA synovial fibroblasts were grown in DMEM containing 10% fetal bovine serum (FBS) (Euroclone, Milan, Italy), with 200 U/mL penicillin and 200 µg/mL streptomycin (P/S) (Sigma–Aldrich, Milan, Italy). The culture medium was changed two times for week. The morphology was examined daily with an inverted microscope (Olympus IMT-2, Tokyo, Japan), and the cells from passages 3 to 6 were employed for the experimental procedures. A cell culture derived from a unique donor was used for each single experiment, for a total of three independent experiments.

4.2. Stimulus of Synovial Cell Cultures

Human OA synovial fibroblasts were transferred and plated in 6-well dishes at a starting density of 1×10^5 cells/well until they became confluent. Human recombinant visfatin (Sigma–Aldrich, Milan, Italy) and human recombinant resistin (BioVendor, Rome, Italy) were dissolved in phosphate buffered saline (PBS) (Euroclone, Milan, Italy), according to the manufacturer's instructions, and then directly diluted in the culture medium for the treatment in order to obtain the final concentration required.

The cells were immersed in DMEM medium enriched with 0.5% FBS and 2% P/S and stimulated for 24 h with visfatin at concentration of 5 and 10 µg/mL or resistin 50 and 100 ng/mL. The concentrations of the adipokines used in our in vitro study were selected according to those used by other authors and in our previous report [16,17,84]; the final concentrations were chosen based on the best results obtained in terms of viability (Figure S1).

After the treatment, the cells were recovered and immediately processed to carry out flow cytometry analysis and quantitative real-time PCR.

In addition, the synovial cells were pre-incubated for 2 h with 1 µM BAY 11-7082 (NF-κB inhibitor, IKKα/β, Sigma–Aldrich, Milan, Italy) and then stimulated 24 h with the selected concentrations of visfatin (5 and 10 µg/mL) and resistin (50 and 100 ng/mL). Then, the gene expression of the target genes (IL-1β, IL-6, IL-17A, TNF-α, MMP-1, MMP-13, Col2a1, BCL2, SOD-2, CAT and NRF2) and miRNA (miR-34a, miR-146a, and miR-181a) was evaluated.

4.3. MTT Assay

The viability of the cells was evaluated, by MTT test, after the treatment of the cells with visfatin and resistin at the tested concentrations.

Chondrocytes were incubated for 3 h at 37 °C in a culture medium containing 10% of 5 mg/mL of MTT (Sigma–Aldrich, Milan, Italy). At the end of this period, the medium was removed and 0.2 mL of dimethyl sulfoxide (DMSO) (Rottapharm Biotech, Monza, Italy) was added to the wells to solubilize the formazan crystals. The absorbance was measured at 570 nm in a microplate reader (BioTek Instruments, Inc., Winooski, VT, USA). A control well without cells was employed for blank measurement.

The percentage of survival cells was evaluated as (absorbance of considered sample) / (absorbance of control) × 100.

The experiments were performed on cell cultures at 80% to 85% of confluence in order to prevent contact inhibition which can alter the results. Data were reported as OD units per 10^4 adherent cells.

4.4. Transfection of Synovial Cells

The cells were grown in 6-well dishes at a starting density of 1×10^5 cells/well until a confluence of 85% in DMEM supplemented with 10% FBS; then, the media were replaced with DMEM 0.5% FBS for 6 h before transfection. Afterwards, synoviocytes were transfected with specific inhibitors of *miR-34a, miR-146a,* and *miR-181a* (Qiagen, Hilden, Germany), at the concentration of 50 nM, or with their relative negative controls siRNA (NC) (Qiagen, Hilden, Germany), at the concentration of 5 nM, in serum-free medium for a period of 24 h. Supernatants were removed and synoviocytes immediately harvested or incubated with visfatin (5 and 10 µg/mL) or resistin (50 and 100 ng/mL) for additional 24 h.

4.5. Quantitative Real-Time PCR of mRNA and miRNA

Synovial fibroblasts were grown in 6-well dishes at a starting density of 1×10^5 cells/well in DMEM supplemented with 10% FBS. Then, the supernatant was removed, and the cells were cultured in DMEM with 0.5% FBS used for the treatment procedure.

Total RNA, including miRNA, was extracted using TriPure Isolation Reagent (Euroclone, Milan, Italy) according to the manufacturer's instructions, and was stored at −80 °C. The concentration, purity, and integrity of RNA were evaluated by measuring the OD at 260 nm and the 260/280 and 260/230 ratios by Nanodrop-1000 (Celbio, Milan, Italy). The quality of RNA was verified by electrophoresis on agarose gel (FlashGel System, Lonza, Rockland, ME, USA). Reverse transcription for miRNA was carried out by the cDNA miScript PCR Reverse Transcription kit (Qiagen, Hilden, Germany), while for target genes the QuantiTect Reverse Transcription kit (Qiagen, Hilden, Germany) was used, according to the manufacturer's instructions.

MiRNA and target genes were examined by real-time PCR using, miScript SYBR Green (Qiagen, Hilden, Germany) and QuantiFast SYBR Green PCR (Qiagen, Hilden, Germany) kits, respectively. A list of the used primers is reported in Table 1.

Table 1. Primers used for RT-qPCR.

miRNA Genes	Cat. No. (Qiagen)
miR-34a	MS00003318
miR-146a	MS00003535
miR-181a	MS00006692
SNORD-25	MS00014007
Target Genes	**Cat. No. (Qiagen)**
IL-1β	QT00021385
IL-6	QT00083720
IL-17A	QT00009233
TNF-α	QT00029162
MMP-1	QT00014581
MMP-13	QT00001764
Col2a1	QT00049518
BCL2	QT00000721
SOD-2	QT01008693
CAT	QT00079674
NRF2	QT00027384
ACTB	QT00095431

Abbreviations: miRNA = microRNA; SNORD-25 = Small Nucleolar RNA, C/D Box 25; IL-1β = interleukin 1β; IL-6 = interleukin 6; IL-17A = interleukin 17A; TNF-α = tumor necrosis factor-α; MMP-1 = matrix metalloproteinase 1; MMP-13 = matrix metalloproteinase 13; Col2a1 = type II collagen alpha 1 chain; BCL2 = B-cell lymphoma; SOD-2 = superoxide dismutase 2; CAT = catalase; NRF2 = nuclear factor erythroid 2 like 2; ACTB = actin beta.

All qPCR reactions were achieved in glass capillaries by a LightCycler 1.0 (Roche Molecular Biochemicals, Mannheim, Germany) with LightCycler Software Version 3.5. The reaction procedure for miRNA consisted of 95 °C for 15 min for HotStart polymerase activation, followed by 40 cycles of 15 s at 95 °C for denaturation, 30 s at 55 °C for annealing, and 30 s at 70 °C for elongation, according to the protocol. Target genes amplification was performed at 5 in at 95 °C, 40 cycles of 15 s at 95 °C, and 30 s at 60 °C. In the final step of both protocols, the temperature was raised from 60 °C to 95 °C at 0.1 °C/step to plot the melting curve.

The analysis of the dissociation curves was performed by visualizing the amplicons lengths in agarose gel to confirm the correct amplification of the resulting PCR products.

For the data analysis, the C_t values of each sample and the efficiency of the primer set were calculated through LinReg Software [85] and then converted into relative quantities and normalized using the Pfaffl model [86].

The normalization was performed considering Small Nucleolar RNA, C/D Box 25 (SNORD-25) for miRNA and Actin Beta (ACTB) for target genes, as the housekeeping genes. The choice of the genes was carried out by using geNorm software version 3.5 [87].

4.6. Apoptosis Detection

Apoptotic cells were evaluated by using Annexin V-FITC and propidium iodide (PI) (ThermoFisher Scientific, Milan, Italy). Human OA synovial fibroblasts were seeded in 12-well plates (8×10^4 cells/well) for 24 h in DMEM with 10% FBS. Then, the medium was discarded, and the cells were cultured in DMEM with 0.5% FBS used for the treatment procedure. Afterwards, the synovial cells were washed and harvested by using trypsin, collected into cytometry tubes, and centrifuged at 1500 rpm for 10 min. The supernatant was replaced, and the pellet was resuspended in 100 µL of 1× Annexin-binding buffer, 5 µL of Alexa Fluor 488 annexin-V conjugated to fluorescein (green fluorescence) and 1 µL of 100 µg/mL PI working solution. Markers were added to 100 µL of cell suspension. Cells were incubated at room temperature for 15 min in the dark. Then, 600 µL of 1× Annexin-binding buffer were added before the analysis at flow cytometer. A total of 10,000 events (1×10^4 cells per assay) were measured by the instrument. The obtained results were analyzed with Cell Quest software (Version 4.0, Becton Dickinson, San Jose, CA, USA). The evaluation of apoptosis was carried out considering staining cells simultaneously with Alexa Fluor 488 annexin-V and PI; a discrimination of intact cells (annexin-V and PI-negative), early apoptosis (annexin-V-positive and PI-negative), and late apoptosis (annexin-V and PI-positive) is allowed [88].

The results were expressed as percentage of positive cells to each dye (total apoptosis), and the data were represented as the mean of three independent experiments (mean ± SD).

4.7. Mitochondrial Superoxide Anion ($\bullet O_2$-) Production

Human OA synovial fibroblasts were seeded in a density of 8×10^4 cells per well in 12 multi-plates for 24 h in DMEM with 10% FCS. Then, the medium was eliminated, and the cells were cultured in DMEM with 0.5% FBS used for the treatment procedure. Then, the cells were incubated in Hanks' Balanced Salt Solution (HBSS) and MitoSOX Red for 15 min at 37°C in dark, to assess mitochondrial superoxide anion ($\bullet O_2$-) production. MitoSOX was dissolved in DMSO, at a final concentration of 5 µM. Cells were then harvested by trypsin and collected into cytometry tubes and centrifuged at 1500 rpm for 10 min. Besides, cells were suspended in saline solution before being analyzed by flow cytometry. A density of 1×10^4 cells per assay (a total of 10,000 events) were measured by flow cytometry and data were analyzed with CellQuest software (Version 4.0, Becton Dickinson, San Jose, CA, USA). Results were collected as median of fluorescence (AU) and represented the mean of three independent experiments (mean ± SD).

4.8. Immunofluorescence Analysis

Human OA synovial fibroblasts were plated in coverslips in Petri dishes (35×10 mm) at a starting low density of 4×10^4 cells/chamber, to prevent possible cell overlapping, and re-suspended in 2 mL of culture medium until 80% of confluence. The cells were processed after 2 h of stimulus with adipokines to evaluate the potential activation of the NF-κB pathway. The synovial cells were washed in PBS and then fixed in 4% paraformaldehyde (ThermoFisher Scientific, Milan, Italy) (pH 7.4) for 10 min at room temperature. Afterwards, the cells were permeabilized with a blocking solution (PBS, 1% bovine serum albumin (BSA) (Sigma–Aldrich, Milan, Italy) and 0.2% Triton X-100 (ThermoFisher Scientific, Milan, Italy) for 20 min at room temperature, and then incubated overnight at 4 °C with mouse monoclonal anti-p50 subunit primary antibody (Santa Cruz Biotechnology, Italy) diluted at 1:100 in PBS, 1% BSA and 0.05% Triton X-100. Three washes in PBS of the coverslips were followed by 1 h incubation with goat anti-mouse IgG-Texas Red conjugated antibody (Southern Biotechnology, Italy) diluted at 1:100 in PBS, 1% BSA and 0.05% Triton X-100. Finally, the coverslips were washed three times in PBS and submitted to nuclear counterstain by 4,6-diamidino-2-phenylindole (DAPI), and then mounted with Vecta shield (Vector Labs). Fluorescence was examined under an AxioPlan (Zeiss, Oberkochen, Germany) light microscope equipped with epifluorescence at 200× and 400× magnification. The negative controls were obtained by omitting the primary antibody. Immunoreactivity of p50 was semi-quantified as the mean densitometric area of p50 signal into the nucleus and into the cytoplasm, by AxioVision 4.6 software measure program [89]. At least 100 synovial cells from each group were evaluated.

4.9. Statistical Analysis

Three independent experiments were carried out and the results were expressed as the mean ± SD of triplicate values for each experiment. Data normal distribution was evaluated by Shapiro–Wilk, D'Agostino and Pearson, and Kolmogorov–Smirnov tests.

Data from real-time PCR were evaluated by one-way ANOVA with a Tukey's post-hoc test using $2^{-\Delta\Delta CT}$ values for each sample. Flow cytometry results were analyzed by ANOVA with Bonferroni post-hoc test.

All analyses were performed through the SAS System (SAS Institute Inc., Cary, NC, USA) and GraphPad Prism 6.1. A significant value was defined with a p-value < 0.05.

5. Conclusions

Growing evidence supports the relevance of synovitis in OA pathophysiology. Among the various factor involved in synovial membrane inflammation and in cartilage degradation during the development and the progression of OA, adipokines, miRNA, and oxidative stress play a crucial role. These findings induced us to deeper investigate the possible link between adipokines and some miRNA in oxidative stress regulation in human OA synovial cultures.

We firstly demonstrated the ability of visfatin and resistin to induce the gene expression of a pattern of pro-inflammatory cytokines (*IL-1β, IL-6, IL-17A* and *TNF-α*), MMPs (*MMP-1, MMP-13*), anti-oxidant enzymes (*SOD-2, CAT* and *NRF2*), as well as *miR-34a, miR-146a,* and *miR-181a*. Furthermore, they caused apoptosis and superoxide anion production, down-regulated the transcriptional levels of Col2a1 and the anti-apoptotic marker BCL2 and increased the p50 NF-κB activation.

Furthermore, we investigated the implication of *miR-34a, miR-146a,* and *miR-181a* as possible regulators of adipokines effects on the modulation of oxidative stress.

Finally, the use of NF-κB specific inhibitor points out the involvement of the pathway in adipokines-mediated effects.

In conclusion, altogether, these results confirm the role of visfatin and resistin in the induction of inflammation and cartilage degradation, and contribute to elucidate the existing crosstalk among adipokines, miRNA and oxidative stress.

However, further studies are required to deeper investigate this complex network and how this evidence can be useful to identify new possible therapeutic targets to reduce synovitis and cartilage degradation in OA.

Author Contributions: The authors declare to have participated in the drafting of this paper as specified below: Conceptualization, S.T. and A.F.; Data curation, S.C., I.G. and M.B.; Funding acquisition, S.G.; Investigation, S.C.; Methodology, S.C., I.G. and M.B.; Project administration, A.F.; Resources, A.F.; Supervision, A.F.; Visualization, S.T.; Writing – original draft, S.C. and A.F.; Writing – review & editing, S.C., S.G., N.M., A.G., S.T. and A.F. SC, ST and AF contributed equally to this work.

Acknowledgments: We thank the University of Siena for economical contribution.

References

1. Vos, T.; Allen, C.; Arora, M.; Barber, R.M.; Bhutta, Z.A.; Brown, A.; Cater, A.; Casey, C.C.; Charlson, J.F.; Chen, Z.A.; et al. Global, regional, and national incidence, prevalence, and years lived with disability for 310 diseases and injuries, 1990-2015: A systematic analysis for the Global Burden of Disease Study 2015. *Lancet* **2016**, *388*, 1545–1602. [CrossRef]

2. Poulet, B.; Staines, K.A. New developments in osteoarthritis and cartilage biology. *Curr. Opin. Pharmacol.* **2016**, *28*, 8–13. [CrossRef] [PubMed]

3. Sellam, J.; Berenbaum, F. The role of synovitis in pathophysiology and clinical symptoms of osteoarthritis. *Nat. Rev. Rheumatol.* **2010**, *6*, 625–635. [CrossRef] [PubMed]

4. Scanzello, C.R.; Goldring, S.R. The role of synovitis in osteoarthritis pathogenesis. *Bone* **2012**, *51*, 249–257. [CrossRef]

5. Sachdeva, M.; Aggarwal, A.; Sharma, R.; Randhawa, A.; Sahni, D.; Jacob, J.; Sharma, V.; Aggarwal, A. Chronic inflammation during osteoarthritis is associated with an increased expression of CD161 during advanced stage. *Scand J. Immunol* **2019**, e12770. [CrossRef]

6. Neumann, E.; Junker, S.; Schett, G.; Frommer, K.; Müller-Ladner, U. Adipokines in bone disease. *Nat. Rev. Rheumatol.* **2016**, *12*, 296–302. [CrossRef]

7. Azamar-Llamas, D.; Hernández-Molina, G.; Ramos-Ávalos, B.; Furuzawa-Carballeda, J. Adipokine Contribution to the Pathogenesis of Osteoarthritis. *Mediators Inflamm.* **2017**, *2017*, 5468023. [CrossRef]

8. Tenti, S.; Palmitesta, P.; Giordano, N.; Galeazzi, M.; Fioravanti, A. Increased serum leptin and visfatin levels in patients with diffuse idiopathic skeletal hyperostosis: A comparative study. *Scand J. Rheumatol* **2017**, *46*, 156–158. [CrossRef]

9. Fioravanti, A.; Cheleschi, S.; De Palma, A.; Addimanda, O.; Mancarella, L.; Pignotti, E.; Pulsatelli, L.; Galezzi, M.; Meliconi, R. Can adipokines serum levels be used as biomarkers of hand osteoarthritis? *Biomarkers* **2018**, *23*, 265–270. [CrossRef]

10. Carrión, M.; Frommer, K.W.; Pérez-García, S.; Müller-Ladner, U.; Gomariz, R.P.; Neumann, E. The Adipokine Network in Rheumatic Joint Diseases. *Int. J. Mol. Sci.* **2019**, *20*, 4091. [CrossRef]

11. Sun, Z.; Lei, H.; Zhang, Z. Pre-B cell colony enhancing factor (PBEF), a cytokine with multiple physiological functions. *Cytokine Growth F. R.* **2013**, *24*, 433–442. [CrossRef] [PubMed]

12. Senolt, L.; Housa, D.; Vernerová, Z.; Jirásek, T.; Svobodová, R.; Veigl, D.; Anderlova, K.; Muller-Ladner, U.; Pavelka, K.; Haluzik, M. Resistin in rheumatoid arthritis synovial tissue, synovial fluid and serum. *Ann. Rheum. Dis.* **2007**, *66*, 458–463. [CrossRef] [PubMed]

13. Fioravanti, A.; Giannitti, C.; Cheleschi, S.; Simpatico, A.; Pascarelli, N.A.; Galeazzi, M. Circulating levels of adiponectin, resistin, and visfatin after mud-bath therapy in patients with bilateral knee osteoarthritis. *Int. J. Biometeorol.* **2015**, *59*, 1691–1700. [CrossRef] [PubMed]

14. Liao, L.; Chen, Y.; Wang, W. The current progress in understanding the molecular functions and mechanisms of visfatin in osteoarthritis. *J. Bone Miner. Metab.* **2016**, *34*, 485–490. [CrossRef]

15. Calvet, J.; Orellana, C.; Gratacós, J.; Berenguer-Llergo, A.; Caixàs, A.; Chillarón, J.J.; Pedro-Botet, J.; Garcia-Manrique, M.; Navarro, N.; Larrosa, M. Synovial fluid adipokines are associated with clinical severity in knee osteoarthritis: A cross-sectional study in female patients with joint effusion. *Arthritis Res. Ther.* **2016**, *18*, 207. [CrossRef]

16. Zhang, Z.; Xing, X.; Hensley, G.; Chang, L.W.; Liao, W.; Abu-Amer, Y.; Sandell, L.J. Resistin induces expression of proinflammatory cytokines and chemokines in human articular chondrocytes via transcription and messenger RNA stabilization. *Arthritis Rheum.* **2010**, *62*, 1993–2003. [CrossRef]

17. Cheleschi, S.; Giordano, N.; Volpi, N.; Tenti, S.; Gallo, I.; Di Meglio, M.; Giannotti, S.; Fioravanta, A. A Complex Relationship between Visfatin and Resistin and microRNA: An In Vitro Study on Human Chondrocyte Cultures. *Int. J. Mol. Sci.* **2018**, *19*, 3909. [CrossRef]

18. Wu, M.H.; Tsai, C.H.; Huang, Y.L.; Fong, Y.C.; Tang, C.H. Visfatin Promotes IL-6 and TNF-α Production in Human Synovial Fibroblasts by Repressing miR-199a-5p through ERK, p38 and JNK Signaling Pathways. *Int. J. Mol. Sci.* **2018**, *19*, 190. [CrossRef]

19. Chen, W.C.; Wang, S.W.; Lin, C.Y.; Tsai, C.H.; Fong, Y.C.; Lin, T.Y.; Wang, S.L.; Huang, H.D.; Liao, K.W.; Tang, C.H. Resistin Enhances Monocyte Chemoattractant Protein-1 Production in Human Synovial Fibroblasts and Facilitates Monocyte Migration. *Cell Physiol. Biochem.* **2019**, *52*, 408–420. [CrossRef]

20. Gerin, I.; Bommer, G.T.; McCoin, C.S.; Sousa, K.M.; Krishnan, V.; MacDougald, O.A. (2010) Roles for miRNA-378/378* in adipocyte gene expression and lipogenesis. *Am. J. Physiol. Endocrinol. Metab.* **2010**, *299*, E198–E206. [CrossRef]

21. Maurizi, G.; Babini, L.; Della Guardia, L. Potential role of microRNAs in the regulation of adipocytes liposecretion and adipose tissue physiology. *J. Cell Physiol.* **2018**, *233*, 9077–9086. [CrossRef] [PubMed]

22. Malemud, C.J. MicroRNAs and Osteoarthritis. *Cells* **2018**, *7*, 92. [CrossRef] [PubMed]

23. Díaz-Prado, S.; Cicione, C.; Muiños-López, E.; Hermida-Gómez, T.; Oreiro, N.; Fernández-López, C.; Blanco, F.J. Characterization of microRNA expression profiles in normal and osteoarthritic human chondrocytes. *BMC Musculoskelet. Disord.* **2012**, *13*, 144. [CrossRef] [PubMed]

24. De Palma, A.; Cheleschi, S.; Pascarelli, N.A.; Tenti, S.; Galeazzi, M.; Fioravanti, A. Do MicroRNAs have a key epigenetic role in osteoarthritis and in mechanotransduction? *Clin. Exp. Rheumatol.* **2017**, *35*, 518–526.

25. Cheleschi, S.; De Palma, A.; Pecorelli, A.; Pascarelli, N.A.; Valacchi, G.; Belmonte, G.; Carta, S.; Galeazzi, M.; Fioravanti, A. Hydrostatic Pressure Regulates MicroRNA Expression Levels in Osteoarthritic Chondrocyte Cultures via the Wnt/β-Catenin Pathway. *Int. J. Mol. Sci.* **2017**, *18*, 133. [CrossRef]

26. Fathollahi, A.; Aslani, S.; Jamshidi, A.; Mahmoudi, M. Epigenetics in osteoarthritis: Novel spotlight. *J. Cell Physiol.* **2019**, *234*, 12309–12324. [CrossRef]

27. Cheleschi, S.; De Palma, A.; Pascarelli, N.A.; Giordano, N.; Galeazzi, M.; Tenti, S.; Fioravanti, A. Could Oxidative Stress Regulate the Expression of MicroRNA-146a and MicroRNA-34a in Human Osteoarthritic Chondrocyte Cultures? *Int. J. Mol. Sci.* **2017**, *18*, 2660. [CrossRef]

28. Bu, H.; Wedel, S.; Cavinato, M.; Jansen-Dürr, P. MicroRNA Regulation of Oxidative Stress-Induced Cellular Senescence. *Oxid. Med. Cell Longev.* **2017**, *2017*, 2398696. [CrossRef]

29. Machado, C.R.L.; Resende, G.G.; Macedo, R.B.V.; do Nascimento, V.C.; Branco, A.S.; Kakehasi, A.M.; Andrade, M.V. Fibroblast-like synoviocytes from fluid and synovial membrane from primary osteoarthritis demonstrate similar production of interleukin 6, and metalloproteinases 1 and 3. *Clin. Exp. Rheumatol.* **2019**, *37*, 306–309.

30. Ioan-Facsinay, A.; Kloppenburg, M. An emerging player in knee osteoarthritis: The infrapatellar fat pad. *Arthritis Res. Ther.* **2013**, *15*, 225. [CrossRef]

31. Tilg, H.; Moschen, A.R. Adipocytokines: Mediators linking adipose tissue, inflammation and immunity. *Nat. Rev. Immunol.* **2006**, *6*, 772–783. [CrossRef] [PubMed]

32. Francin, P.J.; Abot, A.; Guillaume, C.; Moulin, D.; Bianchi, A.; Gegout-Pottie, P.; Jouzeau, J.-Y.; Mainard, D.; Presle, D. Association between adiponectin and cartilage degradation in human osteoarthritis. *Osteoarthr. Cartilage* **2014**, *22*, 519–526. [CrossRef] [PubMed]

33. Su, Y.P.; Chen, C.N.; Chang, H.I.; Huang, K.C.; Cheng, C.C.; Chiu, F.Y.; Lee, K.C.; Lo, C.M.; Chang, S.F. Low Shear Stress Attenuates COX-2 Expression Induced by Resistin in Human Osteoarthritic Chondrocytes. *J. Cell Physiol.* **2017**, *232*, 1448–1457. [CrossRef] [PubMed]

34. Laiguillon, M.C.; Houard, X.; Bougault, C.; Gosset, M.; Nourissat, G.; Sautet, A.; Jacques, C.; Berenbaum, F.; Sellam, J. Expression and function of visfatin (Nampt), an adipokine-enzyme involved in inflammatory pathways of osteoarthritis. *Arthritis Res. Ther.* **2014**, *16*, R38. [CrossRef]

35. Wang, T.; He, C. Pro-inflammatory cytokines: The link between obesity and osteoarthritis. *Cytokine Growth F. R.* **2018**, *44*, 38–50. [CrossRef]

36. Pelletier, J.P.; McCollum, R.; Cloutier, J.M.; Martel-Pelletier, J. Synthesis of metalloproteases and interleukin 6 (IL-6) in human osteoarthritic synovial membrane is an IL-1 mediated process. *J. Rheumatol. Suppl.* **1995**, *43*, 109–114.

37. Altobelli, E.; Angeletti, P.M.; Piccolo, D.; De Angelis, R. Synovial Fluid and Serum Concentrations of Inflammatory Markers in Rheumatoid Arthritis, Psoriatic Arthritis and Osteoarthitis: A Systematic Review. *Curr. Rheumatol. Rev.* **2017**, *13*, 170–179. [CrossRef]

38. Sato, H.; Muraoka, S.; Kusunoki, N.; Masuoka, S.; Yamada, S.; Ogasawara, H.; Imai, T.; Akasaka, Y.; Tochigi, N.; Takahashi, H.; et al. Resistin upregulates chemokine production by fibroblast-like synoviocytes from patients with rheumatoid arthritis. *Arthritis Res. Ther.* **2017**, *19*, 263. [CrossRef]

39. Cawston, T.E.; Young, D.A. Proteinases involved in matrix turnover during cartilage and bone breakdown. *Cell Tissue Res.* **2010**, *339*, 221–235. [CrossRef]

40. Troeberg, L.; Nagase, H. Proteases involved in cartilage matrix degradation in osteoarthritis. *Biochim Biophys Acta* **2012**, *1824*, 133–145. [CrossRef]

41. Pérez-García, S.; Carrión, M.; Jimeno, R.; Ortiz, A.M.; González-Álvaro, I.; Fernández, J.; Gomariz, R.P.; Juarranz, Y. Urokinase plasminogen activator system in synovial fibroblasts from osteoarthritis patients: Modulation by inflammatory mediators and neuropeptides. *J. Mol. Neurosci.* **2014**, *52*, 18–27. [CrossRef] [PubMed]

42. Pérez-García, S.; Gutiérrez-Cañas, I.; Seoane, I.V.; Fernández, J.; Mellado, M.; Leceta, J.; Tío, L.; Villanueva-Romero, R.; Juarranz, Y.; Gomariz, R.P. Healthy and Osteoarthritic Synovial Fibroblasts Produce a Disintegrin and Metalloproteinase with Thrombospondin Motifs 4, 5, 7, and 12: Induction by IL-1β and Fibronectin and Contribution to Cartilage Damage. *Am. J. Pathol.* **2016**, *186*, 2449–2461. [CrossRef] [PubMed]

43. Goldring, M.B.; Otero, M. Inflammation in osteoarthritis. *Curr. Opin. Rheumatol.* **2011**, *23*, 471–478. [CrossRef] [PubMed]

44. Meier, F.M.; Frommer, K.W.; Peters, M.A.; Brentano, F.; Lefèvre, S.; Schröder, D.; Kyburz, D.; Steinmeyer, J.; Rehart, S.; Gay, S.; et al. Visfatin/pre-B-cell colony-enhancing factor (PBEF), a proinflammatory and cell motility-changing factor in rheumatoid arthritis. *J. Biol. Chem.* **2012**, *287*, 28378–28385. [CrossRef] [PubMed]

45. Hwang, H.S.; Kim, H.A. Chondrocyte Apoptosis in the Pathogenesis of Osteoarthritis. *Int. J. Mol. Sci.* **2015**, *16*, 26035–26054. [CrossRef] [PubMed]

46. Wang, N.; Lu, H.S.; Guan, Z.P.; Sun, T.Z.; Chen, Y.Y.; Ruan, G.R.; Chen, Z.K.; Jiang, J.; Bai, C.J. Involvement of PDCD5 in the regulation of apoptosis in fibroblast-like synoviocytes of rheumatoid arthritis. *Apoptosis* **2007**, *12*, 1433–1441. [CrossRef]

47. Feng, L.; Precht, P.; Balakir, R.; Horton, W.E., Jr. Evidence of a direct role for Bcl-2 in the regulation of articular chondrocyte apoptosis under the conditions of serum withdrawal and retinoic acid treatment. *J. Cell Biochem.* **1998**, *71*, 302–309. [CrossRef]

48. Jiao, Y.; Ding, H.; Huang, S.; Liu, Y.; Sun, X.; Wei, W.; Ma, J.; Zheng, F. Bcl-XL and Mcl-1 upregulation by calreticulin promotes apoptosis resistance of fibroblast-like synoviocytes via activation of PI3K/Akt and STAT3 pathways in rheumatoid arthritis. *Clin. Exp. Rheumatol.* **2018**, *36*, 841–849.

49. Sun, L.; Chen, S.; Gao, H.; Ren, L.; Song, G. Visfatin induces the apoptosis of endothelial progenitor cells via the induction of pro-inflammatory mediators through the NF-κB pathway. *Int. J. Mol. Med.* **2017**, *40*, 637–646. [CrossRef]

50. Marchev, A.S.; Dimitrova, P.A.; Burns, A.J.; Kostov, R.V.; Dinkova-Kostova, A.T.; Georgiev, M.I. Oxidative stress and chronic inflammation in osteoarthritis: Can NRF2 counteract these partners in crime? *Ann. N Y Acad Sci* **2017**, *1401*, 114–135. [CrossRef]

51. Marampon, F.; Codenotti, S.; Megiorni, F.; Del Fattore, A.; Camero, S.; Gravina, G.L.; Festuccia, C.; Musio, D.; Felice, F.D.; Nardone, V.; et al. NRF2 orchestrates the redox regulation induced by radiation therapy, sustaining embryonal and alveolar rhabdomyosarcoma cells radioresistance. *J. Cancer Res. Clin. Oncol.* **2019**, *145*, 881–893. [CrossRef]

52. Huang, M.L.; Chiang, S.; Kalinowski, D.S.; Bae, D.H.; Sahni, S.; Richardson, D.R. The Role of the Antioxidant Response in Mitochondrial Dysfunction in Degenerative Diseases: Cross-Talk between Antioxidant Defense, Autophagy, and Apoptosis. *Oxid. Med. Cell Longev.* **2019**, *2019*, 6392763. [CrossRef]

53. Raghuraman, G.; Zuniga, M.C.; Yuan, H.; Zhou, W. PKCε mediates resistin-induced NADPH oxidase activation and inflammation leading to smooth muscle cell dysfunction and intimal hyperplasia. *Atherosclerosis* **2016**, *253*, 29–37. [CrossRef]

54. Teixeira, T.M.; da Costa, D.C.; Resende, A.C.; Soulage, C.O.; Bezerra, F.F.; Daleprane, J.B. Activation of Nrf2-Antioxidant Signaling by 1,25-Dihydroxycholecalciferol Prevents Leptin-Induced Oxidative Stress and Inflammation in Human Endothelial Cells. *J. Nutr.* **2017**, *147*, 506–513. [CrossRef]

55. Lin, Y.T.; Chen, L.K.; Jian, D.Y.; Hsu, T.C.; Huang, W.C.; Kuan, T.T.; Wu, S.Y.; Kwok, C.F.; Ho, L.T.; Juan, C.C. Visfatin Promotes Monocyte Adhesion by Upregulating ICAM-1 and VCAM-1 Expression in Endothelial Cells via Activation of p38-PI3K-Akt Signaling and Subsequent ROS Production and IKK/NF-κB Activation. *Cell Physiol. Biochem.* **2019**, *52*, 1398–1411. [CrossRef]

56. Cheleschi, S.; Tenti, S.; Mondanelli, N.; Corallo, C.; Barbarino, M.; Giannotti, S.; Gallo, I.; Giordano, A.; Fioravanti, A. MicroRNA-34a and MicroRNA-181a Mediate Visfatin-Induced Apoptosis and Oxidative Stress via NF-κB Pathway in Human Osteoarthritic Chondrocytes. *Cells* **2019**, *8*, 874. [CrossRef]

57. Lin, Y.H. MicroRNA Networks Modulate Oxidative Stress in Cancer. *Int. J. Mol. Sci.* **2019**, *20*, 4497. [CrossRef]

58. Yan, S.; Wang, M.; Zhao, J.; Zhang, H.; Zhou, C.; Jin, L.; Zhang, Y.; Qiu, X.; Ma, B.; Fan, Q. MicroRNA-34a affects chondrocyte apoptosis and proliferation by targeting the SIRT1/p53 signaling pathway during the pathogenesis of osteoarthritis. *Int. J. Mol. Med.* **2016**, *38*, 201–209. [CrossRef]

59. Abouheif, M.M.; Nakasa, T.; Shibuya, H.; Niimoto, T.; Kongcharoensombat, W.; Ochi, M. Silencing microRNA-34a inhibits chondrocyte apoptosis in a rat osteoarthritis model in vitro. *Rheumatology* **2010**, *49*, 2054–2060. [CrossRef]

60. Niederer, F.; Trenkmann, M.; Ospelt, C.; Karouzakis, E.; Neidhart, M.; Stanczyk, J.; Kolling, C.; Gay, R.E.; Detmar, M.; Gay, S.; et al. Down-regulation of microRNA-34a* in rheumatoid arthritis synovial fibroblasts promotes apoptosis resistance. *Arthritis Rheum.* **2012**, *64*, 1771–1779. [CrossRef]

61. Zhong, X.; Li, P.; Li, J.; He, R.; Cheng, G.; Li, Y. Downregulation of microRNA-34a inhibits oxidized low-density lipoprotein-induced apoptosis and oxidative stress in human umbilical vein endothelial cells. *Int. J. Mol. Med.* **2018**, *42*, 1134–1144. [CrossRef]

62. Okuhara, A.; Nakasa, T.; Shibuya, H.; Niimoto, T.; Adachi, N.; Deie, M.; Ochi, M. Changes in microRNA expression in peripheral mononuclear cells according to the progression of osteoarthritis. *Mod. Rheumatol.* **2012**, *22*, 446–457. [CrossRef]

63. Zheng, H.; Liu, J.; Tycksen, E.; Nunley, R.; McAlinden, A. MicroRNA-181a/b-1 over-expression enhances osteogenesis by modulating PTEN/PI3K/AKT signaling and mitochondrial metabolism. *Bone* **2019**, *123*, 92–102. [CrossRef]

64. Chen, K.L.; Fu, Y.Y.; Shi, M.Y.; Li, H.X. Down-regulation of miR-181a can reduce heat stress damage in PBMCs of Holstein cows. *In Vitro Cell Dev. Biol. Anim.* **2016**, *52*, 864–871. [CrossRef]

65. Feng, X.; Zhang, C.; Yang, Y.; Hou, D.; Zhu, A. Role of miR-181a in the process of apoptosis of multiple malignant tumors: A literature review. *Adv. Clin. Exp. Med.* **2018**, *27*, 263–270. [CrossRef]

66. Yamasaki, K.; Nakasa, T.; Miyaki, S.; Ishikawa, M.; Deie, M.; Adachi, N.; Yasunaga, Y.; Asahara, H.; Ochi, M. Expression of microRNA-146a in osteoarthritis cartilage. *Arthritis Rheumatol.* **2009**, *60*, 1035–1041. [CrossRef]

67. Li, L.; Chen, X.P.; Li, Y.J. MicroRNA-146a and human disease. *Scand. J. Immunol.* **2010**, *71*, 227–231. [CrossRef]

68. Xie, Y.; Chu, A.; Feng, Y.; Chen, L.; Shao, Y.; Luo, Q.; Deng, X.; Wu, M.; Shi, X.; Chen, Y. MicroRNA-146a: A Comprehensive Indicator of Inflammation and Oxidative Stress Status Induced in the Brain of Chronic T2DM Rats. *Front. Pharmacol.* **2018**, *9*, 478. [CrossRef]

69. Li, J.; Huang, J.; Dai, L.; Yu, D.; Chen, Q.; Zhang, X.; Dai, K. miR-146a, an IL-1β responsive miRNA, induces vascular endothelial growth factor and chondrocyte apoptosis by targeting Smad4. *Arthritis Res. Ther.* **2012**, *14*, R75. [CrossRef]

70. Wang, J.H.; Shih, K.S.; Wu, Y.W.; Wang, A.W.; Yang, C.R. Histone deacetylase inhibitors increase microRNA-146a expression and enhance negative regulation of interleukin-1β signaling in osteoarthritis fibroblast-like synoviocytes. *Osteoarthr. Cartilage* **2013**, *21*, 1987–1996. [CrossRef]

71. Wen, F.; Li, B.; Huang, C.; Wei, Z.; Zhou, Y.; Liu, J.; Zhang, H. MiR-34a is Involved in the Decrease of ATP Contents Induced by Resistin Through Target on ATP5S in HepG2 Cells. *Biochem. Genet.* **2015**, *53*, 301–309. [CrossRef]

72. Gong, Y.Y.; Luo, J.Y.; Wang, L.; Huang, Y. MicroRNAs regulating reactive oxygen species in cardiovascular diseases. *Antioxid. Redox Signal.* **2018**, *29*, 1092–1107. [CrossRef]

73. D'Adamo, S.; Cetrullo, S.; Guidotti, S.; Borzì, R.M.; Flamigni, F. Hydroxytyrosol modulates the levels of microRNA-9 and its target sirtuin-1 thereby counteracting oxidative stress-induced chondrocyte death. *Osteoarthr. Cartilage* **2017**, *25*, 600–610. [CrossRef]

74. Ji, G.; Lv, K.; Chen, H.; Wang, T.; Wang, Y.; Zhao, D.; Qu, L.; Li, Y. MiR-146a regulates SOD2 expression in H2O2 stimulated PC12 cells. *PLoS ONE* **2013**, *8*, e69351. [CrossRef]

75. Wang, L.; Huang, H.; Fan, Y.; Kong, B.; Hu, H.; Hu, K.; Guo, J.; Mei, Y.; Liu, W.L. Effects of downregulation of microRNA-181a on H2O2-induced H9c2 cell apoptosis via the mitochondrial apoptotic pathway. *Oxid. Med. Cell Longev.* **2014**, *2014*, 960362. [CrossRef]

76. Baker, J.R.; Vuppusetty, C.; Colley, T.; Papaioannou, A.I.; Fenwick, P.; Donnelly, L.; Ito, K.; Barnes, P.J. Oxidative stress dependent microRNA-34a activation via PI3Kα reduces the expression of sirtuin-1 and sirtuin-6 in epithelial cells. *Sci. Rep.* **2016**, *6*, 35871. [CrossRef]

77. Huang, X.; Gao, Y.; Qin, J.; Lu, S. The role of miR-34a in the hepatoprotective effect of hydrogen sulfide on ischemia/reperfusion injury in young and old rats. *PLoS ONE* **2014**, *9*, e113305. [CrossRef]

78. Rigoglou, S.; Papavassiliou, A.G. The NF-κB signaling pathway in osteoarthritis. *Int. J. Biochem. Cell Biol.* **2013**, *45*, 2580–2584. [CrossRef]

79. Zhang, Q.; Lenardo, M.J.; Baltimore, D. 30 Years of NF-κB: A blossoming of relevance to human pathobiology. *Cell* **2017**, *168*, 37–57. [CrossRef]

80. Zhang, Z.; Zhang, Z.; Kang, Y.; Hou, C.; Duan, X.; Sheng, P.; Sandell, L.J.; Liao, W. Resistin stimulates expression of chemokine genes in chondrocytes via combinatorial regulation of C/EBPβ and NF-κB. *Int. J. Mol. Sci.* **2014**, *15*, 17242–17255. [CrossRef]

81. Aslani, M.R.; Keyhanmanesh, R.; Alipour, M.R. Increased Visfatin Expression Is Associated with Nuclear Factor-κB in Obese Ovalbumin-Sensitized Male Wistar Rat Tracheae. *Med. Princ. Pract.* **2017**, *26*, 351–358. [CrossRef]

82. Xu, B.; Li, Y.Y.; Ma, J.; Pei, F.X. Roles of microRNA and signaling pathway in osteoarthritis pathogenesis. *J. Zhejiang Univ. Sci. B* **2016**, *17*, 200–208. [CrossRef]

83. Altman, R.; Asch, E.; Bloch, D.; Bole, G.; Borenstein, D.; Brandt, K.; Christy, W.; Cooke, T.D.; Greenwald, R.; Hochberg, M.; et al. Development of criteria for the classification and reporting of osteoarthritis. Classification of osteoarthritis of the knee. Diagnostic and Therapeutic Criteria Committee of the American Rheumatism Association. *Arthritis Rheum.* **1986**, *29*, 1039–1049. [CrossRef]

84. Gosset, M.; Berenbaum, F.; Salvat, C.; Sautet, A.; Pigenet, A.; Tahiri, K.; Jacques, C. Crucial role of visfatin/pre-B cell colony-enhancing factor in matrix degradation and prostaglandin E2 synthesis in chondrocytes: Possible influence on osteoarthritis. *Arthritis Rheum.* **2008**, *58*, 1399–1409. [CrossRef]

85. Ramakers, C.; Ruijter, J.M.; Deprez, R.H.; Moorman, A.F. Assumption-free analysis of quantitative real-time polymerase chain reaction (PCR) data. *Neurosci. Lett* **2003**, *339*, 62–66. [CrossRef]

86. Pfaffl, M.W. A new mathematical model for relative quantification in real RT-PCR. *Nucleic Acid Res.* **2001**, *29*, e45. [CrossRef]

87. Vandesompele, J.; de Preter, K.; Pattyn, F.; Poppe, B.; van Roy, N.; de Paepe, A.; Speleman, F. Accurate normalization of real-time quantitative RT-PCR data by geometric averaging of multiple internal control genes. *Genome Biol.* **2002**, *3*, research0034.1. [CrossRef]

88. Cheleschi, S.; Calamia, V.; Fernandez-Moreno, M.; Biava, M.; Giordani, A.; Fioravanti, A.; Anzini, M.; Blanco, F. In vitro comprehensive analysis of VA692 a new chemical entity for the treatment of osteoarthritis. *Int. Immunopharmacol.* **2018**, *64*, 86–100. [CrossRef]

89. Cheleschi, S.; Fioravanti, A.; De Palma, A.; Corallo, C.; Franci, D.; Volpi, N.; Bedogni, G.; Giannotti, S.; Giordano, N. Methylsulfonylmethane and mobilee prevent negative effect of IL-1β in human chondrocyte cultures via NF-κB signaling pathway. *Int. Immunopharmacol.* **2018**, *65*, 129–139. [CrossRef]

Permissions

List of Contributors

Anna De Palma
Rheumatology Unit, Azienda Ospedaliera Universitaria Senese, Policlinico Le Scotte, Viale Bracci 1, 53100 Siena, Italy
Department of Medical Biotechnologies, University of Siena, Policlinico Le Scotte, Viale Bracci 1, 53100 Siena, Italy

Nicola Antonio Pascarelli and Mauro Galeazzi
Rheumatology Unit, Azienda Ospedaliera Universitaria Senese, Policlinico Le Scotte, Viale Bracci 1, 53100 Siena, Italy

Nicola Giordano
Department of Medicine, Surgery and Neurosciences, Scleroderma Unit, University of Siena, Policlinico Le Scotte, Viale Bracci 1, 53100 Siena, Italy

Sara Tenti
Department of Medicine, Surgery and Neuroscience, Rheumatology Unit, University of Siena, Policlinico Le Scotte, Viale Bracci 1, 53100 Siena, Italy

Ewa Ostrycharz, UrszulaWasik, Agnieszka Kempinska-Podhorodecka and Malgorzata Milkiewicz
Department of Medical Biology, Pomeranian Medical University, 71-111 Szczecin, Poland

Jesus M. Banales
Department of Liver and Gastrointestinal Diseases, Biodonostia Health Research Institute-Donostia University Hospital-Ikerbasque, CIBERehd, University of the Basque Country (UPV/EHU), 20014 San Sebastian, Spain

Piotr Milkiewicz
Translational Medicine Group, Pomeranian Medical University, 71-210 Szczecin, Poland
Liver and Internal Medicine Unit, Medical University of Warsaw, 02-097 Warsaw, Poland

Mirza Muhammad Fahd Qadir
Diabetes Research Institute, University of Miami Miller School of Medicine, Miami, FL 33136, USA
Department of Cell Biology and Anatomy, University of Miami Miller School of Medicine, Miami, FL 33136, USA

Dagmar Klein and Silvia Álvarez-Cubela
Diabetes Research Institute, University of Miami Miller School of Medicine, Miami, FL 33136, USA

Juan Domínguez-Bendala
Diabetes Research Institute, University of Miami Miller School of Medicine, Miami, FL 33136, USA
Department of Cell Biology and Anatomy, University of Miami Miller School of Medicine, Miami, FL 33136, USA
Department of Surgery, University of Miami Miller School of Medicine, Miami, FL 33136, USA

Ricardo Luis Pastori
Diabetes Research Institute, University of Miami Miller School of Medicine, Miami, FL 33136, USA
Department of Medicine, Division of Metabolism, Endocrinology and Diabetes, University of Miami Miller School of Medicine, Miami, FL 33136, USA

Chang-Youh Tsai and Hsien-Tzung Liao
Division of Allergy, Immunology & Rheumatology, Taipei Veterans General Hospital & National Yang-Ming University, #201 Sec.2, Shih-Pai Road, Taipei 11217, Taiwan

Song-Chou Hsieh, Ko-Jen Li and Chia-Li Yu
Department of Internal Medicine, National Taiwan University Hospital, #7 Chung-Shan South Road, Taipei 10002, Taiwan

Cheng-Shiun Lu, Cheng-Han Wu, Yu-Min Kuo and Chieh-Yu Shen
Department of Internal Medicine, National Taiwan University Hospital, #7 Chung-Shan South Road, Taipei 10002, Taiwan
Institute of Clinical Medicine, National Taiwan University College of Medicine, #7 Chung-Shan South Road, Taipei 10002, Taiwan

Tsai-Hung Wu
Division of Nephrology, Taipei Veterans General Hospital & National Yang-Ming University, #201 Sec. 2, Shih-Pai Road, Taipei 11217, Taiwan

Hui-Ting Lee
Section of Allergy, Immunology & Rheumatology, Mackay Memorial Hospital, #92 Sec. 2, Chung-Shan North Road, Taipei 10449, Taiwan

Wen Cai Zhang
Burnett School of Biomedical Sciences, College of Medicine, University of Central Florida, 6900 Lake Nona Blvd, Orlando, FL 32827, USA

Kamesh R. Babu
Cancer Science Institute of Singapore, National
University of Singapore, Singapore 117599, Singapore

Yvonne Tay
Cancer Science Institute of Singapore, National
University of Singapore, Singapore 117599, Singapore
Department of Biochemistry, Yong Loo Lin School of
Medicine, National University of Singapore, Singapore
117597, Singapore

**Bonita Shin, Riley Feser, Braydon Nault, Stephanie
Hunter, Sujit Maiti, Kingsley Chukwunonso
Ugwuagbo and Mousumi Majumder**
Department of Biology, Brandon University, 3rd
Floor, John R. Brodie Science Centre, 270 – 18th Street,
Brandon, MB R7A6A9, Canada

Yang-Hsiang Lin
Liver Research Center, Chang Gung Memorial
Hospital, Linkou, Taoyuan 333, Taiwan

**Jinwei Zhang, Wanling Qiu, Jideng Ma, Yujie Wang,
Zihui Hu, Keren Long, XunWang, Long Jin, Qianzi
Tang, Guoqing Tang, Li Zhu, Xuewei Li, Surong
Shuai and Mingzhou Li**
Institute of Animal Genetics and Breeding, College of
Animal Science and Technology, Sichuan Agricultural
University, Chengdu 611130, Sichuan, China
Farm Animal Genetic Resource Exploration and
Innovation Key Laboratory of Sichuan Province,
Sichuan Agricultural University, Chengdu 611130,
Sichuan, China

**Keiko Yamakawa, Yuko Nakano-Narusawa, Nozomi
Hashimoto, Masanao Yokohira and Yoko Matsuda**
Oncology Pathology, Department of Pathology and Host-
Defense, Faculty of Medicine, Kagawa University, 1750-1
Ikenobe, Miki-cho, Kita-gun, Kagawa 761-0793, Japan

Giulia Cosentino and Ilaria Plantamura
Molecular Targeting Unit, Research Department,
Fondazione IRCCS Istituto Nazionale dei Tumori,
20133 Milan, Italy

Alessandra Cataldo and Marilena V. Iorio
Molecular Targeting Unit, Research Department,
Fondazione IRCCS Istituto Nazionale dei Tumori,
20133 Milan, Italy
IFOM Istituto FIRC di Oncologia Molecolare, 20139
Milan, Italy

**Sara Cheleschi, Ines Gallo, Sara Tenti and Antonella
Fioravanti**
Department of Medicine, Surgery and Neuroscience,
Rheumatology Unit, Azienda Ospedaliera Universitaria
senese, Policlinico Le Scotte, 53100 Siena, Italy

Marcella Barbarino and Antonio Giordano
Sbarro Institute for Cancer Research and Molecular
Medicine, Department of Biology, College of Science
and Technology, Temple University, Philadelphia, PA
19122, USA

Stefano Giannotti and Nicola Mondanelli
Department of Medicine, Surgery and Neurosciences,
Section of Orthopedics and Traumatology, University
of Siena, Policlinico Le Scotte, 53100 Siena, Italy

Index